Bioinorganic Chemistry

# Bioinorganic Chemistry

Dieter Rehder

University of Hamburg

OXFORD
UNIVERSITY PRESS

# OXFORD
## UNIVERSITY PRESS

Great Clarendon Street, Oxford, OX2 6DP,
United Kingdom

Oxford University Press is a department of the University of Oxford.
It furthers the University's objective of excellence in research, scholarship,
and education by publishing worldwide. Oxford is a registered trade mark of
Oxford University Press in the UK and in certain other countries

British Library Cataloguing in Publication Data

Data available

Library of Congress Control Number: 2013948265

ISBN 978-0-19-965519-9

Printed in Great Britain by

Ashford Colour Press Ltd, Gosport, Hampshire

# Contents

# List of sidebars

# Introduction

Bioinorganic chemistry explores the role of 'inorganics' in biological and biomimetic processes. Inorganics are metal ions and metal compounds of variable complexity: anions and cations such as sulfate, phosphate, and the ammonium ion; and neutral molecules involved in biological processes such as dioxygen and ozone, and the oxides of carbon, nitrogen, and sulfur. All of these inorganic ions and molecules can play an intrinsic role in a range of physiological functions. They are therefore essential for sustaining life—as long as homeostasis is not disrupted. Conversely, the dysfunction of life processes can be attributed to an undersupply or oversupply of inorganics, either through malnutrition or as a result of disturbances caused by genetic and/or effects related to environmental and dietary imbalance, or illness.

An estimated one-third of the proteins, and one-half of all enzymes, which collectively perform a variety of tasks, contain metal ions. Many of these enzymatic processes concern redox reactions on the one hand, and processes involving the making and breaking of bonds in respect to the formation and disruption of organic (and inorganic) molecules by non-redox processes on the other. Esterification and hydrolysis are typical reactions in this respect. In order for metal ions to become resorbed, to 'locate' their target molecules, and to be desorbed, organisms require suitable recognition and transport systems—transport systems that have available coordination sites that match the metal ion and its oxidation state.

Life processes involve (i) metabolism—the chemical transformation of matter coupled with energy production or consumption, (ii) replication/reproduction (coupled to the storage and transfer of information), and (iii) evolutionary modulation—the optimization of the chances of survival in a given habitat and the adjustment to changing environmental conditions. Evolutionary adaption is achieved by mutation within the individual gene pool, and by horizontal gene transfer between different (bacterial) species.

Inorganic processes, intrinsically linked to biochemical reactions, participate at all these levels. In this context, during this book we also focus on environmental and medicinal concerns. Examples of 'inorganics' in medicine include: the treatment of certain types of cancer with platinum prescriptions; and the use of paramagnetic gadolinium and radioactive technetium compounds in diagnosis.

The present book arises from a lecture course called 'Bioinorganic Chemistry', delivered by the author during the past two decades at the University of Hamburg, Germany, and, in 2008, at the University of Lund, Sweden. The idea for the book was initiated by Ebbe Nordlander, Lund University, who also had been involved in developing the original concept of this book.

Structuring the wealth of material has been a particularly challenging task. For an inorganic chemist, it is tempting to arrange the material according to elements. In biology, however, the focus is often on function. A combination of these two approaches has been pursued in the present book: elements that have a pivotal role in (almost) all biological functions (alkaline and alkaline earth metals, iron, zinc, and sulfur) are addressed in separate chapters, while other chapters are dedicated to functional processes. Examples of the latter are oxidoreductases, enzymes of the nitrogen cycle, and photosynthesis, while the general role of nickel is covered in the context of methanogenic/methanotrophic prokarya. Two extra chapters are dedicated to the metal–carbon bond and to inorganics in medicine, respectively.

The book is written for students with a basic knowledge of general, inorganic, and organic chemistry (plus basics in biochemistry), including analytical methods. Irrespective of this general caveat, introductory chapters provide a general review of the bio-elements (Chapter 1) and 'Life' (Chapter 2). Additional supporting information is provided in sidebars, which are distributed throughout the book. These briefings encompass analytical methods frequently employed for the characterization of bioinorganic species, overviews of concepts (such as symmetry and radioactivity), and overviews of functional regimes, for example copper enzymes and organic redox systems.

Finally, I would like to thank Jonathan Crowe from Oxford University Press for his constructive comments in the process of the final shaping of the book chapters.

Dieter Rehder
Hamburg, June 2013

# Bio-elements in the periodic table

In this introductory chapter, we will provide a brief overview of those chemical elements that have biological and medicinal functions. For the latter, we will consider metals that have a direct impact on physiological activity; metals employed, for example, in supports or as substitutes for joints (such as titanium alloys) are thus excluded. On the other hand, we include toxic compounds based on mercury, lead, and arsenic. In the second part of the chapter, we provide overview of the main ligands and ligand functions available for metal ions in biological systems. The term 'metal ion' is used here in a broader sense, including metalloids (such as Si, As, Sb, and Se). Ligands do not only mediate the transport and storage of metal ions, but also fine-tune the metal's physiological actions.

In the periodic table of the bio-elements in Fig. 1.1, elements of biological relevance are classified according to four categories. The elements C, H, O, N, and S (in black) account for the main part of organic matter and thus for 'biomass'. In addition, many elements commonly considered 'inorganic' play an important role in a biological and, more specifically, physiological context. Some of these elements, in light grey, are present in (almost) all organisms. The alkaline metals Na and K, the alkaline earth metals Mg and Ca, the transition metals Mn, Fe, Co, Cu, and Zn, and the non-metals P, Se, F, Cl, and I belong to this category. Other elements, shown in dark grey (the metals V, W, Ni, and Cd, the half-metal Si, and the non-metal B) are important in a restricted number of organisms only.

Elements shown in dark blue in Fig. 1.1 are used in medicinal therapy or diagnosis (see Sections 14.3 and 14.4), and may thus be classified as medicinal elements. There are additional elements that affect living organisms either by their direct toxicity in very low doses and/or by their destructive radioactive potential. Of the toxic elements, we explore As, Pb, and Hg (in light blue) in Section 13.2. Toxic effects exerted by overloads of otherwise beneficial elements, in particular Fe and Cu, shall be considered in the respective chapters dedicated to these elements and their functions.

Specific chapters providing surveys on single elements or groups of elements are dedicated to the alkaline and alkaline earth metals (Chapter 3), iron (Chapter 4), and zinc (Chapter 12). The nitrogen and sulfur cycles are discussed in some detail in Chapter 8 (sulfur) and Chapter 9 (nitrogen), respectively.

For elements highlighted in blue and grey shades in Fig. 1.1, we provide a brief summary of their main biological function and/or medicinal application in the following pages, including, where appropriate, links to the respective chapter, section, or sidebar. The numbering of sidebars matches the numbering of chapters. For elements primarily present as free ions or in ionic compounds, the charge is denoted in Arabic numerals (such as $Mg^{2+}$); for elements chiefly present in covalent compounds, Roman numerals are utilized—for example,

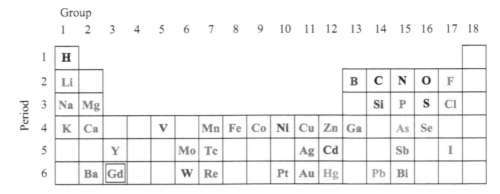

Figure 1.1 Periodic table of the bio-elements. Black: elements which build up biomass; light grey: additional generally essential elements; dark grey: essential for some groups of organisms only; dark blue: medicinally important elements (in therapy and/or diagnosis); light blue: elements addressed in the context of their toxicity. Gadolinium (Gd) is framed because it is a member of the lanthanoid subfamily.

$Fe^{II}$ and $Se^{-II}$. Boron and silicon are included in this overview, including key references, but not treated in extra chapters.

$Li^+$ is used in the treatment of bipolar disorder (manic depression) and hypertension (14.3.4).

$Na^+$ and $K^+$ are the most important 'free' intra- and extracellular cations. They are responsible for, for example, the regulation of the osmotic pressure, membrane potentials, enzyme activity, and signalling (3.3).

$Mg^{2+}$ is the central metal ion in chlorophyll (11.2). Mg is further involved in anaerobic energy metabolism (adenosine triphosphate $\rightarrow$ adenosine diphosphate + inorganic phosphate), and the activation of kinases and phosphatases (3.4), and thus triggers activation paths.

$Ca^{2+}$ plays a pivotal role in signalling, muscle contraction, and enzyme regulation (3.5). $Ca^{2+}$ can be a cofactor in hydrolases, and can play a role in determining the structure of biological molecules, for example in thermolysin (12.2.2). Ca is also constituent of the photosynthetic oxygen-evolving centre (11.2), and plays a role as a second messenger and in the activation of enzymes (3.5). Calcium, in the form of partially fluorinated hydroxyapatite $Ca_5(PO_4)_3(OH,F)$, is the main inorganic part of the endoskeletons (bones, teeth, enamel) of vertebrates (3.5). Exoskeletons of, for example, mussels, shells, corals, and sea urchins are built up of aragonite and calcite, $CaCO_3$ (3.5).

$Gd^{III}$ is the most common paramagnetic metal centre in contrast agents in magnetic resonance imaging (14.4).

$V^{III/IV/V}$ constitutes the active centre of vanadate-dependent haloperoxidases (7.2) and vanadium nitrogenase (9.1). $V^{3+}$ and $VO^{2+}$ are accumulated by ascidians, $V^{IV}$ (in the form of amavadin) by *Amanita* mushrooms (7.2).

$Mo^{IV/VI}$ is constituent of molybdo-pyranopterins, and thus in a component of the active centre of a variety of oxidoreductases and in acetylene hydratase (7.1). Molybdenum is further a constituent of the FeMo-cofactor in molybdenum-nitrogenases (9.1).

$W^{IV/VI}$-based tungsto-pyranopterins (analogues of the corresponding molybdenum cofactors) are present in several oxidoreductases, mainly of thermophilic archaea (7.1).

$Mn^{II/III/IV}$ constitutes the basis of the $\{CaMn_4O_5\}$ cluster of the oxygen evolving complex in photosynthetic water oxidation (11.2). Ribonucleotide reductases can contain one or two Mn centres (6.1), and Mn can also be the active metal ion in superoxide dismutases (6.2).

$^{99m}Tc$ is a metastable $\gamma$-emitter ($t_{1/2}=6\,h$). Its coordination compounds are employed in radio diagnostics of, for example, bone cancer and infarct risk (14.4).

$Fe^{II/III/(IV/V)}$: this multi-functional and omnipresent element is stored and 'operated' by proteins (ferritins, Dps proteins, and frataxins) (4.2). Bio-mineralization of iron compounds leads to the minerals ferrihydrite, goethite, magnetite, and greigite (4.2). The transport protein transferrin (4.2) regulates iron transport; pathological dysfunction can cause iron overload and deficiency (14.2.1). Biological functions mediated by iron include the oxygen transport by haemoglobin (5.1) and haemerythrin (5.2), and electron transfer (redox) reactions. Fe-based electron transfer proteins can depend on iron–sulfur clusters (9.2, sidebar 5.1), haem-type iron (Chapter 5), and dinuclear and mononuclear non-sulfur and non-haem iron proteins (Chapter 6). Additional examples of iron-based enzymes are the oxygenase $P_{450}$ (6.3), methane monooxygenase (10.3), ribonucleotide reductase (7.1), iron-only hydrogenases (10.2), and NO reductase (9.2). Carbonyl and cyano complexes of iron play a role in nickel–iron and iron-only hydrogenases (13.1).

$Co^{I,II,III}$ is the central ion in synthases and isomerases of the cobalamine family. An example is vitamin $B_{12}$ (13.1), the methyl form of which is also employed in the methylation of organic and inorganic substrates, for example in the context of methanogenesis (10.2).

$Ni^{I/II/III}$ is a main metal in methanogenesis, where a NiFe hydrogenase and the so-called factor $F_{430}$ with an interim $Ni–CH_3$ centre (10.2) are active. Carbonyl–nickel intermediates are formed in the course of the activities of NiFe-CO-dehydrogenase and acetyl coenzyme-A synthase (13.1). Additional examples of processes catalysed by Ni-dependent enzymes include the hydrolysis of urea and the dismutation of superoxide (10.4).

$Pt^{II/IV}$-based complexes are used in the chemotherapy of cancer (mainly of the ovaria and testes). A prominent example is cisplatin $cis$-$[Pt(NH_3)_2Cl_2]$ (14.3.3).

$Cu^{I/II}$ mediates oxygen transport by haemocyanin (5.2). Active centres containing 1–7 Cu ions are involved in electron transport enzymes such as plastocyanin (11.2), nitrite and NO reductases (9.3), catechol oxidase and galactose oxidase (6.3), and in oxygenases (tyrosinase) and dismutases (6.3). Copper possibly also plays an active role in Alzheimer's disease (14.2.2).

$Au^{I/III}$ compounds are considered in the context of the treatment of arthritis (14.3.2).

$Zn^{2+}$ is in the active centre of enzymes including hydrolases, carboanhydrase, and alcohol dehydrogenase (12.2). Other zinc dependent functions are manifest in genetic transcription (zinc fingers), in the stabilization of tertiary and quaternary structures of peptides (12.3), and in DNA repair proteins (12.3). Low molecular mass proteins rich in zinc, so-called thioneins, store zinc and regulate zinc levels, but can also act as scavengers for toxic $Cd^{2+}$ and $Hg^{2+}$ (12.4).

$Cd^{2+}$ is a zinc antagonist and therefore toxic, because it binds more effectively to cysteinate residues and thus inhibits the activity of zinc enzymes. By way of exception, $Cd^{2+}$ can replace $Zn^{2+}$ in the carboanhydrase of marine diatoms (12.2.1).

$Hg^{I/II}$: mercurous ($Hg_2^{2+}$) and mercuric ($Hg^{2+}$) compounds are particularly toxic because they denature proteins by the formation of insoluble HgS or HgSe when reacting with cystine and cysteine, or selenocysteine. In mammals, $Hg^{2+}$ is metabolized to methylmercury, $CH_3Hg^+$ (13.2.1).

$B^{III}$ is a constituent of a few naturally occurring antibiotics (such as boromycin). In the form of borate, it can be employed as a stabilizing component of herbal cell walls (see also [1]).

$Si^{IV}$, in the form of silicates, is involved in the build-up of bones. Silica ($SiO_2$) and silica-gels ($SiO_2 \cdot xH_2O$) are employed as a stabilizing support in monocotyledonous plants (such as grass) and *Equisetum*, and constitute the shells of diatoms ($\rightarrow$ kieselgur). Dietary silicon is likely beneficial to bone and connective tissue health [2].

$P^V$ in phosphates, ($H_nPO_4^{(3-n)-}$), is a constituent in hydroxy- and fluorapatite $Ca_5(PO_4)_3(OH/F)$ of the bone and enamel. Esters of mono-, di- and triphosphate are further involved in energy metabolism (ATP/ADP/AMP, *c*-GMP), in the activation of reductants such as NADPH (side-bar 12.1), and in the activation of organic substrates in metabolic and catabolic pathways. Phospholipids—lipids containing a phosphoester unit—in cell membranes, and other phosphate esters, including DNA and RNA, are indispensable and thus common in all organisms.

$As^{III/V}$: toxic $As_2O_3$ (arsenic) is metabolized to methyl arsenates (13.2.4); arsenate ($HAsO_4^{2-}$) is a life-threatening antagonist for phosphate.

$Sb^{III}$, for example in the form of $Sb_2O_3$ or $Sb_2S_3$, has sporadically been utilized as a 'disin-fectant', for example in the treatment of inflammatory skin pimples such as acne (14.3.1). Antimony compounds are toxic.

$Bi^{III}$-based prescriptions are used in the treatment of gastritis (14.3.4) such as caused by *Helicobacter pylori*.

$Se^{-II}$ is the key constituent in selenocysteine, an essential amino acid present in specific enzymes, for example in glutathione peroxidase, and in some representatives of the molyb-dopterin cofactor of oxidoreductases (7.1).

$F^-$ (fluoride) partly replaces OH in apatite (3.5). The teeth of sharks are almost completely fluorapatite $Ca_5(PO_4)_3F$.

$Cl^-$ is, along with hydrogencarbonate, the most important free anion in physiological liquids (Table 14.1). Its functions range from the regulation of ion homeostasis to the regulation of electrical excitability [3].

$I^-$ is an essential constituent of thyroid hormones such as thyroxin. These hormones stimulate diverse metabolic activities in tissues, and are also involved in genetic transcription [3].

The cations of transition metals are commonly not present in a free form, but are rather coordinated to (complexed by) ligands. In particular, this applies to metal ions in the active centres of enzymes, or else integrated into peptides and proteins, mostly as structure stabilizing factors. Representative ligands are listed in Fig. 1.2: **N**-functional ligands can be provided by the amide linkage of the peptide moiety, by porphinogens, histidine N$\delta$ or N$\epsilon$, lysine and arginine; **O**-functional ligands by the peptide amide, tyrosinate, serinate, glutamate, and aspartate; **S**-functional ligands by cysteinate and methionine, and the **Se**-function by seleno-cysteinate. O, S, and Se can also be present as doubly bonded, dianionic ligands (oxido, sulfide and selenido ligand), or as singly bonded $OH^-$, $SH^-$, and $SeH^-$. In addition to the organic peptide/protein and haem-type ligands, simple inorganic ligands are also often employed; see row (5) in Fig. 1.2.

Ligands coordinating via sulfur, such as cysteinate and sulfide, are classified as soft (in the sense of deformable) ligands, ligands coordinating via oxygen donors as hard ligands. Nitrogen donors fall in-between. This soft/hard concept goes back to Pearson; see also sidebar 4.1.

Hard metal ions preferentially bind to hard ligands, soft metal ions to soft ligands. There are, however, many exceptions to this simplified generalization. Alkaline and alkaline earth

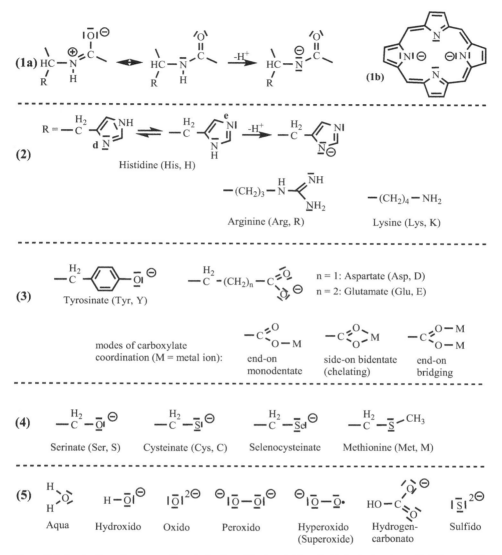

**Figure 1.2** A selection of common ligands for (transition) metal ions in biological systems. (**1a**) The peptide function of the protein backbone. Note that the peptide-N can only coordinate out of its deprotonated form, or the neutral mesomeric resonance hybrid, where N is sp³ hybridized and thus has a free electron pair available. Coordination via the carbonyl-O is also feasible. (**1b**) Porphinogenic ligands, such as the haem-type centre (of cytochromes) shown here, are tetradentate. (**2–4**) Functional groups provided by amino acid residues in peptides and proteins. The three-letter and one-letter codes of the amino acids are given in parentheses. (**5**) Frequently employed inorganic co-ligands.[1]

---

[1] Note that *ligands* such as $O^{2-}$, $OH^-$, $O_2^{2-}$, and $S^{2-}$ are often termed—not quite correctly—oxo, hydroxo, peroxo, and thio ligands. The nomenclature used throughout this book follows IUPAC recommendations: oxido, hydroxido, peroxido, sulfido, etc.

metal ions are considered hard, and thus are commonly found in a coordination sphere dominated by oxygen-functional ligands. An exception is $Mg^{2+}$ in chlorophyll, where the ion is in a porphyrinogenic environment (3.4). Hard ligands are also commonly targeted by *early* transition metals in their high oxidation states: $V^{IV/V}$, $Mo^{IV/VI}$, $W^{IV/VI}$; see, for example, vanadate-dependent haloperoxidases (7.2), molybdo- and tungsto-pyranopterins (7.1). Manganese also favours oxido functionalities, such as in the oxygen evolving centre in photosynthesis (Chapter 11) where it runs through the oxidation states II, III, and IV. Ferrous ($Fe^{II}$) and ferric ($Fe^{III}$) iron, the common oxidation states of iron in nature, are rather unselective: Iron ions bind to hard, soft, and intermediate ligands. Porphinogens readily coordinate Fe, Co, or Ni. Examples are cytochromes (Fe-dependent transporters for $O_2$ and electrons; Chapter 5), vitamin $B_{12}$ (containing Co; 13.1), and the factor $F_{430}$, a Ni-based cofactor in methanogenesis; 10.2).

The late transition metal ions $Cu^{+/2+}$ and $Zn^{2+}$ tend to prefer intermediate to soft ligands. Examples of zinc ions exclusively coordinating to thiolate are thioneins (12.4) and the *structural* zinc centre in alcohol dehydrogenase (12.2.3). The *functional* centre in the zinc enzyme alcohol dehydrogenase (12.2.3) exemplifies $Zn^{2+}$ coordination to a mixed N/S ligand set, while in carboanhydrase, the enzyme responsible for the hydration of $CO_2$ and the dehydration of $H_2CO_3$ (12.2.1), $Zn^{2+}$ exclusively coordinates to histidines. Mixed thiolate/histidine coordination of $Cu^{+/2+}$ in copper enzymes is exemplified by plastocyanin (in photosynthesis; Chapter 11) or cytochrome-c oxidase, the catalyst in the final step of the electron transfer to $O_2$ in the respiratory chain (5.3). Nitrite reductase, which contains two functionally coupled copper centres (9.3), is an example for an enzyme harbouring Cu both in an exclusive N-donor environment *and* in a mixed N/S coordination sphere.

## Suggested reading

Waldron KJ, Rutherford JC, Ford D, et al. Metalloproteins and metal sensing. *Nature* 2009; 460: 823–830.

The article provides a clue as to how metal sensors in proteins distinguish between different metals and thus select the right metal ion for a specific function, including the delivery of metal ions to functional metalloproteins by metal transporters (metallo-chaperones).

## References

1. Miwa K, Kamiya T, Fujiwara T. Homeostasis of the structurally important micronutrients, B and Si. *Curr. Opin. Plant Biol.* 2009; 12: 307–311.

2. Jugdaohsingh R. Silicon and bone health. *J. Nutr. Health Aging* 2007; 11: 99–110.

3. Pearce EN, Andersson M, Zimmermann MB. Global iodine nutrition: where do we stand in 2013? *Thyroid* 2013; 23: 523–528.

# 2 Pre-life and early life forms; extremophiles

Early life forms began to develop on our home planet ca. 3.6 Ga ago (1 Ga $= 10^9$ years). 'Life' is accompanied by, and depends on, the metabolism of simple and complex organic molecules as well as on inorganics, viz. transition metal ions and non-metals, phosphorus (in the form of phosphate) in particular. The development of protocells from the inventory available in the primordial atmosphere and aquatic sanctuaries, and in minerals present in Earth's crust, as well as the successive evolutionary development of these protocells into primitive bacterial and archaean cellular organisms, took place in an anoxic environment. Successors of these organisms still thrive in niches deprived of oxygen. With the Great Oxygen Event 2.4 Ga ago, probably initiated by photosynthetic cyanobacteria, adaption to an oxic environment became an essential precondition for survival for those unicellular organisms which were no longer confined, or could not retreat to, oxygen-free habitats.

For the more complex organisms (eukarya) evolving ca. two billion years ago (2 Ga), oxygen became *the* element of life—along with environmental conditions (temperature, pressure, pH, atmospheric and aquatic compositions), which we usually term 'normal'. In addition, however, many bacterial and archaean species—but some eukaryan algae as well—have adapted to extremes and are therefore referred to as extremophiles. This includes extremes in temperature, pressure, pH, salt and toxic metal concentrations, resistance towards UV and $\gamma$ radiation and, of particular interest in the view of bioinorganic chemistry, to carbon sources other than $CO_2$, electron sources other than ($O^{2-}$ in) water, and energy sources other than light.

In this chapter, we will go back in time to assess briefly the 'chemistry', in the very early days of our home planet, with respect to the scenarios which might have sparked Life. We will then describe the stages of development as the chemical reality and thus the conditions for life to thrive changed through the eons. Finally, adaption to extreme situations, as manifested in extremophiles, will be addressed.

## 2.1 What is Life, and how did Life evolve?

The study of bioinorganic chemistry can help us probe questions such as (i) 'How did life evolve on Earth?'; (ii) 'Are life forms resembling ours also feasible on other planets/moons in our Solar System, or on exoplanets?'; and (iii) 'Can alternative (e.g. non-carbon and non-water) life forms exist?'. The first question in particular can be explored via pre-life scenarios, such as those provided by the Miller–Urey and related experiments, 'clay-organisms', as originally proposed by Cairns-Smith, or 'pioneer organisms' in a primordial iron–sulfur world (Wächtershäuser), as explained in Sidebar 2.1. The comparatively high concentrations of $K^+$,

$Zn^{2+}$, $Mn^{2+}$, and phosphate in modern cells, exceeding the respective concentrations in common aqueous habitats (such as oceans, lakes, and rivers), suggests that primordial volcanic ponds of condensed geothermal vapour, nestling between silicateous and sulfidic rock, acted as hatcheries for the protocells.[1]

In addition to considering primordial scenarios for the development of the first cells and their adaption to an alleged inhospitable environment, the study of 'modern' extremophiles on our planet can help to answer questions such as those set out above. Commonly, extremophiles (for more details see below) are bacteria and archaea, subsumed under the term prokarya (or prokaryota) as opposed to eukarya, which thrive under conditions where 'normal' microorganisms cannot exist. In rare cases, horizontal gene transfer from prokarya to algae can also provide eukarya with the ability to live in extreme conditions.

Scheme 2.1 provides an approximate time line for the development of terrestrial life. Our planet accreted from the presolar nebula 4.56 Ga ago (1 Ga = 10^9 years). Igneous crystallization, providing an initial, still fragile crust, started about 4.53 Ga ago; the oldest minerals found on Earth, micron-sized zircon ($ZrSiO_4$) crystals, date back to 4.4 Ga. The primordial atmosphere consisted of molecular species such as $CO_2$, $N_2$, $NH_3$, $H_2$, and $H_2O$ (and likely trace amounts of $O_2$). Simple inorganic, organic, and transient molecules and ions derived from this primordial atmosphere, such as $H_2O$, $HCN$, $CO/CO_2$, $H_2CO$, and $CH_3SH$, provide,

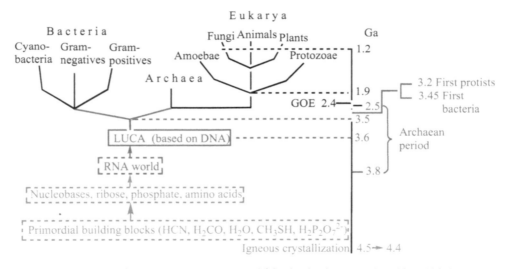

Scheme 2.1 A time line (in billion years, Ga; not to scale) for the development of pre-life and life forms on Earth, including a simplified phylogenetic tree. Colour code: Building blocks for life molecules in light blue, Archaean period in blue, the three trees of life (bacteria, archaea, eukarya)[2] in black. LUCA = last uniform common ancestor. RNA and DNA = ribo- and deoxyribonucleic acid. GOE = Great Oxygen Event. Contrasting prokarya (bacteria and archaea), eukaryotic cells possess a nucleus and differentiated cell organelles (such as mitochondria and chloroplasts). Bacteria and archaea are distinguished by the lipids employed in the cell walls (ester lipids in the case of bacteria; terpenoid ether lipids in the case of archaea).

[1]  For details, see Suggested reading (Mulkidjanian et al.).
[2]  An organism dubbed 'Myoyin parakaryote', with intermediate features between prokarya and eukarya, has recently been reported [4].

together with $HPO_4^{2-}$, the basic units for the synthesis of nucleobases, sugars, and amino acids for the assemblage of biomolecules (nucleotides, peptides, lipids, carbohydrates). These basic building blocks will have been generated under primordial terrestrial conditions; and/ or they might have been provided by meteorites, interplanetary, and interstellar dust.

The Miller–Urey and Wächtershäuser experiments (Sidebar 2.1) provide, at least in part, appropriate scenarios for the formation of simple amino acids (Eq. (2.1)), thioacetic acid ester (Eq. (2.2)), and nucleobases, resorting to energy sources such as discharge (lightning) and the oxidation of FeS to pyrite $FeS_2$. 'Clay organisms', i.e. aluminosilicate or silicate

---

### Sidebar 2.1    Scenarios for the primordial supply of basic life molecules

**The Miller–Urey scenario** [2]: Miller and Urey had based their experiments on a gas mix in the primordial atmosphere consisting of $CH_4$, $NH_3$, $H_2O$, $N_2$, and some $CO_2/CO$, i.e. completely devoid of oxygen or potentially oxidizing constituents. Their presumed energy source was essentially electric discharge (as naturally provided by lightning and corona discharges from pointed surface objects). In their primordial broth they could identify, among others, formic acid, acetic acid, and lactic acid, the biogenic (i.e. biologically essential) amino acids glycine and α-alanine, and the non-biogenic amino acids β-alanine and α-amino-$n$-butyric acid. The proposed mechanism is the formation, by spark, of aldehydes and HCN in the gas phase, which further react in the aqueous phase to amino nitriles and amino acids (Strecker synthesis); i.e. according to the following non-stoichiometric reaction sequence:

$$CH_4, \ N_2, \ H_2O \rightarrow\rightarrow HCN, \ HCHO$$
$$HCN + HCHO \rightarrow NH_2CH_2CN$$
$$NH_2CH_2CN + H_2O \rightarrow NH_2CH_2CO_2H$$

**The Wächtershäuser scenario** [3]: Wächtershäuser adapted the atmosphere to fit our current understanding of the composition of the primitive atmosphere ($CO_2$, $N_2$, $H_2O$, traces of $H_2$ and $O_2$), i.e. to an atmosphere which contained trace amounts of oxygen, stemming from the splitting of water by spark discharge, cosmic rays (mainly high-speed protons), and high-energy UV. The energy source for bond formation and thus the generation of more complex molecules is the oxidation of nickel and ferrous sulfide, e.g. troilite FeS (with $S^{-II}$) to disulfide, e.g. pyrite $FeS_2$ (containing $S^{-I}$).

An example for C–C bond formation, the generation of 'activated acetic acid' (the methyl ester of thioacetic acid) is:

$$FeS + HS^- \rightarrow FeS_2 + H^+ + 2e^-    \Delta G = -118 \, kJ \, mol^{-1}$$
$$2CH_3SH + CO_2 + FeS \rightarrow CH_3C(O)SCH_3 + H_2O + FeS_2$$

Amorphous iron sulfides, initially formed in hot, aqueous, $H_2S$-rich environments, transform to the thermodynamically stable crystalline sulfides with a microscopic 'honeycomb' texture, providing a large reactive surface area as well as protective pockets for the inventory of basic organic molecules.

**Clay organisms** [5]: Clays are phyllosilicates that 'breathe' by absorbing and releasing water (depending on the respective water vapour pressure). They can catalyse condensation reactions, including those which result in the formation of peptides from amino acids, and oligonucleosides from nucleosides. A typical clay in this respect is montmorillonite, composed of alternate layers of interconnected anionic $\{SiO_4\}$ and cationic $\{Al(O/OH)_6\}$ units, and interstitial water and hydrated alkaline and alkaline earth metal ions. The amino acids and nucleosides are activated by coordinating to vacant sites of aluminium, and by hydrogen-bonding to bridging OH.

$$2NH_2CH_2CO_2H \rightarrow NH_2CH_2C(O)NHCH_2CO_2H + H_2O$$

surfaces and interlayers, have been shown to promote the formation of peptide bonds, of sugars, nucleotides, and even polynucleotides, up to oligomers of 50 units, the lower limit of oligomerization believed to enable autocatalytic reproduction. For details on these scenarios, see [1].

$$2CO + N_2 + H_2 + discharge\ energy \rightarrow \rightarrow H_2N-CH_2-CO_2H \tag{2.1}$$

$$2CH_3SH + CO_2 + FeS \rightarrow CH_3-C(O)SCH_3 + H_2O + FeS_2 \tag{2.2}$$

The first pre-life or pseudo-life forms, a world based on ribonucleic acid (RNA) [6], may have arisen 3.8 Ga ago, followed by LUCA, the Last Uniform Common Ancestor 3.6 Ga ago. LUCA already employed the hydrolytically more stable deoxyribonucleic acid (DNA), the information storage pool common to all contemporary life forms. Nonetheless, RNA is still indispensable for information *transfer* in present life forms, and as a catalyst (so-called 'ribozymes') for certain biochemical processes. Many viruses use RNA instead of DNA for information storage; however, viruses are not considered life forms in a narrow sense, since they lack an independent metabolism.

There have been many LUCAs, but only one survived [6]. Starting from LUCA, the three phyla—bacteria, archaea, and eukarya—were formed (Fig. 2.1). Eukarya are uni- and multicellular organisms, the cells of which contain a differentiated nucleus, while cells of bacteria and archaea lack such a nucleus.

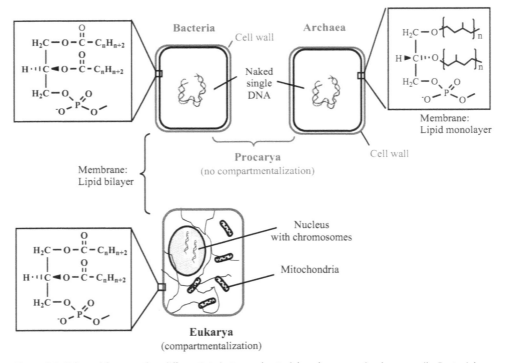

Figure 2.1 Selected features that differentiate between bacterial, archaean, and eukaryan cells. Bacterial and eukaryan cell membranes contain a lipid ester bilayer with glycerol in the D configuration, archaean cells a lipid ether monolayer with glycerol in the L configuration. Some eukaryan cells also contain cell walls.

Figure 2.2 Left: Dome-shaped stromatolites from the Shark Bay, Western Australia. From http://www. heritage.gov.au/ahpi. Right: Cross section of a fraction of a typical stromatolite of cyanobacterial origin, showing light calcareous layers along with dark, silica-enriched carbonaceous layers. Courtesy of Didier Descouens. This file is licensed under the Creative Commons Attribution-Share Alike 3.0 Unported license. *Reproduced in full colour in Plate 1*.

The time span encompassing 3.8 Ga (RNA world) to 2.5 Ga (first eukarya) is referred to as the Archaean Period. The oldest microfossils, found in stromatolites, date back to 3.45 Ga. Stromatolites are often dome-shaped, rocky constructions (Fig. 2.2) formed from remains of biofilms of microorganisms (commonly dominated by cyanobacteria), cemented together with sediments and calcareous minerals [7]. Recent findings suggest that the first protists (uni- and oligocellular eukarya), so-called acritarchs, linking to the eukarya of today, are as old as 3.2 Ga [8]. Further, there is a chance that life might have developed on our neighbour planet Mars even earlier, perhaps 3.9 Ga ago. The evidence for such a setting is provided by the Martian meteorite ALH84001, found in Alan Hills, Antarctica, in 1984. This meteorite, with an igneous crystallization age of 4.1 Ga, contains particularly pure magnetite ($Fe_3O_4$) crystallites embedded in carbonates, which are considered to be 'biomarkers' [9] (cf. also Section 4.2): Terrestrial magnetotactic bacteria—bacteria that orient along Earth's magnetic field lines—employ magnetite (and sometimes greigite, the sulfur analogue of $Fe_3O_4$), assembled in magnetosomes, for orientation in Earth's magnetic field; early Mars also had available a permanent magnetic field, still imprinted in today's Martian ferric rocks.

With the Great Oxygen Event 2.4 Ga ago, cyanobacteria started to provide oxygen, by photosynthesis (i.e. the oxidation of water, energetically driven by light; Chapter 11), in amounts that could no longer be completely coped with by oxidative conversion of inorganic and organic matter. The emergence of photosynthesis transformed life on our planet from anaerobic to essentially aerobic processing.

We can consider an entity to represent 'Life' if an organizational unit (a cellular system) fulfils the following criteria:

1. It exhibits a metabolism (production and transformation of chemical species) and arrangement of cellular components, also named autopoiesis.

2. There is feedback between the cellular system and the environment via a membrane. 'Environment' here refers to (i) the chemical status of the cellular surroundings (the

respective feedback is termed 'cognition'), and (ii) to cellular individuals of the same and of other provenience ('communication'), while 'membrane' relates to a protective, partially permeable wall defining the geometry of the cell.

3. There is information storage, commonly by DNA.
4. The cellular system can reproduce; reproduction encompasses complete information transfer.
5. Information transfer includes minor 'imperfections' (mutations within the DNA) to allow for adaptation to environmental changes, and thus to evolution. Evolution is additionally carried forth by *horizontal* gene transfer, i.e. by genetic alteration from the direct uptake of exogenous DNA provided by cellular organisms of different species and genera.

In order to sustain metabolism and reproduction, and to allow for evolution, prokarya are reliant on an energy source, an electron source (reduction equivalents) for respiration, and a carbon source for synthesizing organics. The energy source is usually light energy (photons) or chemical energy (derived from oxidation reactions); accordingly, one distinguishes between phototrophs, who generate energy from light, and chemotrophs, who generate energy from oxidation reactions. Examples for the latter are Eqs. (2.3)–(2.7).

$$Fe^{2+} \rightarrow Fe^{3+} + e^- \tag{2.3}$$

$$H_2 \rightarrow 2H^+ + 2e^- \tag{2.4}$$

$$HS^- + 4H_2O \rightarrow SO_4^{2-} + 9H^+ + 8e^- \tag{2.5}$$

$$Mn^{2+} + 3H_2O \rightarrow MnO(OH)_2 + 4H^+ + 2e^- \tag{2.6}$$

$$NH_3 + 3H_2O \rightarrow NO_3^- + 9H^+ + 8e^- \tag{2.7}$$

The reduction equivalents may be provided by organic sources (organotrophs) or inorganic sources (lithotrophs). Finally, the carbon source can be an organic molecule (heterotrophs) or carbon dioxide (autotrophs) [10]. Scheme 2.2 provides an overview. As an example,

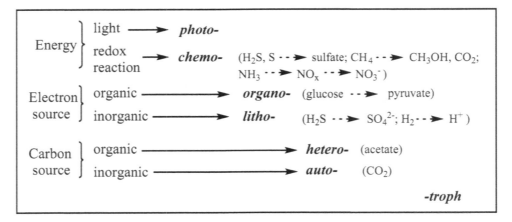

Scheme 2.2 Classification of prokarya according to energy, electron, and carbon sources employed in metabolic pathways. Examples for reactions/substrates are given in parentheses.

a chemo-litho-heterotrophic bacterium is a bacterium which resorts to a redox process (such as the oxidation of $HS^-$ to $SO_4^{2-}$; Eq. (2.5)) as an energy and electron source, and acetate as a carbon source.

## 2.2 Extremophiles

Most of the life forms we are familiar with thrive under moderate conditions, where 'moderate' refers to temperatures around 15 °C (with seasonal and geographical variations), a pressure range of $10^3$–$10^7$ Pa (10 mbar to 100 bar), an approximately neutral pH, salt concentrations not exceeding 3.5% (ca. 0.3 M), low background γ- and X-ray flux (around 2 mGy $a^{-1}$)[3], and the absence of appreciable amounts of toxic elements such as Hg, Cd, and As. Further, supply of nutrients and water is afforded. There are, however, many prokarya that thrive under extreme conditions. These so-called *extremophiles* can cope with a variety of drastic situations. Thermophiles may tolerate temperatures up to 120 °C. But there are also halophilic organisms adapted to permafrost, with temperatures as low as −15 °C; reproduction rates are, however, very low (around 50 days) in these frosty environments. Halophiles tolerate brines with salt concentrations up to 5 M ($NaCl/NaHCO_3$; $MgCl_2/NaCl/CaCl_2/KCl$ in the Dead Sea). Halophiles living in soda lakes are also alkaliphiles; they tolerate pH values up to 11. Highly acidic (down to pH ≈ 0) habitats of volcanic (*solfataras*) or industrial origin house acidophiles.[4] Further, there are baryophiles (also termed piezophiles) that have been found in rocky micro-niches almost devoid of any water and carbon supply, at depths down to 10 km.

As noted above, horizontal gene transfer from extremophilic archaea and bacteria can occasionally provide *eukarya* with the ability to thrive under extreme conditions. An example is the red alga *Galdieria sulphuraria*. This photosynthetic alga lives in hot (up to 56 °C), acidic (pH 0–4) environments rich in toxic metals such as cadmium, mercury, and arsenic [11].

In all of these cases, the organisms have developed special mechanisms to protect themselves against the 'hostile' conditions including—as in the case of baryophiles and microorganisms retrieved from subglacial brine pools—extremely low reproduction rates, covering several hundred or even thousand years. By studying these mechanisms, we can derive potential scenarios for life forms in extraterrestrial habitats. A bodacious event in this context is the existence of terrestrial bacterial populations at the interface between liquid $CO_2$-$CH_4$ and solid clathrates $CO_2 \subset (H_2O)_6$ in deep sea carbon dioxide lakes in the Okinawa Trough between Japan and Taiwan [12]. Also noteworthy in this regard is the presence of stromatolites in nitrate-rich lakes in the high Andes [13] with a very high UV flux.

With the appearance of free oxygen, $O_2$, about 2.4 Ga ago, dubbed the 'Great Oxygen Event' or 'Oxygen Crisis' (in view of the survival problems for hitherto exclusively anaerobic life forms), novel strategies for survival in an aerobic world had to be developed by those

---

[3] The unit Gy (Gray; 1 Gy=1 J/kg) measures the absorbed dose of radiation energy, while the unit Sv (Sievert, 1 Sv=1 J/kg) indicates the dose equivalent radiation for *tissue*. Absorbed dose (in Gy)×weighting factor $f$=equivalent dose to a tissue (in Sv). Depending on the nature of the radiation, $f$ can vary between 1 and ca. 100: $f$=1 ($e^-$, hν, muons, neutrons n with energies $E < 10$ keV/μm); $f$=2 (protons, $\pi^+$, $\pi^-$), 20 (α), 100 (n with $E$=100 keV/μm).

[4] For zinc-dependent hydrolases isolated from thermophiles and acidophiles, see Subsection 12.2.2.

organisms which did not retreat into anaerobic niches. Free atmospheric oxygen, and excess oxygen dissolved in the oceans, became available as soon as oxygen consuming systems, such as ferrous iron and decaying organics, could no longer chemically absorb all of the oxygen produced in the course of photosynthesis (Chapter 11). In the context of survival strategies, those metabolic processes developed by early terrestrial microorganisms *prior* to the Great Oxygen Event may have been helpful, by which 'hidden' oxygen was used to oxidize readily available organic substrates. An example is the oxidation of methane to methanol (and further to $CO_2$) with oxygen derived from nitrite [14], Eq. (2.8). The conversion of nitrite to $N_2$ resembles the common microbial denitrification process (Section 9.3), except that the intermediate $N_2O$ is omitted.

$$2NO_2^- \longrightarrow \longrightarrow 2NO \longrightarrow N_2 + O_2; O_2 + CH_4 \longrightarrow H_2O + CH_3OH(\longrightarrow \longrightarrow CO_2) \tag{2.8}$$

As the availability of atmospheric oxygen increased, oxidative precipitation of water-soluble components, as well as oxidative weathering of continental minerals, became a progressively prominent process. Examples include (i) the oxidation of ferrous to ferric iron, and (ii) the oxidative formation of molybdate(VI). Iron is an indispensable element for all living organisms. The conversion of soluble iron, present in water of pH 7 in the form of $[Fe(H_2O)_6]^{2+}$ and $[Fe(H_2O)_5OH]^+$, to insoluble ferric hydroxides ($Fe(OH)_3 \cdot nH_2O$) and ferric oxides afforded evolutionary adaption of the organisms in order to ascertain the supply of iron from (mineralized) ferric sources. Molybdenum is another element essential for most organisms. The solubilization of this element from molybdenum containing minerals by oxidative conversion of $Mo^{II-IV}$ to $Mo^{VI}$ delivered $MoO_4^{2-}$ into the oceans. Nowadays, molybdate, at an average concentration of ca. 100 nM, is the most abundant transition metal in sea water.

The availability of molybdate allowed nitrogen-fixing microorganisms to switch from vanadium and iron-only nitrogenases to the more efficient molybdenum nitrogenase, which appeared about 2 Ga ago [15]. Nitrogenases catalyse the reduction of inert $N_2$ to biologically utilizable $NH_4^+$. This process, referred to as nitrogen fixation (section 9.1), secures the supply of exploitable nitrogen, and thus thriving growth and reproduction.

 ## Summary

Our home planet accreted from the presolar nebula 4.56 Ga (1 Ga = $10^9$ years) ago. Pre-life forms began to develop with the start of the Archaean period (3.8–2.5 Ga), resorting to molecules available in the primordial broth, such as $H_2O$, HCN, CO, $CO_2$, $CH_3SH$, and phosphate, and to energy sources provided by spark discharges (Miller–Urey) or the oxidation of FeS (to $FeS_2$; Wächtershäuser). The first bacteria appeared 3.45 Ga ago, the first protists date back to 3.2 Ga. An excess of oxygen became available 2.4 Ga ago (the 'Great Oxygen Event'), allowing for the evolution of aerobic organisms. However, even prior to this process, some bacteria made use of 'hidden' oxygen ($NO_2^-$, NO) to oxidize organic substrates. The supply of $O_2$ also resulted in the oxidation of inorganic substrates. Prominent examples of the prime importance for an ongoing evolution are the conversion of soluble ferrous compounds to insoluble ferric minerals, and the oxidation of insoluble molybdenum compounds to soluble molybdate, $MoO_4^{2-}$, enabling evolution of the particularly effective molybdenum-containing nitrogenase enzyme.

Life relies on the availability of an energy source (photo- vs. chemotrophs), reduction equivalents (organo- vs. lithotrophs) and a carbon source (auto- vs. heterotrophs) and, commonly, moderate environmental conditions. Many prokarya (bacteria and archaea) have adapted, however, to extreme conditions, viz. temperatures as high as 120 °C, pH values up to 11 and down to 0, salt concentrations in aqueous habitats up to 5 M, high UV and $\gamma$ flux, high pressure, and the liquid $CO_2$–water interface.

## Suggested reading

**David LA and Alm EJ. Rapid evolutionary innovation during an archaean genetic expansion.** *Nature* **2011; 469: 93–96.**
This article addresses the genetic innovation within and after the Archaean eon (cf. Scheme 2.1), showing that genes involved in electron-transport and respiratory pathways developed during the Archaean expansion, while genes arising thereafter with the Great Oxygen Event are increasingly involved in processes using oxygen and redox-active transition metals.

**Mulkidjanian AY, Bychkov AYu, Dibrova DV, et al. Origin of first cells at terrestrial, anoxic geothermal fields.** *Proc. Natl. Acad. Sci. USA* **2012; 10.1073/pnas.1117774109.**
The authors suggest that the ion composition in modern cells (high potassium, manganese, zinc, and phosphate) reflects the ion composition of the habitats for primordial protocells.

**Rampelotto PH. Resistance of microorganisms to extreme environmental conditions and its contribution to astrobiology.** *Sustainability* **2010; 2: 1602–1623.**
Bacterial life thriving at (and even depending on) extreme terrestrial conditions is summarized, and the possibility of primitive extraterrestrial life and its save through-space travel to Earth is addressed.

## References

1. Rehder D. *Chemistry in space*. Weinheim: Wiley-VCH, 2010, ch.7.

2. Miller SL and Urey HC. Organic compound synthesis on the primitive Earth. *Science* 1959; 130: 245–251.

3. Wächtershäuser G. On the chemistry and evolution of the pioneer organism. *Chem. Biodivers.* 2007; 4: 584–602.

4. Yamaguchi M, Mori Y, Kozuba Y, et al. *J. Electron Microsc.* 2012; 61: 423–431.

5. (a) Fitz D, Reiner H, Rode BM. Chemical evolution towards the origin of life. *Pure Appl. Chem.* 2007; 79: 2101–2117; (b) Joshi PC, Aldersley MF, Delano JW, et al. Mechanism of montmorillonite catalysis in the formation of RNA oligomers. *J. Am. Chem. Soc.* 2009; 131: 13369–13374.

6. Plaxco KW and Gross M. *Astrobiology*, 2nd ed. Baltimore, MD: The Hopkins University Press, 2011.

7. Gottschalk G. *Discover the world of microbes*. Weinheim: Wiley-Blackwell (Wiley-VCH), 2012, chs.6–7.

8. Buick R. Ancient acritarchs. *Nature* 2010; 463: 885–886.

9. Thomas-Keprta KL, Clemett SJ, McKay DS, et al. Origins of the magnetic nanocrystals in Martian meteorite ALH84001. *Geochim. Cosmochim. Acta* 2009; 73: 6631–6677.

10. Berg IA, Kockelkorn D, Ramos-Vera WH, et al. Autotrophic carbon fixation in archaea. *Nature Rev. Microbiol.* 2010; 8: 447–460.

11. Schönknecht G, Chen W-H, Ternes CM, et al. Gene transfer from bacteria and archaea facilitated evolution of an extremophilic eukaryote. *Science* 2013; 339: 1207–1210.

12. Nealson K. Lakes of liquid $CO_2$ in the deep sea. *Proc. Natl. Acad. Sci. USA* 2006; 38: 13903–13904.

13. Newman DK. Feasting on minerals. *Science* 2010; 324: 793–794.

14. (a) Ettwig KF, Butler MK, Le Paslier D, et al. Nitrite-driven anaerobic methane oxidation by oxygenic bacteria. *Nature* 2010; 464: 543–548; (b) Oremland RS. NO connection with methane. *Nature* 2010; 464: 500–501.

15. Boyd ES, Anbar AD, Miller S, et al. A late methanogen origin of molybdenum-dependent nitrogenase. *Geobiology* 2011; 9: 221–231.

# 3 The alkaline and alkaline earth metals

Alkaline metal ions ($Na^+$, $K^+$) and alkaline earth metal ions ($Mg^{2+}$, $Ca^{2+}$), the most abundant metal ions in organisms, play a pivotal role in a plethora of physiological processes. Among these four ions, $Mg^{2+}$ adopts a special role due to its rather high charge density (the ratio of charge to ionic radius), allowing for the formation of stable, $Mg^{2+}$-based complexes, such as the chlorophylls, and for the activation of hydrolytic processes involving phosphoester bonds. $Ca^{2+}$ is distinct from the other ions because it forms sparingly soluble compounds with anions such as carbonate and phosphate. $Ca^{2+}$ thus attains functions not only in its ionic form (an example is triggering of the muscle contraction), but also in scaffolds (an example is calcium phosphate in bone structures).

The balance of the intra- and extracellular concentration levels of $Ca^{2+}$, $K^+$ (the dominant cytosolic cation) and $Na^+$ (the main extracellular cation), necessary for the proper functioning of a variety of regulatory processes on the physiological level, is typically guaranteed by ion pumps. These specific ion pumps are trans-membrane proteins forming channels which are gated by mechanical, electrical, or chemical stimuli, and powered—in the case of ion transport against a concentration gradient—by the hydrolysis of phosphoester bonds.

This chapter will thus address the physiological role of the four cations regarding their extrinsic functions as related to their intrinsic disparity. In this context, we will discuss the mechanisms that are responsible for steering and control of the ion concentrations in the extracellular space and the cytosolic cell compartments. Further, biomineralization of calcium will be scrutinized with special reference to its impact on the stabilization of bone structures through the formation of hydroxyapatite.

## 3.1 Overview

Physiologically relevant alkaline and alkaline earth metal ions are $Na^+$, $K^+$, $Mg^{2+}$, and $Ca^{2+}$. $Li^+$, when carefully applied in physiological doses, is of therapeutic interest in the treatment of mood disorder (depressive disorder and bipolar disorder) and hypertension, to be dealt with in some detail in Section 14.3.4. The chemical similarity between potassium and caesium accounts for their physiological indistinguishability, i.e. $Cs^+$ is taken up and distributed in the body along with $K^+$. In a similar manner, the organism does not distinguish between chemically closely related strontium and calcium. Given the low abundance of Cs and Sr, this lack of differentiation does not cause health problems. The situation changes, however, as radio-active isotopes are eventually released into the environment by atomic reactors. The main radio-toxins in this respect are $^{137}$Cs (half-life $t_{1/2}=30.2$ a, $\beta^-$ decay)[1] and $^{90}$Sr ($t_{1/2}=28.8$ a; $\beta^-$ decay). Further, as a consequence of the similar ionic radii $r$ of Li$^+$ ($r=76$ pm) and Mg$^{2+}$ ($r=72$ pm), there is no straightforward physiological distinction between these two ions. Li$^+$ can thus act as an antagonist of Mg$^{2+}$ and is therefore basically toxic.

Average amounts of the four essential alkaline and alkaline earth metals are (per 70 kg body mass): Na 105 g, K 140 g, Mg 35 g, and Ca 1050 g. The particularly high content of cal-cium reflects the fact that this metal is a main constituent of the inorganic part of the bone structure. The daily requirements are: Na 1.1–3.3, K 2.0–5.0, Mg 0.3–0.4, and Ca 0.8–1.2 g. These amounts are readily provided by a normal diet.

With the exception of magnesium, extra- and intracellular concentrations of the cati-ons are strikingly different, causing concentration gradients to be established. Many of the physiological actions of Na$^+$, K$^+$, and Ca$^{2+}$ depend on these concentration gradients. Control and maintenance of the appropriate concentration gradients between the cytosol and the extracellular space is of prime importance to ascertain the specific function of these ions, such as control of the osmotic pressure and cell membrane potentials, triggering and cut-ting off of signal transduction, and activation of enzymes. Table 3.1 summarizes intra- and extracellular concentrations of the four cations, along with the respective concentrations of the main anionic counterparts. The concentration of these ions in sea water, commonly considered the cradle of life, is included for comparison.

Note that the intracellular concentration of K$^+$ is about 40 times that of the extracellular K$^+$ concentration, while for Na$^+$, the settings are just the other way round.

The following survey summarizes some of the central functions of the alkaline and alkaline earth metal ions:

- Support in endo- and exo-skeletons: Ca (and Mg)
- Stabilization of cellular membranes by cross-linking membrane proteins and polysaccharides: Mg (and Ca)
- Regulation of the osmotic pressure: Na, K

Table 3.1 Mean concentrations (in mM) of selected cations and anions in the intra- and extracellular compartments (averaged for all human cell types) and in blood (erythrocytes and plasma). For comparison, the mean ion concentrations in sea water have been included.

| | K$^+$ | Na$^+$ | Ca$^{2+}$ | Mg$^{2+}$ | Cl$^-$ | HCO$_3^-$ | HPO$_4^{2-}$ | SO$_4^{2-}$ |
|---|---|---|---|---|---|---|---|---|
| Extracellular, all cell types | 4 | 142 | 2.5 | 0.9 | 120 | 27 | 1 | 10 |
| Intracellular, all cell types | 155 | 10 | 0.001 | 15 | 8 | 10 | 65 | 0.5 |
| Sea water | 10 | 460 | 10 | 52 | 550 | 30 | 0.002 | 29 |
| Erythrocytes | 92 | 11 | 0.0001 | 2.5 | 50 | 15 | 3 | – |
| Blood plasma | 4 | 140 | 1.2 | 0.8 | 100 | 25 | 1.2 | 0.3 |

[1] The half-life of a radioactive nucleus is the time it takes for half of a given number of nuclei to decay. The decay of a neutron into a proton, an electron ($\beta$ particle), and an antineutrino is referred to as a $\beta^-$ decay.

- Regulation of membrane potentials: Na, K
- Information transfer by migration along concentration and/or electrochemical gradients: Na, K, Mg, Ca
- Signal transduction, e.g. in neuro-transmission: Ca, K
- Activation and regulation of enzymes: Ca (and Mg, K)
- Co-activation of phosphate-dependent metabolic and catabolic paths: Mg
- Stabilization/activation of the hydrolysis of adenosine triphosphate: Mg
- Stabilization and/or functionalization of (enzyme) cofactors, e.g. Mg in chlorophyll, Ca in the oxygen evolving centre of photosystem II and in calmodulin.

For the physiological functions, the charge density $CD = q/r$ ($q$ is the ionic charge, $r$ the ionic radius) is of central importance:

| | $Li^+$ | $Na^+$ | $K^+$ | $Mg^{2+}$ | $Ca^{2+}$ | $Mn^{2+}$ |
|---|---|---|---|---|---|---|
| $r/Å^*$ | 0.76 | 1.02 | 1.38 | 0.72 | 1.00 | 0.83 |
| CD | 1.32 | 0.98 | 0.72 | 2.78 | 2.0 | 2.41 |

* For coordination number 6. 1 Å corresponds to 100 pm.

The larger the charge density, the higher is the ability of the ion to polarize a molecule.[2] The charge density of $Mg^{2+}$ is particularly high: Contrasting the alkaline metal ions, but in accordance with transition metal ions such as $Mn^{2+}$, $Mg^{2+}$—and to some extent $Ca^{2+}$ as well— also form stable complexes with N- and O-functional ligands. Examples are chlorophyll-a and -b with $Mg^{2+}$ in a porphinogenic ligand sphere (Chapter 11), and calmodulin with $Ca^{2+}$ in a coordination sphere of oxygen-functional ligands (Section 3.5).

Alkaline and alkaline earth metal ions are rather mobile and, typically, their complexes are characterized by small stability constants, meaning that they are weak. In aqueous media, and in the absence of other ligands, the ions are present in hydrated form, $[M(H_2O)_x]^{n+} \equiv 'M^{n+} \cdot aq'$, where $x \approx 10 \pm 2$ (cf. also Sidebar 3.1). Unlike aqua complexes of transition metal ions, the number of water molecules in the hydration sphere is hardly defined, and the interactions are weak, i.e. ion–dipole interactions rather than covalent bonds prevail. The rate constants for the exchange of water in the hydration sphere and surrounding water, i.e. for the equilibrium Eq. (3.1), are in the order of magnitude of $10^{-10}$–$10^{-7} s^{-1}$ for $Na^+$, $K^+$, and $Ca^{2+}$, and $10^{-7}$–$10^{-5} s^{-1}$ for $Mg^{2+}$, reflecting the stronger interaction between $Mg^{2+}$ and its hydration sphere (that is, stronger as compared to $Na^+$, $K^+$, and $Ca^{2+}$). Generalizing this issue, complexes of $Mg^{2+}$ are not only thermodynamically but also kinetically more stable than those of the other three ions.

$$[M(H_2O)_x]^{n+} + H_2O^* \rightleftharpoons [M(H_2O)_{x-1}(H_2O^*)]^{n+} + H_2O$$
$$(M = Na, K : n = 1; \ M = Mg, Ca : n = 2) \tag{3.1}$$

---

[2] The volume charge density (of an ion) is the charge per unit volume. The smaller the ion, the larger is its charge density. As a cation comes into electrostatic contact with a molecule, the molecule's electron shell becomes dislocated with respect to the undisturbed situation: it becomes polarized, i.e. a dipole moment is induced. The extent of polarization is a function of the charge density of the ion.

## Sidebar 3.1    Exchange kinetics

For any ligand L, the exchange between water in the coordination sphere of a metal ion $M^{n+}$, $[M(H_2O)_x]^{n+}$ (where $x$ indicates the quantity of water ligands) can be expressed by the equilibrium

$$[M(H_2O)_x]^{n+} + L \rightleftharpoons [M(H_2O)_{x-1}L]^{n+} + H_2O$$

assuming that L is any neutral monodentate ligand. The reaction rates for the reaction proceeding from left to right (forward reaction), $v_\rightarrow$, and for the reverse reaction, $v_\leftarrow$, are

$$v_\rightarrow = k_\rightarrow c([M(H_2O)_x]^{n+})c(L)$$

and

$$v_\leftarrow = k_\leftarrow c([M(H_2O)_{x-1}L]^{n+})c(H_2O)$$

where $c$ indicates concentration, and $k_\rightarrow$ and $k_\leftarrow$ are the rate constants for the forward and the reverse reaction, respectively. In the case of equilibrium, $v_\rightarrow = v_\leftarrow$, and $k_\leftarrow/k_\rightarrow = K$. $K$ is the equilibrium constant, connected to the Gibbs free energy $\Delta G$ by $\Delta G = -RT \ln K$ ($R$ is the gas constant, and $T$ the temperature).

In the special case of water exchange, i.e. $L = H_2O^*$, the above equilibrium and the forward reaction read

$$[M(H_2O)_x]^{n+} + H_2O^* \rightleftharpoons [M(H_2O)_{x-1}H_2O^*]^{n+} + H_2O$$

and

$$v_\rightarrow = k_\rightarrow c([M(H_2O)_x]^{n+})c(H_2O^*)$$

Incorporating the constant concentration of water into $k_\rightarrow$, viz. $k_\rightarrow c(H_2O^*) = k_1$, we arrive at a (pseudo) first-order law for the rate of water exchange:

$$v_\rightarrow = k_1 c([M(H_2O)_x]^{n+})$$

$k_1$ is related to the half-life $t_{1/2}$ (the time interval for half of one of the complexed $H_2O$ molecules to become exchanged for $H_2O^*$) by $t_{1/2} = \ln 2/k_1$. The half-life $t_{1/2}$ for $M = Mg^{2+}$ is typically around $10^{-6}$ s for $Na^+$, $K^+$, and $Ca^{2+}$ $10^{-8}$ to $10^{-9}$ s.

The maintenance of a 'correct' balance of ion concentrations in the intra- and extracellular space is pivotal for the operation of correct physiological functions to be maintained. To achieve this, efficient and specific transport facilities for cations and anions have to be made available. There are four main transportation routes for alkaline and alkaline earth metal ions: (1) ion exchange across a membrane (e.g. $Na^+$ for $H^+$); (2) passive, entropy-driven diffusion down a concentration gradient via gated or open (for $K^+$ only) channels; (3) active transport against a concentration gradient ('ion pumping'), coupled to the hydrolysis of adenosine triphosphate; and (4) trans-membrane transport with the help of vehicles, so-called ionophores, also referred to as hydrophobic transport.[3] The ion transport across ion channels

---

[3] Hydrophilic transport along ion channels will be covered in Section 3.2, while active and passive trans-membrane transport will be dealt with in sections 3.3 ($Na^+$, $K^+$) and 3.5 ($Ca^{2+}$).

will be detailed in Section 3.2; see also Fig. 3.1 in Section 3.2 for a summary. The task of ion channels can also be adopted by carrier proteins which, to some extent, function in the same way as ion channels, but are significantly shorter. The transport by carriers is comparatively slow, since it involves coordination of the ions to functional groups of the carrier molecule, migration of the carrier across the cell membrane, and finally detachment of the metal ions off the coordinating functions of the carrier.

## 3.2  Ion channels

Ion channels are trans-membrane proteins, the inner surface of which is lined with carboxylate (of Glu and Asp) or/and carbonyl groups of the peptide backbone, allowing for the transport of ions such as $Na^+$ and $K^+$. These ions are usually partly or completely deprived of their hydration shell while transported through weak coordinative interaction with the oxygen functionalities in the channel. The following types of ion channels are distinguished; (cf. also Fig. 3.1):

- Leak channels; only for $K^+$: they are always open
- Gated channels; these are channels with 'gates' (or locks) which are usually closed, but can be opened by a stimulus when required. Depending on the stimulus, there are mainly four different qualities of steerage:
  - Voltage gated: the channel opens and closes as a consequence of a change in the electrochemical membrane potential
  - Light-gated: illumination of channel rhodopsin induces a flow of cations across the membrane. Rhodopsin, or visual purple, is a pigment that acts as a photoreceptor.

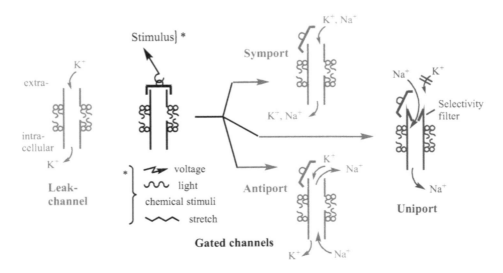

Figure 3.1 The main types of trans-membrane ion channels. See text for details.

- Ligand gated: a chemical stimulus communicates the option to close or open the channel. Examples for chemical stimuli are neurotransmitters (acetylcholine, dopa, glutamate, NO), toxins (nicotine), $Ca^{2+}$, and polyunsaturated fatty acids.
- Stretch gated: a mechanical stimulus, e.g. a physical change in the membrane as a consequence of strain, provoking the opening/closure mechanism.

Channels (or carrier proteins) which transport one specific ion only are termed uniporters. The synchronous transport of two or more types of ions is referred to as symport, if the direction of the transport is the same, and as antiport in the case where the ions are transported in opposite directions (into the cell vs. out of the cell). In many cases, however, ion channels are uniporters, guaranteed by a 'selectivity filter' at the channel entrance.

Ion channels are expressed in almost all cell types, where they govern a variety of operations, such as the control of the membrane potential, regulation of hormone secretion (e.g. insulin), regulation of salt and water transport in the epithelium of the kidneys, control of cell proliferation, regulation of the myocardial contraction, and the electrical excitability of neurons. Most of the voltage-gated $K^+$ channels consist of four α subunits arranged in such a way that a channel forms (Fig. 3.2a and b). In addition, there are four β subunits which can be associated either with the cytosol or the membrane. The β subunits modulate the activity of the channel pores. A typical amino acid sequence at the pore domain is Thr–Leu/Ile/Ala–Gly–Tyr–Gly–Asp. Intermittent binding of the ion (partly or completely deprived of its hydration shell) to the entrance pore and during transport across the channel is electrostatic, commonly achieved by the carbonyl oxygen functionalities stemming from the protein backbone, and negatively charged side-chain groups (carboxylate, alkoxide, phenolate) plus, eventually (and in particular in the case of $Na^+$), intervening $H_2O$.

The selectivity of ion channels for $K^+$ vs. $Na^+$ can be implemented by geometrical factors, as shown in Fig. 3.2, for a $K^+$ selective channel. The coordination cage of the selectivity

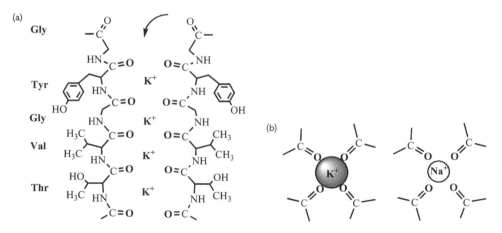

Figure 3.2 Selectivity filter (channel entry) of a $K^+$ selective ion channel. (a) Side view (only two of the four α subunits of the channel are shown) [2]. (b) Top view (only one of the two overlaid planes of four carbonyl functions is shown). In addition to backbone carbonyls, side-chain carboxylates [3] can come in as binders for the alkaline metal ion. The selectivity for $K^+$ vs. $Na^+$ is overestimated here, assuming rigidity of the entrance port. See text for details.

filter—two planes of four oxo functions each—provides an optimal fit for $K^+$, while the $Na^+$ cation is too small to fit snugly into the opening. This is, however, a somewhat simplified view [1], since it does not take into consideration the overall spectrum of interactions, which comprise, along with the attractive interactions between the ion and the functionalities of the selectivity filter, the repulsive interactions between the functionalities within the filter and between the ions, and the interactions between these functionalities and the ions, respectively, with the environment. In addition, the flexibility (conformational changes) of the protein residues in the entry gate has to be taken into account.

Geometrical factors are also responsible for the activation of potassium ion channels that are gated by $Ca^{2+}$. As intracellular $Ca^{2+}$ ions bind to the four subunits of the so-called gating ring, which is located on top of the $K^+$ channel in the cytoplasm, the gating ring opens like the petals of a flower, enabling the trans-membrane transport of $K^+$ [4].

Organometallic and inorganic systems have been developed which model, to some extent, the transport of alkaline and alkaline earth metal ions. Prominent examples are copper-based metal-organic polyhedra (MOPs) in lipid membranes (Fig. 3.3a) [5], and molybdenum- [6] or tungsten-based [7] polyoxidometalates (POMs) in aqueous solution (Fig. 3.3b). MOPs such as MOP-18, are assemblies consisting of 24 $Cu^{2+}$ linked by an equal number of a dodecoxyl benzoates.[4] They contain an intrinsic hydrophilic cavity (13.8 Å in diameter), accessible to ions such as $H^+$ and alkaline metals through triangular and square 'windows'. Embedded in lipid double layers, they transport ions across the lipid membrane.

POMs of the so-called Keplerate type are built up of polyhedral $Mo^{V/VI}O_n$ units linked by, for example, sulfate or acetate. They are about 30 Å in diameter, contain a hydrophilic cavity, and have 20 partially flexible pores with an inner aperture of 1.2 Å, allowing for the exchange of alkaline metal ions and $Ca^{2+}$.

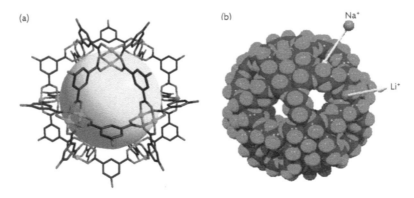

Figure 3.3 Organometallic/inorganic models for cation transporters. (a) MOP-18, viewed down one of the openings of threefold symmetry [5]. Reprinted from [5]; copyright © 2008 WILEY-VCH Verlag GmbH Co. KGaA, Weinheim. Part (a) reproduced with kind permission from Dr Kimoon Kim. (b) Space-filling representation of the polyoxidomolybdate $[Mo_{132}O_{372}(SO_4)_{30}]^{72-}$ [6]. The counter transport of $Na^+/Li^+$ across one of the 20 channels of threefold symmetry connecting to the (water-filled) cavity is shown. From [6]; copyright © 2006 WILEY-VCH Verlag GmbH & Co. KGaA, Weinheim. Part (b) kindly supplied by Dr Achim Müller. *Also reproduced in colour in Plate 2.*

---

[4] 'Dodecoxyl benzoate' is 1,3-dicarboxybenzene carrying, in position 5, the $OC_{12}H_{25}$ (dodecoxy) substituent.

## 3.3 Sodium and potassium

There is a roughly 15-fold excess of $Na^+$ over $K^+$ in the *extra*cellular space, $Na^+(ex)$ and $K^+(ex)$, while this ratio is reversed in the *intra*cellular medium, $Na^+(in)$ and $K^+(in)$ (Table 3.1). The maintenance of this imbalance is of prime importance for correct cellular function. Along with diffusion, and transport of the ions across the lipophilic membrane through ion channels lined with hydrophilic groups (as described in the preceding chapter), a very efficient transport mechanism is provided through active transport by means of $Na^+,K^+$-specific ion pumps [8]. These membrane-bound pumps are $Na^+,K^+$-ATPases $E$ ($E$, for enzyme), where ATP—adenosine triphosphate (Fig. 3.7 in Section 3.4)—is linked to two trans-membrane glycoprotein subunits, $\alpha$ and $\beta$. The larger $\alpha$ subunit is the transport protein, while the $\beta$-subunit provides the anchoring in the cell membrane. Binding is provided by, for example, side chain hydroxides (Thr, Ser) and carboxylates (Asp, Glu), and (less efficiently) main chain carbonyls [8].

The energy necessary for the transport of $Na^+(in) \rightarrow Na^+(ex)$ // $K^+(ex) \rightarrow K^+(in)$ *against* a concentration gradient is provided by the hydrolysis of magnesium-activated ATP, MgATP: $MgATP + H_2O \rightarrow MgADP + P_i$, where $P_i = HPO_4^{2-}/H_2PO_4^-$. The Gibbs free energy for this process is $\approx -35$ kJ mol$^{-1}$.[5] In the course of the ion transport, the $\alpha$ subunit switches between the conformations $E_1$ (which is sensitive for $Na^+$) and $E_2$ ($K^+$ sensitive) coupled to phosphorylation/dephosphorylation of $E$. For each MgATP becoming hydrolysed, two $K^+$ are translocated into the intracellular space, while three $Na^+$ are transported outside. The charge imbalance thus created is partially balanced by a $Na^+,Ca^{2+}$-ATPase (Section 3.5), and partially by passive transport. Details of the overall process are pictured in Scheme 3.1; the individual events in the process of potassium import and sodium export are summarized in Fig. 3.4.

Translocation of alkaline metal ions across cell membranes can also be carried out by ionophores. Ionophores, which are predominantly synthesized in bacteria and fungi,

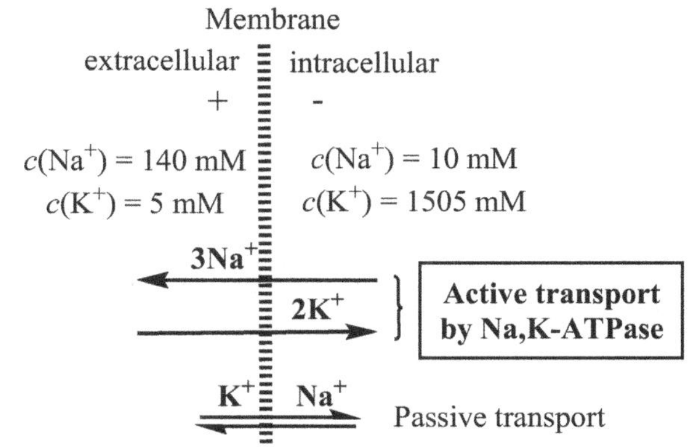

Scheme 3.1 A schematic view of the overall $Na^+/K^+$ trans-membrane transport. Details are shown in Figs. 3.4a–c.

[5] For details concerning MgATP, see also Section 3.4.

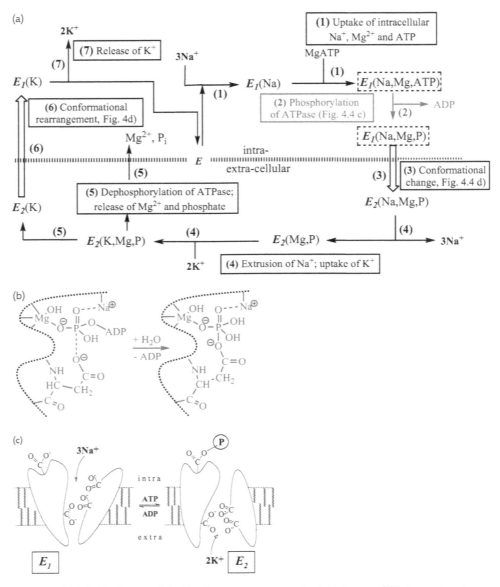

Figure 3.4 (a) Individual events of the Na+/K+ counter transport. The initiating step (1) is the uptake of Na+, Mg$^{2+}$ and HPO$_4^{2-}$ by $E_1$; $E$ is short for ATPase. Phosphorylation/dephosphorylation of the ATPase is highlighted in blue; for additional details see (b), where the pentavalent transition state and the direct linkage of phosphate to a carboxylate of an amino acid side-chain (Asp) are shown. (c) A symbolical depiction of the conformational change in the course of the trans-membrane transport of Na+ and K+.

are macrocyclic compounds with mostly exclusively O-functional groups for the specific coordination of Na+ or K+. The coordination by the multi-dentate ionophores is entropy-driven, comparable to the 'chelate effect' in transition metal coordination chemistry. The entropy ('disorder') increases as the hydrate shell of the alkaline metal ion is replaced by

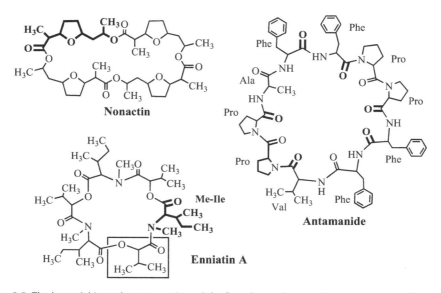

**Figure 3.5** The bacterial ionophore nonactin and the fungal ionophore enniatin-A coordinate $K^+$, while the fungal ionophore antamanide specifically takes up $Na^+$. Antamanide is a cyclic decapeptide, which coordinates $Na^+$ through six of the peptidic keto functions, highlighted in bold. Nonactin, with all together eight alternating ether and ester functions (one of the four subunits is highlighted) binds $K^+$ via the eight oxygen-functional groups in the 32-membered ring. Enniatin-A is a mixed ester-peptide, composed of four amino acids ($N$-methylisoleucine, Me-Ile; in bold) and four hydroxy acids ($\alpha$-hydroxyisovaleric acid; framed). For the coordination of $K^+$, the six carbonyl functions are employed. The hydrophobic periphery is provided by the methyl groups+the furan backbones (nonactin), the phenyl, methyl, and isopropyl substituents (antamanide), and the methyl, isopropyl, and isobutyl substituents (enniatin-A).

the multidentate ligand; Eq. (3.2). Examples for fungal and bacterial ionophores are shown in Fig. 3.5.

$$[K(H_2O)_n]^+ + L^- \rightarrow [K(L)] + nH_2O \quad (n \approx 10) \tag{3.2}$$

During transport, the metal ion is coordinated by the hydrophilic oxido functions and thus encapsulated by the ionophore. The ionophores form a more or less globular sphere, with the lipophilic periphery provided by alkyl and phenyl groups, which allows for the transport through the lipophilic membrane. Artificial macrocyclic compounds such as those shown in Fig. 3.6–crown ethers, cryptands, and calixarenes–are models for ionophores.

## 3.4 Magnesium

Magnesium plays a crucial role in phosphate (and hence energy) metabolism, in as far as it coordinates to the phosphate groups of adenosine triphosphate (ATP) and adenosine diphosphate (ADP). Depending on the mode of coordination, $Mg^{2+}$ can stabilize as well as destabilize ATP and ADP against hydrolysis. Stabilization of ATP is achieved by coordination

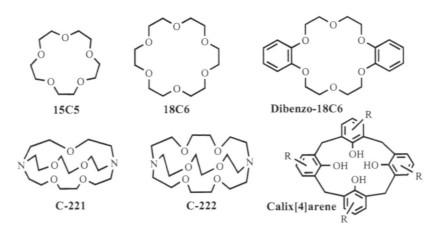

Figure 3.6 Model systems for ionophores suited for the coordination of Na$^+$ and K$^+$: crown ethers (top row), cryptands (bottom row, left and centre), and calixarenes (bottom right). 18C6 = 18-crown-6 (18-membered ring, 6 O-functions); C-221 = cryptand-221 (221 denotes the number of O-functions in the three bridges connecting the amine-nitrogens). The symbol [n] (here: n = 4) in the calixarenes indicates the number of phenolic units.

of Mg$^{2+}$ to the oxygens of the γ and β phosphates, or to all three phosphates [9]. In addition to the phosphate oxygen(s), Mg$^{2+}$ is coordinated to water/OH$^-$ (depending on the pH), and to carboxylate function(s) of Asp or Glu in the protein matrix. Hydrolytic removal of the terminal phosphate (the γ-phosphate) occurs as Mg$^{2+}$ moves to the P$_\gamma$, concomitantly directing a water molecule to this phosphate. An activated pentavalent transition state thus forms, from which phosphate H$_n$PO$_4^{(3-n)-}$ ($n = 1, 2$) is hydrolytically removed. This is illustrated in Fig. 3.7. The common abbreviation for inorganic phosphate used in a biochemical context is P$_i$; at pH 7.2, HPO$_4^{2-}$ and H$_2$PO$_4^-$ are present in equal amounts. The complex formation constant for Mg$^{2+}$ATP (Mg$^{2+}$ + ATP$^{3-}$ ⇌ [MgATP]$^-$) is about 10$^4$ M$^{-1}$ (corresponding to a dissociation constant of 0.1 mM); it is therefore a comparatively weak interaction.

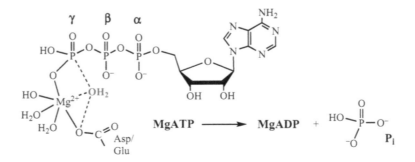

Figure 3.7 Activation of the hydrolysis of adenosine triphosphate (embedded in a protein matrix; not shown; for details, cf. [10]) by magnesium ions: Mg$^{2+}$ directs the water molecule for hydrolysis (in blue) to the terminal (γ) phosphate.

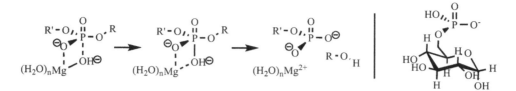

**Scheme 3.2** Mediation of the hydrolysis of a phosphoester bond by $Mg^{2+}$. Note the activated pentagonal–bipyramidal transition state (centre), formed intermittently in the course of the interaction of $[Mg(H_2O)_nOH]^+$ with the phosphoester. In the successive step of the reaction sequence, the ester bond is broken, alcohol ROH is released, and the tetrahedral configuration at the phosphorus centre is restored. R' is H or a second ester residue. An example for a phosphoester is glucose-6-phosphate, shown at the right.

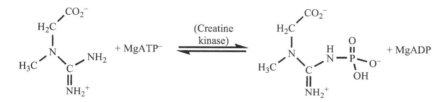

**Scheme 3.3** Phosphorylation/dephosphorylation of creatine.

In a similar manner, $Mg^{2+}$ also mediates the hydrolysis of phosphoester bonds in phosphorylated substrates, such as glucose-6-phosphate and DNA, as well as in phosphatases by minimizing the activation barrier for the formation of the trigonal–bipyramidal transition state, as illustrated in the reaction in Scheme 3.2.

The phosphate group generated by ATP hydrolysis can be transferred to suitable substrates which thus become activated. An example is the phosphorylation of glucose to glucose-6-phosphate (Scheme 3.2, right), the starting point of glucose degradation in the glucose cycle. The daily turnover of ATP at rest corresponds to about half the body weight. In cells with a high turnover rate for ATP, e.g. muscle cells in the course of sporting activity, phosphate is transferred to creatine, a process catalysed by creatine kinase (Scheme 3.3). Creatine phosphate serves as a source for rapid regeneration of ATP.

We explore the prominent role of $Mg^{2+}$ as a constituent of chlorophyll and thus in photosynthesis in Chapter 11.

## 3.5 Calcium

Sparingly soluble ionic calcium compounds, such as carbonates (aragonite, calcite, vaterite), phosphates (apatite), and fluorides (fluorapatite), can act as scaffold when incorporated into exo- and endo-skeletons. Our bones (and those of other vertebrates) are composite materials comprising about 50% collagen (a fibrous protein) and 50% hydroxyapatite $Ca_5(PO_4)_3(OH)_{1-x}F_x$ ($x \leq 0.01$). In dental enamel, about 95% of the overall material is apatite, with a somewhat larger content of fluoride ($x \approx 0.1\%$) than in bone. A person of 70 kg contains roughly 1.1 kg of calcium, only about 10 g of which is not confined to bony materials. These

10 g are used in a variety of functions, including the regulation of cell functions, muscle contraction/relaxation, blood clotting, enzyme regulations, and the gating of $K^+$ channels. Aerobic metabolism and cell survival greatly depend on mitochondrial $Ca^{2+}$ homeostasis. More generally, $Ca^{2+}$ can act as a second messenger, that is to say by activating, regulating and reinforcing signals. To some extent, $Ca^{2+}$ resembles $Zn^{2+}$: Calcium ions can take over the role of a cofactor in hydrolases; an example is the $Ca^{2+}/Zn^{2+}$ driven hydrolysis of the phosphodiester bond in nucleases. Like $Zn^{2+}$, $Ca^{2+}$ can also stabilize the tertiary structure of proteins (see below). Overall $Ca^{2+}$ concentrations are low: Cytosolic concentrations are typically in the order of magnitude of 1 μM, extracellular concentrations are larger by a factor of $10^3$ to $10^4$.

As in the case of $Na^+$ and $K^+$, the trans-membrane transport of $Ca^{2+}$ is accomplished by $Ca^{2+}$-specific ion channels [11], ATP-driven calcium ion pumps [12], and $Na^+/Ca^{2+}$ counter-transport. The latter is a transporter that extrudes intracellular $Ca^{2+}$ across the cell membrane against its concentration gradient by exploiting the downhill gradient of $Na^+$ [13].

Malfunction of calcium metabolism can result in the precipitation of sparingly soluble calcium deposits such as oxalate, phosphate, and steroids in the blood vessels (where they cause calcification responsible for cardiovascular diseases) and in secretory organs where these deposits are responsible for stones in gall, kidney, and bladder. Dysfunction of the calcification process in bones (apatite crystallization) results in osteomalacia (a softening of the bones) and osteoporosis.

Calcium ions have a prominent role in muscle contraction and relaxation. Muscle cells contain protein filaments, termed myofibrils, which are embedded within the sarcoplasmatic reticulum (SR); Fig. 3.8. The SR is an extended and interconnected intracellular network containing cisternal spaces, or vesicles *ves*, which store $Ca^{2+}$ at concentrations of 1–5 mM. The $Ca^{2+}$ storage is provided by calsequesterin, a protein of ca. 50 kDa containing a plethora of acidic residues (Asp and Glu) which bind up to 50 $Ca^{2+}$. Muscle contraction is achieved by release—following the concentration gradient—of $Ca^{2+}$ into the cytoplasm, *cyt*, across the membrane of the SR. The return transport from the cytosol into the vesicles of the SR, *against*

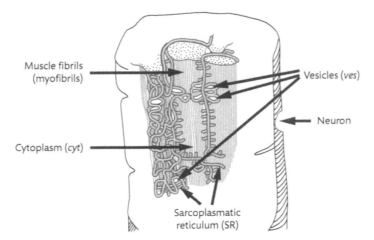

Figure 3.8 Components of a muscle cell involved in $Ca^{2+}$-driven muscle contraction and relaxation.

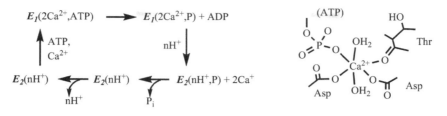

Scheme 3.4 Left: The cycling of $Ca^{2+}$ between the vesicles of the sarcoplasmatic reticulum (SR) and the cytosol. The transport path *against* the concentration gradient (i.e. from the cytosol into the vesicles of the SR, going along with *relaxation* of the muscle) is powered by ATP in the $E_1$ conformation of the $Ca^{2+}$-ATPase. Right: The $Ca^{2+}$ binding site in $E_1(Ca^{2+},ATP)$, adapted from [12].

the concentration gradient, is powered by an ATPase, coupled to the counter transport of, e.g. protons. As in the case of $K^+,Na^+$-ATPase, the enzyme switches between two conformations, $E_1$ and $E_2$ (Section 3.3). The overall process is summarized in Eq. (3.3), and details are depicted in Scheme 3.4, where $P_i$ again stands for inorganic phosphate. Scheme 3.4 also contains an illustration of the coordination environment of $Ca^{2+}$ bound to the $\gamma$-phosphate of ATP, water, a main chain carbonyl, and side-chain carboxylates of the protein.

$$Ca^{2+}(cyt) + ATP + E_1 \rightleftharpoons Ca^{2+}(ves) + ADP + P_i + E_2 \tag{3.3}$$

The activation of $Ca^{2+}$-dependent enzymes is promoted by proteins of the calmodulin family, where calmodulin stands for *cal*cium *modul*ating prot*ein*. Calmodulin is a small protein: 148 amino acids (including unusual trimethylammonium lysine) are arranged in four helix-loop structural domains, Fig. 3.9a, each of which can coordinate a calcium ion. $Ca^{2+}$ is in a seven- to eight-coordinate environment, dominated by acidic side chain residues of Asp and Glu [14]; Fig. 3.9b. Additional ligands are water, a main-chain carbonyl, and a hydroxide of a side-chain Ser, Thr, or Tyr.

$Ca^{2+}$ binding to calmodulin [15] results in a conformational change, accompanied by the exposure of hydrophobic residues, and particularly methyl residues from Met, Ile, Val, and Leu. This conformational change allows for coupling to an enzyme, the substrate of which becomes activated; Fig. 3.9c. Examples for enzymes activated by $Ca^{2+}$·calmodulin are adenylate cyclase, NAD kinases, NO-synthase (Section 9.4), and Ca-ATPases.

The coordination environment of $Ca^{2+}$ shown in Fig. 3.9b is also typical for other calcium containing proteins, including those where the function of $Ca^{2+}$ is restricted to a structural role. An example is thermolysin, a thermo-stable 34.6 kDa protein produced by the thermophile *Bacillus thermoproteolyticus*.[6] Thermolysin catalyses the hydrolysis of peptide bonds joining bulky hydrophobic amino acids. There is one $Zn^{2+}$ in the catalytic centre (Section 12.2.2), while four structural $Ca^{2+}$ ions ascertain a protein folding in such a way that the enzyme is protected against autoproteolysis.

As noted at the start of this chapter, calcium also plays a pivotal role in the structures of endo- and exo-skeletons. Shell fragments, egg shells, the spines of sea urchins, and corals are examples of structural supports provided by calcium carbonates $CaCO_3$ (trigonal calcite,

---

[6] Thermophiles are introduced in Chapter 2.

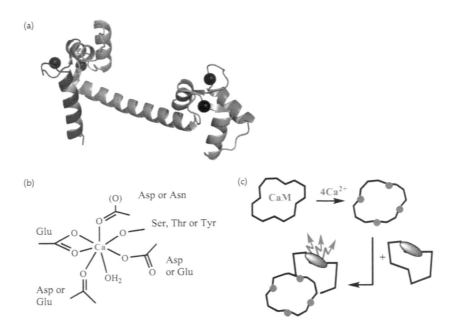

Figure 3.9 Calmodulin (CaM). (a) The four helix-loop domains are shown in light blue, dark blue, light grey, and dark grey. Black balls represent $Ca^{2+}$. Generated from PDB 1CLL by J. Crowe. (b) Coordination environment of the calcium ions. (c) Symbolic representation of the activation of an enzyme by CaM: Loading of CaM with $Ca^{2+}$ leads to a conformational change, triggering binding of $Ca^{2+}$·CaM to the enzyme. The enzyme's substrate (grey) thus becomes activated (blue arrows).

orthorhombic aragonite), while vaterite, a thermodynamically unstable hexagonal variant of $CaCO_3$, is found in pearls of the freshwater shell *Hyriopsis*, and is also generated by soil bacteria belonging to the genus *Myxococcus*. Calcification by coccolithophores (Fig. 3.10a)—micron-sized organisms belonging to the marine phytoplankton—is responsible for roughly one third of calcium carbonate found in the oceans today, and thus plays an important role in marine

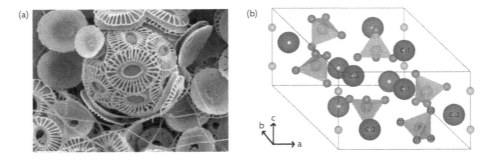

Figure 3.10 (a) The coccolithophore *Emiliania huxleyi*. Coccolithophores are phytoplankton with calcareous ($CaCO_3$) scales. Part (a) reproduced with kind permission from Dr Alex Poulton. (b) Crystal structure (unit cell) of apatite/fluorapatite $Ca_5(PO_4)_3(OH/F)$. Reprinted from [18], © 2013, with permission from Elsevier. Part (b) reproduced with kind permission from Dr Barbara Pavan. **Panel (b) also reproduced in full colour in Plate 3**.

$CO_2$ fixation—and, eventually, release of $CO_2$ as a consequence of increasing anthropogenic $CO_2$ emission/acidification of sea water. Cyanobacteria have been major contributors to calcification, produced via extracellular processes, on the geological timescale. A more recent discovery is the *intracellular* calcification, resulting in the formation of the mineral benstonite. Benstonite is an unusual mineral of composition $(Ba,Sr)_6Ca_6Mg(CO_3)_{13}$, i.e. this amorphous carbonate contains all of the four stable alkaline earth metals [16].

The apical vacuoles of the desmid *Closterium* contain tiny crystallites of calcium sulfate (gypsum, $CaSO_4 \cdot 2H_2O$). Calcium oxalate (in the form of the minerals whewellite $CaC_2O_4 \cdot H_2O$ and weddelite $CaC_2O_4 \cdot 2H_2O$) are examples of the biomineralization of calcium in plants.

$Ca_5(PO_4)_3(OH)_{1-x}F_x$, an apatite[7] with some of the hydroxide positions replaced by fluoride (Fig. 3.10b), is the main inorganic support in bone ($x \approx 0.01$) and dental enamel ($x \approx 0.1$). The teeth of sharks are almost 100% fluorapatite $Ca_5(PO_4)_3F$. Bone is a composite material, mainly made up of apatite and collagen. Collagen is a triple helix protein, which has high glycine and hydroxy-proline content. Collagen fibrils form the scaffold of the matrix for the axially oriented apatite crystals. The formation of calcium phosphate is regulated by proteins such as asporin, which bind to collagen. *Asporin* is a leucine-rich protein with a terminal poly-*asp*artate domain. Calcium ions bind to this domain and further to phosphate ions. Asporin thus enables the biomineralization of an initially amorphous calcium phosphate phase. The amorphous calcium phosphate becomes infiltrated in-between the collagen fibrils, from which nucleation and growth of apatite crystals proceed [17].

Regulation of calcium homeostasis (intake, distribution and discharge of $Ca^{2+}$) is highly dependent on vitamin D, in the form of its metabolic variant calcitriol, or $1,25\text{-}(OH)_2D$, i.e. the di-hydroxylated form of the vitamin. Calcitriol regulates the uptake of dietary $Ca^{2+}$ from the gastro-intestinal tract via $Ca^{2+}$ channels, its delivery into the blood stream by means of a $Ca^{2+}$-ATPase, and its re-absorption from kidney tissue. It further takes part in active phosphate absorption, and is thus intimately involved in the assembly and disassembly of apatite in the bone cells. Insufficient vitamin D supply thus causes demineralization and accordingly softening of the bones, eventually causing rickets and osteomalacia.

Understanding the specific strategies—the interactions and processes—that enable nature to form sophisticated materials via biomineralization on the micro- and nanometre scale is of prime interest in the context of the controlled manufacture of micro- and nano-scaled artificial materials [19], as is the employment of biocompatible nanoparticles for specifically designed purposes, including those in medicinal applications, such as the treatment of osteoporosis.

 ## Summary

The ions $Na^+$, $K^+$, $Mg^{2+}$, and $Ca^{2+}$ are of prime importance in the regulation of a plethora of physiological functions. Therapeutic doses of lithium are employed in a medicinal context. $Na^+$ and $K^+$ are mainly present as hydrates, while $Ca^{2+}$ and $Mg^{2+}$ readily coordinate to O-functional sites and, in the case of $Mg^{2+}$, N-functional sites. Calcium, in the form of insoluble carbonates, oxalates,

[7] The actual formula unit for apatite is $Ca_{10}(PO_4)_6(OH)_2$; here, we are employing $Ca_5(PO_4)_3OH$ for short.

and phosphates, also participates in load bearing. Examples are seashells ($CaCO_3$), *cactaceae* ($CaC_2O_4 \cdot nH_2O$), and bone ($Ca_5(PO_4)_3OH$).

In order to ascertain proper operation, high intracellular $K^+$ and low intracellular $Na^+$ as well as $Ca^{2+}$ concentrations are afforded. Regulation of the intra-/extracellular concentration is achieved by trans-membrane transport of the ions via diffusion, by gated and open (only in case of $K^+$) ion channels, and by active transport through MgATP-driven ion pumps, so-called ATPases (ATP = adenosine triphosphate).

Ion channels are membrane-bound proteins which are gated by chemical, mechanical, electric, or light stimuli. Negatively charged amino acid side-chain functions and protein backbone carbonyls are involved in the shuttle of the ions. Selectivity is implemented by geometrical factors. The free energy afforded for the ion transport by ion pumps is provided by the hydrolysis of $Mg^{2+}$-activated ATP to ADP and inorganic phosphate. Bacteria and fungi also employ cyclic oligopeptides (ionophores) as transport vehicles for $K^+$ and $Na^+$ across lipophilic membranes.

$Ca^{2+}$ plays a prominent role in muscle contraction and relaxation. Relaxation is coupled to $Ca^{2+}$ import, powered by an ATPase specific for $Ca^{2+}$, into the vesicles of the sarcoplasmatic reticulum. $Ca^{2+}$ bound to proteins such as calmodulin also mediates the activation of various enzymes, or simply acts as a structure-stabilizing factor as in the case of thermolysin, a $Zn^{2+}$-dependent hydrolase. In vertebrates, by far the main amount of calcium is locked within the structures of bones and teeth. Bone is a composite material consisting of about equal amounts of the protein collagen and crystalline apatite $Ca_5(PO_4)_3OH$, with a low percentage (<0.01%) of OH replaced by fluoride. Vitamin D attains a central role in the regulation of $Ca^{2+}$ levels.

## Suggested reading

**Kim I and Allen TW. On the selective ion binding hypothesis for potassium channels.** *Proc. Natl. Acad. Sci. USA* **2011; 108: 17963–17968.**
This overview provides a differentiated picture—including thermodynamic and kinetic factors—of the mechanism by which selectivity filters of a potassium channel select for $K^+$ over $Na^+$.

**Wopenka B and Pasteris JD. A mineralogical perspective on the apatite in bone.** *Mater. Sci. Eng. C* **2005; 25: 131–143.**
A mineralogical approach to bone that provides insight into the process of bone development and the body's ability to biochemically control the formation of hydroxyapatite.

## References

1. (a) Andersen OS. Perspectives on: ion selectivity. *J. Gen. Physiol.* 2011; 137: 393–395; (b) Kim I and Allen TW. On the selective ion binding hypothesis of potassium channels. *Proc. Natl. Acad. Sci. USA* 2011; 108: 17963–17968.

2. (a) Fowler PW, Tai K, Sansom MSP. The selectivity of $K^+$ ion channels: testing the hypotheses. *Biophys. J.* 2008; 95: 5062–5072; (b) Cao Y, Jin X, Hung H, et al. Crystal structure of a potassium ion transporter, TrkH. *Nature* 2011; 471: 336–341; (c) Valiyaveetil FI, Leonetti M, Muir TM, et al. Ion selectivity in a semisynthetic $K^+$ channel locked in the conductive conformation. *Science* 2006; 314: 1004–1007.

3. Payandeh J, Scheuer T, Zheng N, et al. The crystal structure of a voltage-gated sodium channel. *Nature* 2011; 475: 353–358.

4. Yuan P, Leonetti MD, Hsiung Y, et al. Open structure of the $Ca^{2+}$ gating ring in the high-conductance $Ca^{2+}$-activated $K^+$ channel. *Nature* 2012; 481: 94–97.

5. Jung M, Kim H, Baek K, et al. Synthetic ion channel based on metal–organic polyhedra. *Angew. Chem. Int. Ed.* 2008; 47: 5755–5757.

6. (a) Rehder D, Haupt ETK, Bögge H, et al. Countercation transport modeled by porous spherical molybdenum oxide based nanocapsules. *Chem. Asian J.* 2006; 1: 76–81;

(b) Rehder D, Haupt ETK, Müller A. Cellular cation transport studied by $^{6,7}$Li and $^{23}$Na NMR in a porous $Mo_{132}$ Keplerate type nanocapsule as model system. *Magn. Reson. Chem.* 2008; 46: 524–529.

7. Mitchell SG, Streb C, Miras HN, et al. Face-directed self-assembly of an electronically active Archimedean polyoxometalate architecture. *Nat. Chem.* 2010; 2: 308–312.

8. Shinoda T, Ogawa H, Cornelius F, et al. Crystal structure of the sodium–potassium pump at 2.4 Å resolution. *Nature* 2009; 459: 446–451.

9. Liao J-C, Sun S, Chandler D, et al. The conformational states of Mg·ATP in water. *Eur. Biophys. J.* 2004; 33: 29-37.

10. (a) Håkansson KO. The structure of Mg-ATPase nucleotide-binding domain at 1.6 Å resolution reveals a unique ATP-binding motif. *Acta Cryst. D* 2009; 65: 1181–1186; (b) Qian X, He Y, Luo Y. Binding of a second magnesium is required for ATPase activity of RadA from *Methanococcus voltae. Biochemistry* 2007; 46: 5855–5863.

11. De Stefani D, Raffaello A, Teardo E, et al. A forty-kilodalton protein of the inner membrane is the mitochondrial calcium uniporter. *Nature* 2011; 476: 336–340.

12. Picard M, Lund Jensen A-M, Sørensen TL-M, et al. $Ca^{2+}$ versus $Mg^{2+}$ coordination at the nucleotide-binding site of the sarcoplasmatic reticulum $Ca^{2+}$-ATPase. *J. Mol. Biol.* 2007; 368: 1-7.

13. Liao J, Li H, Zeng W, et al. Structural insight into the ion-exchange mechanism of the sodium/calcium exchanger. *Science* 2012; 335: 686–690.

14. (a) Babu YS, Bugg CE, Cook WJ. Structure of calmodulin refined at 2.2 Å. *J. Mol. Biol.* 1988; 204: 191–204; (b) Kovacs E, Harmat V, Tóth J, et al. Structure and mechanism of calmodulin binding to a signaling sphingolipid reveal new aspects of lipid–protein interaction. *J. Fed. Am. Soc. Exp. Biol.* 2010; 24: 3829–3839.

15. Tidow H, Poulsen LR, Andreeva A, et al. A bimolecular mechanism of calcium control in eukaryotes. *Nature* 2012; 491: 468–472.

16. Curadeau E, Benzerara K, Gérard E, et al. An early-branching microbialite cyanobacterium forms intracellular carbonates. *Science* 2012; 336: 459–462.

17. (a) Kalamajski S, Aspberg A, Lindblom K, et al. Asporin competes with decorin for collagen binding, binds calcium and promotes osteoblast collagen mineralization. *Biochem. J.* 2009; 423: 53–59; (b) Nudelman F, Pieters K, George A, et al. The role of collagen in bone apatite formation in the presence of hydroxyapatite nucleation inhibitors. *Nat. Mater.* 2010; **9**, 1004–1009.

18. Pavan B, Ceresoli D, Tecklenburg MMJ, et al. First principles NMR study of fluorapatite under pressure. *Solid State Nucl. Magn. Reson.* 2012; 45–46: 59–65.

19. Schwarz K and Epple M. Biomimetic crystallization of apatite in a porous matrix. *Chem. Eur. J.* 1998; 4: 1898–1903.

# 4 Iron: general features of its inorganic chemistry and biochemistry

Among the transition metals, iron is exceptional because of its ubiquity and abundance in all living organisms. The universal dependence of life on iron, a potentially toxic metal, and the extreme insolubility of $Fe^{3+}$ in inorganic resources, presents a significant problem regarding storage, maintenance, and availability of this metal for biological processes.

The biological importance of iron is linked to the ease by which it shuttles electrons between the ferric and ferrous state, and to switch readily between high- and low-spin electronic states, depending on the coordination environment (hard vs. soft ligands; coordination number and geometry).

In this chapter, we consider the highlights transport, delivery, and functioning of iron as connected to its coordination environment. In addition, biological 'mineralization' (storage) of iron will be dealt with. Oxygen transport and functional aspects related to the transmutation of oxygen species will primarily be addressed in Chapters 5, 6, and 7.

## 4.1 General and aqueous chemistry

Iron in the form of ferrous ($Fe^{II}$) and ferric ($Fe^{III}$) iron adopts an overwhelming role in life. Hardly any organism can sustain life, and metabolism in particular, without iron.[1] Iron deficiency, dysfunctions of iron metabolism, and iron overload cause severe health problems (cf. Section 14.2.1). Iron may already have taken a central role in the primordial development of what has been termed 'pioneer organisms' by Wächtershäuser [1]: pseudo-cellular constructs ('honeycombs') of sub-micron-sized iron- and iron–nickel sulfide minerals allowing for chemical reactions between basic constituents in the primordial broth, their stabilization and accumulation by adsorption to the 'cell walls' of the honeycomb compartments, and finally their organization into more complex life-molecules.

The main reactive gaseous constituents in the primordial atmosphere of our planet, about 4 billion years ago, were $CO_2$, $N_2$, $H_2O$, plus traces of CO, $CH_4$, $H_2$ and $O_2$, $H_2S$ and $CH_3SH$, among others. The main iron and iron–nickel sulfides available were pyrite ('fool's gold') $FeS_2$,

---

[1] In the lactic acid bacterium *Lactobacillus plantarum*, the role of iron in, e.g., coping with reactive oxygen species is taken over by manganese (see [2]).

troilite FeS, pyrrhotite $Fe_{1-x}S$, $x \approx 0-0.2$, and pentlandite $(Fe,Ni)_9S_8$, all containing ferrous iron. Pyrite is ferrous disulfide, i.e. the sulfur in the $S_2^{2-}$ anion is in the oxidation state −I. The oxidation of troilite (with sulfur in the oxidation state −II) to pyrite, Eq. (4.1), is an exergonic process, with a redox potential comparable to that of the $Zn/Zn^{2+}$ pair:

$$FeS + HS^- \rightarrow FeS_2 + H^+ + 2e^-; \ \Delta E = -620mV, \Delta G = -118kJmol^{-1} \tag{4.1}$$

The energy released here can be employed for redox-coupling between methyl sulfide and carbon dioxide to form the methyl ester of thioacetic acid $CH_3CO(SCH_3)$, i.e. 'active' acetic acid, Eq. (4.2). The electrons released according to Eq. (4.1) are consumed to reduce carbon in the educts of Eq. (4.2) (with carbon in the mean oxidation state 0) to the mean oxidation state− 2/3 per carbon in the product. Thioacetic acid methyl ester is a simple analogue of the more complex acetyl coenzyme A, acetyl-*CoA*, an energy-rich cofactor which plays a central role in metabolic pathways in living organisms. Alternatively, thioacetic acid methyl ester can also form in a non-redox C-C coupling reaction from methyl sulfide and carbon monoxide, catalysed by pentlandite, Eq. (4.3a). In the presence of aniline, the anilide of acetic acid is obtained, Eq. (4.3b), a compound containing an amide bond, and thus reminiscent of a peptide linkage.

$$2CH_3SH + CO_2 + FeS \longrightarrow CH_3-C\overset{O}{\underset{SCH_3}{}} + H_2O + FeS_2 \tag{4.2}$$

$$2CH_3SH + CO \xrightarrow{\{FeNiS\}} CH_3-C\overset{O}{\underset{SCH_3}{}} + H_2S \tag{4.3a}$$

$$\xrightarrow[C_6H_5NH_2]{} CH_3-C\overset{O}{\underset{\underset{H}{N}}{}}\bigcirc + CH_3SH \tag{4.3b}$$

The prodigious self-assembly of $Fe^{2+/3+}$ and sulfide to cuboid iron–sulfur clusters in the presence of an electron acceptor {EA} (to convert part of the indigenous ferrous to ferric iron) is another example of an inorganic process involving iron playing a pivotal role in life. A representative example of the assembly of such a cluster is provided by Eq. (4.4) where, in biological systems, SH is commonly replaced by cysteinate. A broad variety of iron–sulfur clusters is used in electron transport by practically all known extant life forms. Traces of oxygen present in the primordial atmosphere (stemming from the splitting of $H_2O$ by lightning, UV, γ rays, and solar wind protons) may have acted as electron acceptors.

$$4Fe^{2+} + 4S^{2-} + 4HS^- + \{EA\} \longrightarrow \left[\begin{array}{c} HS_{\diagdown}\ S-Fe^{\diagup SH} \\ Fe{\rightleftharpoons}S \ | \\ HS-|\ Fe-|\diagdown S \\ S-Fe_{\diagdown SH} \end{array}\right]^{2-} + \{EA^{2-}\} \tag{4.4}$$

The central role which iron takes in biological events is due, in part, to its general availability. Iron is abundant and ubiquitous in the geo-sphere; with an abundance of 4.7% by weight, it is the fourth most abundant element in the Earth's crust, only outnumbered by O, Si, and Al.

Cosmologically, iron is—next to O, C, and Ne—the most abundant 'metal' in the Universe (in cosmological phraseology, all elements heavier than helium are metals). In addition, along with its availability, iron exhibits properties which make it particularly valuable in biological processes:

(i)   The ease of change between the oxidation states +II and +III, and the disposability also of the oxidation states +IV and +V, allows for flexibility in redox processes, including those where an intermittently high oxidation state of the catalytic metal centre is required. In aqueous solution and at pH = 7, the redox potential, Eq. (4.5), is −230 mV. Under aerobic conditions, $O_2$ is commonly the oxidizing agent: the redox potential for the couple $(\frac{1}{2}O_2 + 2\,H^+)/H_2O$ at pH 7 is +820 mV. Thus, aerobic weathering of magmatic rock (which generally contains ferrous iron) ends up with ferric iron which, in contrast to $Fe^{2+}$, is essentially insoluble in a neutral aqueous medium, a fact which causes constraints on its availability.

$$Fe^{3+} + e^- \rightleftharpoons Fe^{2+} \tag{4.5}$$

(ii)  High flexibility with respect to the nature of ligand functions (hard and soft, i.e. O- and N-functional ligands on the one hand, and S-functional ligands on the other hand), coordination number (3, 4, 5, and 6) and coordination geometry allow for an unrestricted involvement of iron in a plethora of coordination environments and substrate targets.

(iii) Ease of change between different spin states (spin cross-over between high-spin and low-spin) in medium strength ligand fields, conditions which are fulfilled in (principally octahedral) coordination environments dominated by N-functional ligands. The switching between spin states enables the fine-tuning of changes in the local geometry of the catalytic centre and thus allows for catalytically conducted reactions to be switched on and off. See Sidebar 4.1 for ligand classification, stability constants, and ligand field splitting, and Sidebar 4.3 for magnetic behaviour.

Figure 4.1 provides a generalized overview of the main type of iron centres employed in reactions relying on iron. For details and an extension see Chapter 5. A versatile tool for the characterization of iron centres is Mössbauer spectroscopy, which is sensitive to the nature of the coordination environment, the oxidation state of iron, its spin state and magnetic environment; cf. Sidebar 4.2.

In very acidic aqueous media, ferric ions are present as hexaquairon(III) cations. These are Brønsted acids which gradually form hydroxido complexes, Eqs. (4.6a-c), and finally—in slightly acidic media—essentially insoluble ferric hydroxide $[Fe(H_2O)_3(OH)_3] \equiv Fe(OH)_3 \cdot aq$. The proteolysis reactions, Eqs. (4.6), are accompanied by condensation reactions, i.e. the formation of hydroxido- and oxido-bridged aggregates, clusters, and, eventually, colloids of a composition in between that of the minerals goethite $FeO(OH)$ and ferrihydrite $Fe_{10}O_{14}(OH)_2$, Scheme 4.1

$$[Fe(H_2O)_6]^{3+} + H_2O \rightleftharpoons [Fe(H_2O)_5OH]^{2+} + H_3O^+ \quad (pK_{a1} = 2.2) \tag{4.6a}$$

$$[Fe(H_2O)_5(OH)]^{2+} + H_2O \rightleftharpoons [Fe(H_2O)_4(OH)_2]^+ + H_3O^+ \quad (pK_{a2} = 3.5) \tag{4.6b}$$

$$[Fe(H_2O)_4(OH)_2]^+ + H_2O \rightleftharpoons [Fe(H_2O)_3(OH)_3] + H_3O^+ \quad (pK_{a3} = 6.0) \tag{4.6c}$$

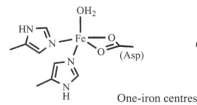

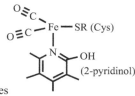

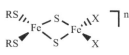

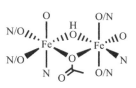

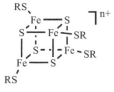

One-iron centres
(e.g. Rieske dioxygenase (left) and [Fe]-hydrogenase (right))

Haeme-type
(e.g. cytochromes, haemoglobin)

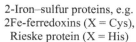

2-Iron–sulfur proteins, e.g.
2Fe-ferredoxins (X = Cys),
Rieske protein (X = His)

Two-iron centres
(e.g. ribonucleotide reductase)

4-Iron–sulfur proteins,
(e.g. 4Fe-ferredoxins)

**Figure 4.1** General classification of the main iron type centres in iron-dependent functional systems. Haem-type iron centres such as in cytochromes and haemoglobin are used in electron transfer and oxygen transfer, iron-sulfur proteins (see also Eq. (4.4)) in electron transfer, and oxido/hydroxido-bridged two-iron centres in reductases and hydrolyses.

---

## Sidebar 4.1   Coordination compounds: definition, stability, ligand classification

Coordination compounds, or complexes, are integral molecular or ionic units comprising a central metal ion (or, sometimes, atom), bonded to a defined number of ligands in a defined geometrical arrangement. The ligands can be simple or composite ions, or molecular (induced) dipoles. Commonly, each ligand provides a free electron pair, i.e. the ligands are Lewis bases, while the metal in the coordination centre is the Lewis acid. The bonding can thus be described in terms of Lewis acid/Lewis base interaction. Alternative notions for the bonding situation are: (i) donor–acceptor bond, and (ii) coordinative covalent bond, often denoted $L \rightarrow M$, where L = ligand (the donor) and M = metal (the acceptor), and $\rightarrow$ refers to the bonding electron pair. Complexes tend to be stable when the overall electron configuration at the metal, i.e. the sum of metal valence electrons plus electron pairs provided by the ligands, is 18 (electron configuration $ns^2np^6(n-1)d^{10}$; $n$, for transition metals, is 4, 5, or 6), or 16 for the late transition metals.

The stability of a complex, formed according to the equilibrium $M + nL \rightleftharpoons (ML_n)^q$ ($n$ = number of ligands, $q$ = charge of the complex) is quantified by $[(ML_n)^q]/[M][L]^n = K$. Square brackets denote equilibrium concentrations. $K$ is the *stability constant* or *complex formation constant* ($pK = -\log K$); its inverse, $K^{-1}$, is termed the *dissociation constant*.

In an undisturbed transition metal (ion), the $d$ orbitals are degenerate. This degeneracy is lifted in part or completely in the complexes $(ML_n)^q$. The extent by which the d-orbital set is split depends on the charge of the metal centre, the amount of ligands and their arrangement (local symmetry; see Sidebar 4.4), and the 'strength' of the ligands. Accordingly, ligands can be arranged in a series of increasing ligand strength:

Halides $\approx \{S\} < \{O\} < \{N\} < CN^- < NO^+ \approx CO$

where {S} etc. symbolize ligands coordinating through S etc. Examples of ligand field splitting by a weak ligand set (high-spin complexes) and a strong ligand set (low-spin complexes) in an octahedrally coordinated complex of $O_h$ symmetry are provided in Sidebar 4.3.

Another mode of classification has been introduced by Pearson. According to the Pearson classification, one distinguishes between soft and hard metal centres on the one hand, and soft and hard ligands on the other hand. Early and transient transition metals in high oxidation states, e.g. $Mo^{6+}$ and $Fe^{3+}$ are hard; they prefer hard ligands, such as oxygen-based donors. Soft metal centres, e.g. $Cu^+$, preferentially coordinate soft ligands, e.g. the thiolate function of cysteinate. There are many exceptions from this 'rule'.

Additional stabilization of a complex can also come about by multidentate ligands, a phenomenon which is referred to as chelate effect. An example is the particularly stable complex formed between ferric iron and the siderophore enterobactin ($ent^{6-}$, a hexadentate ligand; cf. Fig. 4.2):

$$[Fe(H_2O)_6]^{3+} + ent^{6-} \rightarrow [Fe(ent)]^{3-} + 6H_2O$$

The chelate effect is an entropic effect: here, an increase of the particle number (from 2 on the left hand to 7 on the right hand side).

## Sidebar 4.2    Mössbauer spectroscopy

Mössbauer spectroscopy is based on the recoil-free emission and absorption of γ-rays. The γ-rays are associated with transitions between different *nuclear* energy levels. A recoil-free situation is provided in a solid (crystalline) matrix. The experimental set up is sketched below:

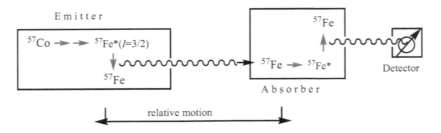

The γ-emitter is mounted on a slide, the relative movement of which with respect to the absorber (representing the sample under investigation) can be regulated—exploiting the Doppler effect (the change in frequency of an electromagnetic wave for a detector moving relative to the source of the wave)—so as to set up resonance with the absorber. The loss of γ-ray intensity by absorption is plotted as a function of the relative absorber velocity and termed isomer shift δ (mm s$^{-1}$), which largely depends on the chemical environment of the nucleus under consideration, just as the chemical shift in NMR spectroscopy.

In biological chemistry, the most common Mössbauer isotope is $^{57}$Fe which, in its stable ground state, has a nuclear spin of $I = \frac{1}{2}$. Absorption of a γ-quantum of the 'correct' energy (resonance case) leads to an exited state $^{57}$Fe* with $I = 3/2$ associated with a quadrupole moment. The emitter source employed in $^{57}$Fe Mössbauer spectroscopy is $^{57}$Co, which decays (half-life $t_{1/2} = 272$ days), via electron caption from the $K$ shell, to $^{57}$Fe*($I = 5/2$) and $^{57}$Fe*($I = 3/2$). The half-life for the decay of $^{57}$Fe*($I = 3/2$) is $1.4 \times 10^{-7}$ s, the energy of the γ radiation accompanying the transition $^{57}$Fe*($I = 3/2$) → $^{57}$Fe is 14.4 keV, corresponding to a wavelength of 86 pm.

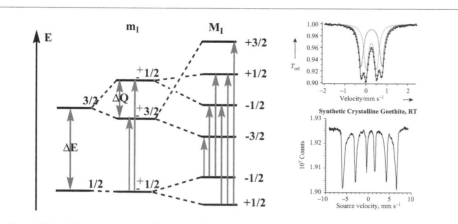

The splitting ΔE between the $I=1/2$ and $I=3/2$ states determines the degree of isomer shift. The $I=3/2$ state is subject to quadrupole splitting ΔQ, giving rise to a quadrupole doublet ($m_I = 1/2$ and 3/2). ΔQ (mm s$^{-1}$) is sensitive to the oxidation number, the spin state (high-spin vs. low-spin) and the molecular symmetry. The spectrum at the upper right shows two overlapping quadrupole doublets (Copyright © 2008 WILEY-VCH Verlag GmbH & Co. KGaA, Weinheim; *Angew. Chem. Int. Ed.* 2008; 47: 9537) for the two different iron sites in a model for a Rieske protein (cf. Fig. 4.1). Each of the doublets represents the two transitions indicated in the $m_I$ part in the centre of the energy diagram. In a magnetic field, such as produced by a magnetic material (an example is magnetite $Fe_3O_4$), additional splitting (Zeeman splitting) into six energy sublevels occurs. With the selection rule ΔM = ±1, six transitions are allowed, leading to a six-line system. The spectrum to the lower right shows the Zeeman sextet for goethite FeO(OH). The six lines reflect, from left to right with increasing energy, the six transitions shown in the $M_I$ part of the energy diagram. See also [3].

## 4.2 Mobilization, transport, delivery, and mineralization of iron

A general view representing some of the aspects of the iron cycle is depicted in Scheme 4.2. As noted, ferric iron in aqueous media, and in the absence of ligating systems, is hardly available and its uptake is limited due to the low solubility of $Fe(OH)_3$ above pH ≈ 6. From the solubility product $K_{sp}$, Eq. (4.7), we find that $Fe^{3+}$ has a solubility at pH 7 ($[OH^-] = 10^{-7}$ mol L$^{-1}$) of $s = [Fe^{3+}] \approx 10^{-18}$ mol L$^{-1}$, corresponding to $s \times N_A \times 10^{-3} = 600$ $Fe^{3+}$ ions per cubic centimetre; $N_A$ is the Avogadro constant. This clearly demonstrates the acutely low availability of iron from deposits of ferric hydroxide and ferric oxide hydrates.

$$K_{sp} = \left[Fe^{3+}\right]\left[OH^-\right]^3 = 10^{-39} \, mol^4 \, L^{-4} \tag{4.7}$$

Bacteria, fungi, aquatic plants, and the root system of many non-aquatic plants can mobilize $Fe^{3+}$ by secreting siderophores. These 'iron carriers' are multifunctional chelating ligands which very efficiently coordinate and thus mobilize $Fe^{3+}$ from its insoluble deposits. Stability constants $K$ of the resulting stable complexes such as that given in Eq. (4.8) (which describes complex formation), may go up to $10^{50}$. A selection of siderophores is shown in Fig. 4.2. The

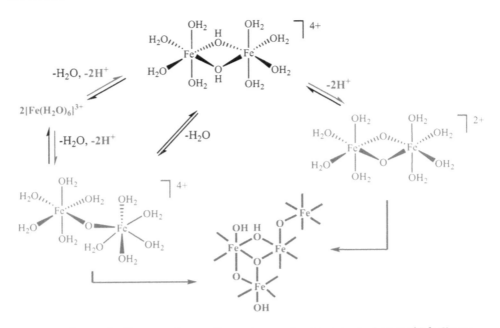

Scheme 4.1 Protonation/deprotonation reactions and aggregation in aqueous solutions of $Fe^{3+}$. Shown are dimers (those resulting from deprotonation in blue) and, in grey, a section of a colloid with the basic FeO(OH) structural unit also present in the minerals goethite α-FeO(OH), lepidokrokite γ-FeO(OH) and ferrihydrite $Fe_{10}O_{14}(OH)_2$, commonly recorded as $5Fe_2O_3·9H_2O$ or $FeO(OH)·0.4H_2O$.

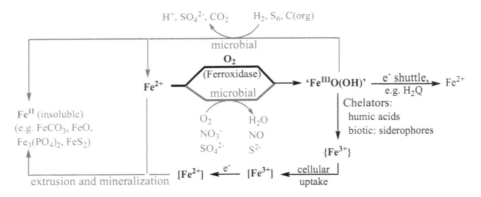

Scheme 4.2 Selected features of the iron cycle: Under oxic conditions, iron is present in the form of insoluble ferric oxides/hydroxides (right). Reductants such as hydroquinone ($H_2Q$) or those shown in sidebar 9.2 can mobilize iron by reducing ferric to ferrous iron. Alternatively, chelators can pick up $Fe^{3+}$ to form water-soluble complexes, {$Fe^{3+}$}, which are transported through an aqueous medium and internalized by cells, [$Fe^{3+}$], where reduction occurs, [$Fe^{2+}$]. Alternatively (upper part), microbial reduction and mobilization can take place, reverting to reductants such as hydrogen, p-sulfur or diverse organics. Once $Fe^{2+}$ is formed (central part), re-oxidation may be effected in an aerobic environment in vitro or in vivo (eventually catalysed by Fe- or Cu-based ferroxidases). In an anoxic ambience, mineralization of ferrous iron can occur, forming minerals such as siderite $FeCO_3$, wüstite FeO, vivianite $Fe_3(PO_4)_2$, pyrite $FeS_2$, and others. Microbially-induced reactions are high-lighted in blue, mineralization processes in grey.

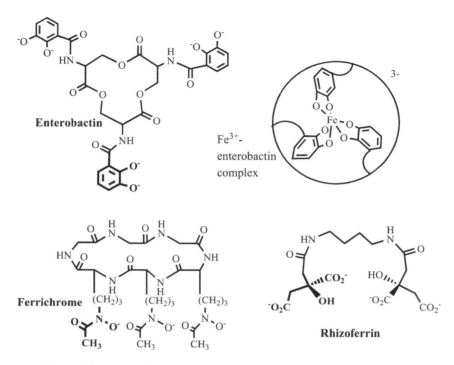

Figure 4.2 A bacterial (enterobactin) and two fungal siderophores. In each of the siderophores, one of the functions coordinating to $Fe^{3+}$ is highlighted in bold.

functional groups can be catecholate moieties, as in the case of enterobactin, hydroxamates (ferrioxamines and ferrichromes), or hydroxycarboxylates (rhizoferrin). The $Fe^{3+}$–siderophore complexes are globular and contain hydrophilic groups in their periphery, promoting their solubility in water. The complexed $Fe^{3+}$ is transported to the cells, the complex taken up by a receptor in the cell wall and translocated into the cytosol, often via endocytosis.[2] Release of iron within the cell occurs either by degradation of the siderophore, or by reductive release of ferrous iron from the siderophore.

$$K = [FeL]^{3-}/[Fe^{3+}][L^{6-}] \approx 10^{50}\,M^{-1} \text{(for L = enterobactin; Fig. 4.2)} \tag{4.8}$$

For the inter-cellular transport of iron, including the transport of $Fe^{3+}$ in the blood plasma of vertebrates, transferrins[3] are employed. Human apotransferrin, $hTfH_2$, (where 'apo' relates to metal-free), is a glycoprotein (carbohydrate fraction 6%) of 80 kDa molecular mass (1 Da = 1 g mol$^{-1}$), which can take up—as shown in Fig. 4.3a—two $Fe^{3+}$, one each in the terminal $N$ and $C$ lobes of the protein. The coordination centres comprise two tyrosines, one

[2] Endocytosis is a transport process for (larger) molecules across the cell membrane: the molecule to be transported is taken up by a receptor at the outer cell membrane, engulfed by the receptor, transported to the inner membrane, and released into the intracellular medium.

[3] For the possible role of transferrin in diseases such as Parkinson's and Alzheimer's, see Section 14.2.1.

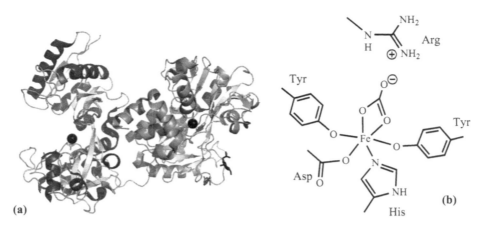

Figure 4.3 Serum transferrin Tf. **(a)** The subunits (lobes) of the two domains of Tf are highlighted in blue and grey. Ferric ions are indicated by black balls. The protein is composed of $\alpha$-helices and $\beta$-sheets. Generated from PDB 3QYT by J. Crowe. **(b)** Coordination environment of the $Fe^{3+}$ ion incorporated into transferrin and additionally coordinated to a synergistic carbonate.

aspartate, and one histidine. In addition, carbonate is taken up as a so-called 'synergistic' ligand as shown in Eq. (4.9), providing an octahedral coordination sphere. Further stabilization comes about by ion-pair contacts to a neighbouring arginine, as shown in Fig. 4.3b; such an interaction is also termed a 'salt interaction'.

$$hTfH_2 + Fe^{3+} + HCO_3^- \rightarrow \left[ (hTf)Fe^{III}(CO_3) \right]^- + 3H^+ \tag{4.9}$$

The stability constant of the complex formed between transferrin and $Fe^{3+}$ at pH 7.4—the pH of blood—is $10^{20.2}$. The iron-loaded transferrin delivers iron to sites of potential use, e.g. incorporation into protoporphyrin IX, a precursor of the haem group of haemoglobin, and frataxins, small iron proteins involved in the biosynthesis of iron–sulfur clusters. Excess iron is stored in iron storage proteins called ferritins; see below. The delivery of iron requires its reduction from the ferric to the ferrous state, Eq. (4.10); the stability constant for the $Fe^{2+}$ transferrin complex goes down to $10^{3.2}$ and thus enables release of ferrous iron into the cytosol. Such reduction can be effectively mediated by an agent such as ascorbic acid (vitamin C).

$$\left[ (hTf)Fe^{III}(CO_3) \right]^- + e^- + 3H^+ \rightarrow hTfH_2 + HCO_3^- + Fe^{2+} \tag{4.10}$$

Iron, when taken up with the food and processed in the mouth (chewing, admixture of saliva), is mostly present in its ferric form. After passing through the gastro-intestinal tract and arriving in the reducing environment of the small intestines, ferric iron is reduced to ferrous iron, a soluble form that allows for its uptake through the intestinal wall. The re-conversion to transportable $Fe^{3+}$, Eq. (4.11), is generally catalysed by Cu-based ferroxidases, {Cu} in Eq. (4.11), such as hephaestin, an oxidoreductase homologous to the copper transporter and ferroxidase ceruloplasmin (Section 14.2).[4] More generally, there are close links between Fe

---

[4] For a more detailed account on (the regulation of) iron intake, see Fig. 14.2 in Section 14.2.

and Cu homeostasis [4], where homeostasis refers to the internal regulation of a system with respect to parameters such as (metal ion) concentration.

$$2Fe^{2+} + \tfrac{1}{2}O_2 + 2H^+ \rightarrow 2Fe^{3+} + H_2O \tag{4.11}$$

Any iron that is not immediately used is stored in ferritins. The immediate embedding into functional units such as haem, mitochondrial frataxins, and iron–sulfur clusters, or the storage in ferritins, is imperative, since free $Fe^{2+/3+}$ is toxic due to its potential to produce reactive oxygen radicals as per the Fenton reaction shown in Eq. (4.12). For more details on oxygen and reactive oxygen species (ROS), see Section 5.1.

$$Fe^{2+} + H_2O_2 \rightarrow Fe^{3+} + \cdot OH + OH^- \tag{4.12a}$$

$$Fe^{3+} + H_2O_2 \rightarrow Fe^{2+} + \cdot OOH + H^+ \tag{4.12b}$$

The iron storage proteins, present in all cell types, consist of a hollow protein sphere, the apoferritin. Human apoferritin is a 450 kDa protein, made up of 24 subunits each of around 170 amino acids, which are arranged in *234* symmetry [5] (Fig. 4.4). The outer diameter is 130 Å, the inner diameter 75 Å. The inner surface of this protein capsule is lined with carboxylate functions (Glu and Asp) for the coordination of the first incoming $Fe^{3+}$ ions. Up to 4500 $Fe^{3+}$ can be incorporated, inter-connected by bridging oxido and hydroxido groups, very much as in the mineral goethite FeO(OH) or ferrihydrite $FeO(OH)\cdot 0.4H_2O$; see also Scheme 4.1. The overall composition of the iron nucleus comes close to $8FeO(OH)\cdot FeO(H_2PO_4)$, i.e. some phosphate is also built in.

Pores and channels of threefold symmetry in the protein envelope, furnished with Asp and Glu residues, allow for an exchange of iron ions between the interior and exterior. For the primary uptake, iron has to be in the oxidation state +II. As noted above, this is achieved by reductive elimination of iron from its transporter transferrin. Uptake through the channels and assembly into the core structure of ferritin is accompanied by re-oxidation to +III.

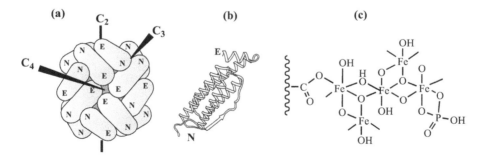

Figure 4.4 The iron storage protein apoferritin. (a) The 24 subunits and the fourfold, threefold, and twofold symmetry axes are shown. The threefold axes serve as ion transport channels. (b) Subunit of ferritin; N and E refer to the N-terminal lobe and the α helical moiety, respectively, of each subunit. (c) Schematic view of a section from the iron core. The phosphate can further link to adenosine.

The oxidant is oxygen, Eq. (4.13). The oxidation proceeds via a dinuclear, peroxido-bridged intermediate. Interestingly, the effective magnetic moment per $Fe^{3+}$ centre in ferritin is only 3.85 Bohr magnetons, and hence clearly less than the 5.92 expected for high-spin ferric iron with five unpaired electrons per iron. This difference points to cooperativity between the iron centres via super-exchange (partial antiferromagnetic coupling) in the highly ordered, pseudo-crystalline ferritin core; cf. Sidebar 4.3 on magnetism.

$$2\,Fe^{2+} + O_2 \longrightarrow Fe^{3+}\!\!\overset{O^{\ominus}}{\underset{\ominus O}{\diagdown\!\diagup}}\!Fe^{3+} \xrightarrow{+\,3H_2O} 2\,Fe\overset{O}{\underset{OH}{\diagup}} + 1/2\,O_2 + 4\,H^+ \tag{4.13}$$

Haemosiderins, sometimes discussed in the older literature as a second form of iron storage proteins, essentially are denatured ferritins. Along with the common iron storage ferritins present in all eukaryotic cells, there is a second ferritin family, termed *Dps*-proteins or *Dps*-ferritins. The *Dps*-proteins[5] were originally discovered in *Escherichia coli* (but are also present in eukarya and archaea) as DNA binding proteins from starved cells: they protect DNA against oxidative stress (the generation of reactive oxygen species as shown in Eq. (4.12)) in cells which are insufficiently supplied with nutrients. They are distinct from the genuine eukaryan ferritins in that they are dodecamers arranged into a hollow sphere of 23 symmetry [6], and bind ferrous iron at ferroxidase sites, Eq. (4.14), prior to oxidation and mineralization. The storage capacity of Dps proteins is ca. 500 $Fe^{3+}$, arranged in ferrihydrite particles with super-paramagnetic behaviour [7].

Human *frataxins* are comparatively small proteins of around 14 kDa (for the apoprotein), which can take up six to seven $Fe^{2+}$. The dissociation constants are in the micromolar range, indicating that frataxins can take over the role of chaperones, mediating the transfer of iron to the nucleation sites for [2Fe–2S] clusters [8]. Binding sites for iron are the dangling carboxylate groups of Glu and Asp, and histidine-N$\epsilon$ (Fig. 4.5). A genetically inherited lack of frataxins, and concomitant undersupply of {FeS} clusters, is known as Friedreich ataxia, and leads to

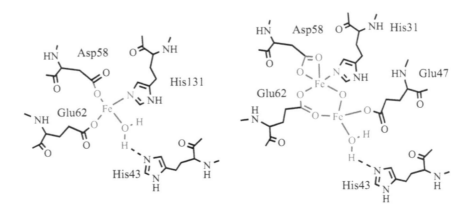

Figure 4.5 Examples for mono- and di-iron ferroxidase sites in bacterial Dps proteins (modified from [8b]).

[5] Also referred to as Dpr proteins for Dps-like peroxide resistance.

the progressive loss of muscle coordination. Frataxin subunits can aggregate to multimeric assemblies which store $Fe^{3+}$ very much like ferritins and Dps proteins. Unlike ferritins and Dps proteins, however, they disassemble as $Fe^{3+}$ is reduced to $Fe^{2+}$.

$$\{Fe_2\}^{2+} + H_2O_2 + H_2O \rightarrow \{Fe_2O_2(OH)\}^- + 3H^+ \tag{4.14}$$

As noted, the assembly of the $\{FeO(OH)\}_n$ core in ferritins resembles that in the mineral ferrihydrite. Most—if not all—of the terrestrial nanocrystalline ferrihydrite may originate from organic sources, and hence will have been produced by *biomineralization*, a term which is employed wherever a mineral is produced in the context of biological activity. Biomineralization is considered to be either biologically *controlled* when it occurs within a cellular framework (as in the case of the formation of ferrihydrite in the cavity of apoferritin), or biologically *induced* when generated from the interaction between (micro)organisms and their surroundings [9].

A second prominent example of biologically controlled biomineralization is the formation of the mixed-valent magnetic materials magnetite $Fe_3O_4$ ($Fe^{II}Fe^{III}_2O_4$) and greigite $Fe_3S_4$ ($Fe^{II}Fe^{III}_2S_4$) by magnetotactic bacteria [10] (Fig. 4.6) and other organisms, including fruit flies, bees, homing pigeons, robins, salamanders, sea turtles, and fish, where it is employed for orientation in the weak, ca. $50\,\mu T$ magnetic field of our planet. While magnetosomes are commonly on a scale of a few tens of nanometres, giant magnetites (Fig. 4.6c), up to a length of $4\,\mu m$, had been produced by magnetotactic bacteria in suboxic zones of aquatic environments with high iron availability in clay-rich sediments deposited about 56 Ma ago [11].

Particularly pure magnetite crystallites are usually considered 'biomarkers', i.e. they point towards a biological origin. Interestingly, ultra-pure magnetite crystallites have also been found in the Martian meteorite ALH84001 [12], picked up in Alan Hills (Antarctica) in 1984. ALH84001 was released from the surface of Mars by an impact event about 15 million years ago, and captured by Earth 13,000 years ago. The rocky matrix of the meteorite would have formed ca. 4.1 billion years ago, while the carbonaceous material embedding the Martian

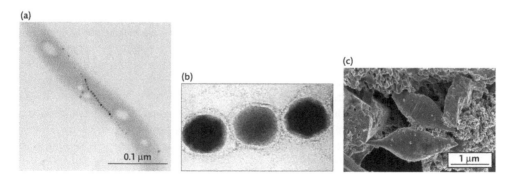

(a)

(b)

(c)

0.1 μm

1 μm

Figure 4.6  (a) A magnetotactic bacterium with magnetosomes (magnetite crystallites in a membrane) aligned to a chain. (b) Magnetite particles enveloped by the magnetosome membrane (c) Giant spearhead magnetites found in clay-rich sediments; cf. text [11]. (a) and (b): Reprinted with permission from Faivre D and Schüler D. *Chem. Rev.* 2008; 108: 4875–4898. Copyright (2013) American Chemical Society. (c) Reprinted with permission from [11]; copyright (2013) National Academy of Sciences, USA. Part (c) kindly supplied by Dr Dirk Schumann.

magnetite would have been 3.9 billion years old, making the Martian magnetite 0.4 billion years older than the oldest fossils on Earth.

The formation of greigite is not restricted to magnetotactic bacteria: The scales of the foot of a gastropod living at the base of black smokers (hot deep sea exhausts) consist of conchiolin (a complex protein) mineralized with pyrite and, in lower proportions, greigite, the latter accounting for the ferrimagnetism (Sidebar 4.3) of the material [13].

The nanocrystalline and, in many cases, mono-domain magnetite produced by magnetotactic bacteria is accommodated in special cell envelopes; its production is therefore biologically *controlled*. The ensemble of magnetite embedded in the lipid bilayer membrane is termed magnetosome. But magnetite can also be generated in a biologically *induced* manner by non-magnetotactic bacteria: Dissimilatory bacteria belonging to genera such as the facultatively anaerobic, psychrotolerant[6] soil bacteria *Geobacter* and *Shewanella* can employ [14] $Fe^{3+}$ as an external electron acceptor under oxygen-free conditions, using lactate, formate, pyruvate, or hydrogen as electron donors. Such processes yield, in the case of several *Shewanella* strains, single-domain magnetite (particle size $> 35\,nm$). Eq. (4.15) exemplifies the reduction of $Fe^{3+}$ to $Fe^{2+}$ with formic acid as the source of reduction equivalents.

*Ferrous* minerals such as siderite $FeCO_3$ can form under appropriate (i.e. non-acidic) conditions in the presence of $CO_2$; Eq. (4.16). Sulfide and elemental sulfur produced by sulfate-reducing bacteria provide the basis for the formation of iron sulfides such as troilite FeS, pyrrhotite $Fe_{1-x}S$ ($x < 0.2$) and pyrite $FeS_2$, see Eq. (4.17) for the latter.

$$HCO_2^- + 2\,FeO(OH) + 4\,H^+ \rightarrow HCO_3^- + 2\,Fe^{2+} + 3\,H_2O \tag{4.15a}$$

$$Fe^{2+} + 2\,FeO(OH) \rightarrow Fe_3O_4 \downarrow + 2\,H^+ \tag{4.15b}$$

$$Fe^{2+} + HCO_3^- + OH^- \rightarrow FeCO_3 \downarrow + H_2O \tag{4.16}$$

$$2\,Fe^{3+} + 3\,HS^- \rightarrow FeS_2 \downarrow + FeS \downarrow + 3\,H^+ \tag{4.17}$$

While Eqs. (4.15a) and (4.17) represent examples of the exogeneous bacterial *reduction* of ferric iron, the extracellular *oxidation* of ferrous iron by photoautotrophic bacteria (Scheme 2.2 in Chapter 2) may also come in as a source of iron biomineralization. Examples are the anaerobic photosynthetic bacterium *Rhodobacter* and the nitrate reducer *Acidovorax* [15]. These bacteria oxidize soluble $Fe^{2+}$ to (nanocrystalline) haematite, ferrihydrite, or goethite. Eqs. (4.18) and (4.19), which depict the formation of goethite, are examples of a photosynthetic and a nitrate-dependent $Fe^{2+}$ oxidation, respectively. The formula fragment $\{CH_2O\}$ in Eq. (4.18) represents a product of $CO_2$ fixation, such as glucose. The periplasm (the space between outer and inner membrane of the bacterial cell), the outer cell surface, or extracellular lipid-polysaccharide fibres associated with the bacterial membrane, act as a template for the nucleation of the mineral crystallites.

$$4\,Fe^{2+} + CO_2 + 7\,H_2O + h\nu \rightarrow 4\,FeO(OH) \downarrow + \{CH_2O\} + 8\,H^+ \tag{4.18}$$

$$5\,Fe^{2+} + NO_3^- + 7\,H_2O \rightarrow 5\,FeO(OH) \downarrow + \tfrac{1}{2}N_2 + 9\,H^+ \tag{4.19}$$

---

[6] Psychro- or cryotolerant bacteria thrive at normal temperatures, but also grow, at somewhat reduced growth and production rates, at low temperatures.

## Sidebar 4.3   Magnetism and spin moment

Electrons orbiting an atomic nucleus are accompanied by a magnetic angular moment, associated with the electron's angular motion. An external magnetic field alters the orbit velocity of the electrons in an atom in such a way that a magnetic dipole moment is induced which is directed opposite to the applied magnetic field, causing a repulsive effect. The material thus influenced is said to exhibit *diamagnetic* behaviour. Diamagnetism is a general property of *all* matter.

In addition, electrons ($e^-$) carry a magnetic spin moment going with the rotation (the spin) around its internal axis. The spin orientation is commonly denoted by an arrow pointing either up or down. When all of the electrons in an atom are paired, these effects are annulled. If one or more unpaired electrons are present in a system, an effective spin moment, or *paramagnetic* behaviour, arises. A paramagnetic probe is attracted by an external magnetic field.

As long as an external field is absent, the spin moments are usually thermally randomly distributed, and the material does not exhibit a resulting bulk magnetism. Under specific conditions, communication between the spins of single atoms or ions in a bulk material is possible, and partial or complete parallel alignment of the individual atomic spins arises, giving rise to a *superparamagnetic* (partial alignment) or *ferromagnetic* (complete alignment) behaviour. If neighbouring spins in a bulk domain are aligned in an antiparallel manner, the material is *antiferromagnetic*. If the overall lattice is built up of two sublattices composed of unequal paramagnetic atoms/ions each, and hence of particles where unequal spins are arranged antipodally, the system is said to be *ferrimagnetic*.

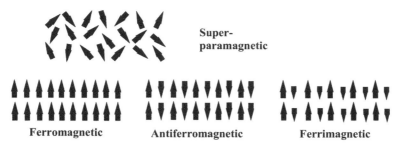

Octahedral $Fe^{II}$ ($d^6$) and $Fe^{III}$ ($d^5$) complexes are typical examples of spin systems giving rise to diamagnetic or paramagnetic behaviour:

- diamagnetic: low-spin $Fe^{II}$, overall spin state $S = 0$
- paramagnetic: (i) high-spin $Fe^{II}$: 4 unpaired $e^-$, $S = 2$; (ii) low-spin $Fe^{III}$, 1 unpaired $e^-$, $S = 1/2$; (iii) high-spin $Fe^{III}$: 5 unpaired $e^-$, $S = 5/2$.

Low-spin (*ls*) complexes are formed with strong ligands such as $CN^-$, high-spin (*hs*) complexes with weak ligands such as $H_2O$, other O-donors, and S-donors. A high oxidation state of the central metal ion favours *ls*; repulsive interaction between negatively charged ligands disfavours *ls* situations.

N-donors, in particular porphinogenic ligands (in haems), can give rise to spin cross-over settings, i.e. situations where minor environmental influences (temperature, solvent and other environmental effects, coordination number, exchange of one axial ligand) can switch the system between *hs* and *ls*, usually via an intermediate situation, such as $S = 3/2$ (3 unpaired $e^-$) in the case of $Fe^{III}$. The figure below displays *ls* ($S = 0$) and *hs* ($S = 1$) states for the active centre of the (reduced, i.e. $Fe^{II}$) oxidase cytochrome-$P_{450}$ (Section 5.3, Sidebar 5.2) with local $C_{4v}$ symmetry. For 'symmetry' see Sidebar 4.4. Spin states intermediate between *hs* and *ls* occur.

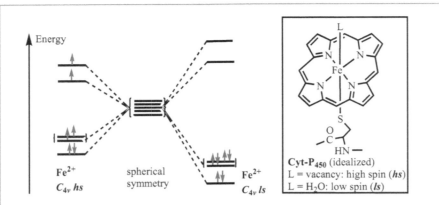

The spin state can be measured with a 'magnetic balance'. The expected overall spin moment $\mu$ of a paramagnetic compound is given by $\mu = \sqrt{n(n+2)}$ in units of the Bohr magneton ($\mu_B$), where $n$ is the number of unpaired electrons. This formula applies reasonably well for the first row transition metals.

In molecular chemistry, the terms ferromagnetic and antiferromagnetic coupling refer to molecules in which two or more paramagnetic metal centres M are linked through a ligand function X. In this multinuclear system, the magnetic moments of the metal ions may be oriented either parallel (ferromagnetic [exchange] coupling) or antiparallel (antiferromagnetic coupling). If ferromagnetic coupling is mediated through the intermediary (anionic and non-magnetic) ligand X, the phenomenon is also referred to as superexchange. In case of ferromagnetic coupling in a molecule containing an uneven number of metal centres, one of the local magnetic moments remains unpaired. This phenomenon is referred to as spin frustration.

Superexchange

---

## Sidebar 4.4   Symmetry

In chemistry, 'symmetry' describes a property of a molecule (or assembly of molecules) which allows for the transformation of the molecule into itself. This imaging in itself is carried out by symmetry operations as shown below for a regular square; examples are cyclobutadiene(2−), $C_4H_4^{2-}$, and tetracynonickelate(2−), $[Ni(CN)_4]^{2-}$. The vertices of the square are—imaginarily—numbered through 1 to 4.

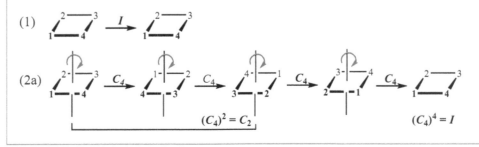

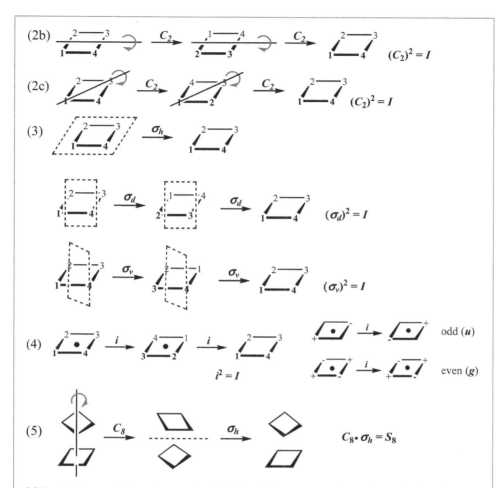

(1) The simplest case is the unit operation, $I$ (for identity), an operation that can be applied to all molecules.

(2) (a) Rotation by 90° around an axis passing through the centre of the square (the axis of symmetry) transforms the square into itself. The operation, denoted $C_4$ ($C$ for cyclic symmetry; the index 4 indicates a rotation by 360/90°, i.e. a quarter turn) can be carried out four times—$(C_4)^4$—to arrive at the numbering represented by the initial situation. Two successive rotations by 90° are identical to $C_2$, a rotation by 180°. The '$C$' descriptor is often omitted when rotational symmetry is indicated. An example is the 24 subunit protein apoferritin (Fig. 4.4a), which has three axes of symmetry, $C_2$, $C_3$ (threefold, rotation by 120°), and $C_4$, an overall symmetry which is simply referred to as 234. (b) and (c) typify twofold rotational symmetry about an axis perpendicular to the $C_4$ axis.

(3) Reflections at a mirror plane are symbolized by $\sigma$ (Greek letter "s", from *Spiegel* – German for mirror). Reflections can be in-plane, i.e. horizontal, $\sigma_h$, or perpendicular to the plane of the square. Such a mirror plane can be positioned dihedrally ($\sigma_d$), or diagonally vertical ($\sigma_v$). Two successive reflections at the same plane restore identity.

(4) Point reflection, or inversion $i \equiv S_2$, can be even (gerade, $g$) or odd (ungerade, $u$). p orbitals are odd with respect to inversion, while d orbitals are even.

(5) A composition of rotation and reflection, rotary reflection or improper rotation, is denoted by $S_n$, where $S$ is the rotation–reflection axis, with the plane of reflection perpendicular to the axis. In the case shown, n=8, i.e. rotation by 45°. Metallocenes are examples for this kind of symmetry.

## Summary

Iron is an essential element for (almost) all living beings, and has possibly already been employed, in the form of FeS, on the primordial Earth for the generation of 'active acetic acid' (the methyl ester of thioacetic acid). The central role in life attributable to iron roots in its general distribution, its flexibility with respect to oxidation states (mainly +II and +III), to the nature of ligands, coordination number and coordination geometry, and its spin state. With respect to its availability, the main problem is the formation of ferric hydroxides under aerobic conditions, which are insoluble at common physiological pH. Many organisms have learnt to cope with this problem by excreting organic transport systems, so-called siderophores.

Once taken up, iron can be bound in its ferric form to the iron transport protein transferrin, which contains two essentially identical binding sites for $Fe^{3+}$, providing two tyrosinates and two histidines. Simultaneously, carbonate is coordinated. Excess iron is mainly stored in ferritins in the form of FeO(OH). Ferritins are hollow globular proteins with an iron storage capacity of up to 4500 $Fe^{3+}$. Along with ferritins, Dps proteins and frataxins can pick up iron. The former protect DNA against oxidative stress, the latter, which can bind up to 7 $Fe^{2+}$, function as iron chaperones. Iron is finally delivered to the active centres of systems responsible for the transport of electrons, the transport and processing of oxygen, and a large variety of redox-active enzymes. The main type of functional iron are haem-type molecules, [2Fe,2S] and [4Fe,4S] centres, and 1Fe and 2Fe centres in a carboxylate/histidine coordination environment.

Biomineralization of iron is achieved by magnetotactic bacteria (magnetosomes containing single domain magnetite $Fe_3O_4$ or greigite $Fe_3S_4$), and by diverse other bacterial activities: dissimilatory reduction of $Fe^{3+}$ in the presence of $CO_2$ yields siderite $FeCO_3$, ferrous sulfides such as troilite FeS and pyrite $FeS_2$. $FeS_2$ and $Fe_3S_4$ have also been found in the scales of a gastropod. Oxidation of $Fe^{2+}$ induced by bacterial activity can yield goethite FeO(OH), ferrihydrite $Fe_{10}O_{14}(OH)_2$ ('FeO(OH)·0.4$H_2O$'), and haematite $Fe_2O_3$.

## Suggested reading

**Taylor AB, Stoj CS, Ziegler L, et al. The copper–iron connection in biology: structure of the metallo-oxidase Fet3p.** *Proc. Natl. Acad. Sci. USA* 2005; 102: 15459–15464.

Based on structure features of a multi-copper ferroxidase (which catalyses the oxidation of $Fe^{2+}$ to $Fe^{3+}$ with concomitant reduction of $O_2$ to $H_2O$), the authors delineate the role of ferroxidases in iron trafficking on the one hand, and linkages between copper and iron homeostasis on the other hand.

**Crichton RR and Declerq J-P. X-ray structures of ferritins and related proteins.** *Biochim. Biophys. Acta* 2010; 1800: 706–718.

This review covers the superfamily of iron storage proteins, including 'classical' ferritins, bacterioferritins (which bind haem), the DNA-binding Dps proteins, and ruberythrins (which contain $FeS_4$ domains).

**Lewin A, Moore GR, Le Brun NE. Formation of protein-coated iron minerals.** *Dalton Trans.* 2005; 3597–3610.

A review in a similar context as that of Chrichton and Declerq, emphasizing mineralization processes in the cavities of ferritins, Dps proteins, and frataxins.

## References

1. Wächtershäuser G. On the chemistry and evolution of the pioneer organism. *Chem. Biodivers.* 2007; 4: 584–602.

2. Archibald F. *Lactobacillus plantarum*, an organism not requiring iron. *Microbiol. Lett.* 1983; 19: 29–32.

3. Gütlich P, Schröder C, Schünemann V. Mössbauer spectroscopy – an indispensable tool in solid state research. *Spectrosc. Eur.* 2012; 24: 21–31.

4. Taylor AB, Stoj CS, Ziegler L, et al. The copper-iron connection in biology: structure of the metallo-oxidase Fet3p. *Proc. Natl. Acad. Sci. USA* 2005; 102: 15459–15464.

5. (a) Ford GC, Harrison POM, Rice DW, et al. Ferritin: design and formation of an iron-storage molecule. *Phil. Trans. R. Soc. Lond. B* 1984; 304: 551–565. (b) Crichton RR and Declerq J-P. X-ray structures of ferritins and related proteins. *Biochim. Biophys. Acta* 2010; 1800: 706–718.

6. (a) Lewin A, Moore GR, Le Brun NE. Formation of protein-coated iron minerals. *Dalton Trans.* 2005; 3597–3610. (b) Haikarainen T, Paturi P, Lindén J, et al. Magnetic properties and structural characterization of iron oxide nanoparticles formed by *Streptococcus suis* DPr and four mutants. *J. Biol. Inorg. Chem.* 2011; 16: 799–807.

7. Zeth K, Offermann S, Essen L-O, et al. Iron-oxo clusters biomineralizing on protein surfaces: structural analysis of *Halobacterium salinarum* DpsA in its low- and high-iron state. *Proc. Natl. Acad. Sci. USA* 2004; 101: 13780–13785.

8. (a) Yoon T and Cowan JA. Iron-sulfur cluster biosynthesis. Characterization of frataxin as an iron donor for assembly of (2Fe–2S) clusters in ISU-type proteins. *J. Am. Chem. Soc.* 2003; 125: 6078–6084. (b) Haikarainen T and Papageorgiou AC. Dps-like proteins: structural and functional insights into a versatile protein family. *Cell. Mol. Life Sci.* 2010; 67: 341–351.

9. Konhausera KO. Bacterial iron mineralisation in nature. *FEMS Microbiol. Rev.* 2006; 20: 315–326.

10. (a) Chen L, Bazylinski DA, Lower BH. Bacteria that synthesize nano-sized compasses to navigate using Earth's geomagnetic field. *Nature Educ. Knowl.* 2012; 334: 1720–1723. (b) Lefèvre CT, Menguy N, Abreu F, et al. A cultered greigeite-producing magnetotactic bacterium in a novel group of sulfate-reducing bacteria. *Science* 2011; 334: 1720–1723.

11. Schumann D, Raub TD, Kopp RE, et al. Gigantism in unique biogenic magnetite at the Paleocene–Eocene thermal maximum. *Proc. Natl. Acad. Sci. USA* 2008; 105: 17648–17653.

12. Thomas-Keprta KL, Clemett SJ, McKay DS, et al. Origins of the magnetic nanocrystals in Martian meteorite ALH84001. *Geochim. Cosmochim. Acta* 2009; 73: 6631–6677.

13. Warén A, Bengtson S, Goffredi SK, et al. A hot-vent gastropod with iron sulfide dermal sclerites. *Science* 2003; 302: 1007.

14. Roh Y, Gao H, Vali H, et al. Metal reduction and iron biomineralization by a psychrotolerant Fe(III)-reducing bacterium, *Shewanella sp.* strain PV-4. *Appl. Environ. Microbiol.* 2006; 72: 3236–3244.

15. Schädler S, Burkhardt C, Hegler F, et al. Formation of cell-iron mineral aggregates by phototrophic and nitrate-reducing anaerobic Fe(II)-oxidizing bacteria. *Geomicrobiol. J.* 2009; 26: 93–103.

# 5 Oxygen transport and the respiratory chain

During the first about two billion years of the existence of our planet, the atmosphere was dominated by $N_2$, $CO_2$, and $H_2O$, plus some methane. Then, ca. 2.4 billion years ago, photosynthesis (Chapter 11) carried out by cyanobacteria became a prevalent life process, rapidly converting the atmospheric composition from anaerobic to aerobic: oxygen (along with $N_2$) became the main atmospheric constituent, an occurrence commonly referred to as 'The Great Oxygen Event'. This event cleared the way for evolutionary processes in the course of which aerobic organisms developed, hence organisms that use the reductive conversion of oxygen (in effect, a biogenic combustion) as a source of energy to power life.

The release of energy through the oxidation of glucose by $O_2$ to form water and carbon dioxide on the cellular level, referred to as 'cellular respiration', is preceded by the uptake of $O_2$ from the outside and its delivery to the tissue cells, coupled to the counter-transport of $CO_2$ to the outside.

Dry air contains 20.95% (by volume) of $O_2$, along with 78.09% of $N_2$ as the main constituent, some argon (0.93%), $CO_2$ (0.039%), and trace amounts of other gases, including ozone $O_3$. The oxygen content of water highly depends on temperature and depth. Sea water at 15 °C and a pressure of 200 kPa (corresponding to a depth of 10 m) contains 16 mg of $O_2$ dissolved in 1 L of water. Despite the diradical character of $O_2$ and the distinctively exothermic character of oxidation processes involving $O_2$, dioxygen is comparatively stable. This stability at ambient temperature is a consequence of the rather high activation barrier for the oxidative conversion of organic matter, protecting organisms against 'spontaneous combustion'.

In order to target potential sites for the oxidation of organic matter in an organism, $O_2$ has to be taken up from air or water by an adequate transporter, channelled into the oxidation site, and activated in such a way that a substrate can be oxidized without becoming oxidatively destructed. To achieve this goal, nature exploits transport systems for $O_2$ based on iron and copper, and complex systems for processing the oxygen and supplying the electrons necessary for its reduction. These systems in turn are based, in many cases, on the redox-active metals iron, copper, manganese, molybdenum, tungsten, and vanadium, often in cooperation with *organic* redox-active cofactors.

In this chapter, the uptake, storage, transport, and reduction of oxygen by iron and copper proteins is covered. Chapter 6 provides an overview of electron transporters employing iron, copper, and manganese, emphasizing cofactors involved in the interconversion of oxygen species. Reductases, oxidases/peroxidases, and dismutases based on V, Mo, and W are introduced in Chapter 7.

## 5.1 Oxygen and oxygen transport by haemoglobin and myoglobin

Neutral, molecular oxygen exists predominantly in the form of diatomic $O_2$.[1] Energy sources such as electrical discharge and UV light (of wavelength < 240 nm) can split molecular oxygen into oxygen atoms. Recombination of an oxygen atom and an oxygen molecule results in the formation of ozone, $O_3$. In the context of atmospheric chemistry, ozone can also be formed from $O_2$ and atomic oxygen generated by, for example, photo-dissociation of $NO_2$.

Charged oxygen species include the radical cation $O_2^+$, the radical anion $O_2^-$ (superoxide[2]) and $O_2^{2-}$ (peroxide). Superoxide, peroxide, and protonated forms thereof ($HO_2^-$, $H_2O_2$) are important intermediates and side-products in oxygen metabolism. These species, along with the hydroxyl radical OH, are termed reactive oxygen species, ROS. ROS can take messenger functions, but they also are co-responsible for damaging DNA and can intervene with other cellular constituents and functions. The oxidative stress caused by the accumulation of ROS is also well known to be associated with senescence. The formation of OH radicals in the reaction of ferrous ions and $H_2O_2$, Eq. (5.1), is known as the 'Fenton reaction'. Scheme 5.1 provides an overview of the Lewis formulae of the various oxygen species, and Table 5.1 contains bond characteristics of bimolecular oxygen species and ozone.

$$Fe^{2+} + H_2O_2 + H^+ \rightarrow Fe^{3+} + H_2O + HO \cdot \tag{5.1}$$

The molecular orbital diagram in Fig. 5.1 provides a more detailed description of the distribution of electrons in oxygen: In its stable ground state, $O_2$ is in the triplet state $^3O_2$, i.e. two of the electrons are unpaired. This situation is achieved because the two highest occupied orbitals, the antibonding $\pi^*$ orbitals, have the same energy (they are doubly degenerate).

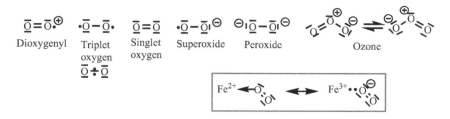

**Scheme 5.1** Lewis formulae of oxygen species, and (boxed) of the two resonance formulae for the Fe·$O_2$ centre in haemoglobin and myoglobin. Note that, for $^3O_2$, one of the valence dash descriptions of the distribution of electrons does not reflect the net bond order of 2. The arrow (←) in the formula for Fe·$O_2$ at the left hand side indicates a dative bond, i.e. both of the bonding electrons in the Fe–O bond of $Fe^{2+}O_2$ are provided by oxygen. In the right hand formula ($Fe^{3+}O_2^-$), each of the bonding partners contributes one electron to the bonding electron pair. In either case, a diamagnetic singlet state results for oxy-Hb and oxy-Mb.

---

[1]  $O_2$ molecules can associate to short-lived dimers ($O_4$) and tetramers ($O_8$).
[2]  More correct—but not in use—is 'hyperoxide'.

Table 5.1 Bond data and vibrational stretching frequencies[a] of dioxygen species and ozone.

| Molecule/molecular ion | $O_2^+$ | $^1O_2$ | $^3O_2$ | $O_2^-$ | $O_2^{2-}$ | $O_3$ |
|---|---|---|---|---|---|---|
| Bond order | 2.5 | 2 | 2 | 1.5 | 1 | 1.5 |
| $d$(O–O), pm | 112 | ca. 125 | 121 | 133 | 149 | 128 |
| $\nu$(O–O), cm$^{-1}$ | 1860 | – | 1555 | 1145 | 770 | 1135(sym), 1089(asym) |

[a] The stretching frequencies correlate with the strength of the O–O bond (larger values for stronger bonds). sym = symmetric, asym = antisymmetric stretch.

Excitation converts triplet-$O_2$ into singlet-$O_2$, $^1O_2$, where the degeneracy is lifted.[3] Singlet oxygen is highly reactive; it is generated as a commonly 'unwanted' ROS in various physiological processes [1], and also forms as a short-lived primary product when $H_2O_2$ disproportionates into water and oxygen, Eq. (5.2).

$$H_2O_2 \rightarrow H_2O + 1/2\,^1O_2 \qquad (5.2a)$$

$$^1O_2 \rightarrow {}^3O_2, \; \Delta H = -94.3\,\text{kJ}\,\text{mol}^{-1} \qquad (5.2b)$$

In vertebrates, and also in some invertebrates (including insects), the transport of oxygen in the blood stream from the respiratory organ to the body tissues is carried out by haemoglobin, Hb, the red 'pigment' present in the red blood cells, the erythrocytes. Hb is an $\alpha_2\beta_2$

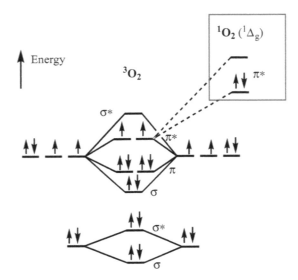

Figure 5.1 Molecular orbital diagram for triplet-$O_2$ ($^3O_2$) and—boxed—the $\pi^*$ state of the $^1\Delta_g$ variant of singlet-$O_2$ ($^1O_2$).

[3] There are actually two forms of singlet-$O_2$. Along with the more stable $^1O_2$ where the two electrons reside in the same $\pi^*$ orbital (see Fig. 5.1), the two non-degenerate $\pi^*$ orbitals can also accommodate one electron each. (The electron spins remain, however, antiparallel.) The two variants of singlet oxygen are referred to as $^1\Delta_g$ (both e$^-$ in the same $\pi^*$) and $^1\Sigma_g$ (each of the $\pi^*$ accommodates one e$^-$).

(adult Hb) or $\alpha_2\gamma_2$ (foetal Hb)[4] tetramer with a molar mass of 64 kDa, and with one iron per subunit. Muscle tissues contain light red myoglobin Mb (molecular mass 17.7 kDa), essentially a monomeric variant of Hb. Due to its higher oxygen affinity, Mb takes over $O_2$ from Hb, delivering the oxygen to the mitochondrial respiratory chain.

The prosthetic group of Hb and Mb is the haem group: a porphyrin ring, consisting of four pyrroles that are bridged by four methynes, with a ferrous ion coordinated to the four pyrrole nitrogens (Fig. 5.2). This system is commonly referred to as protoporphyrin IX. The $Fe^{2+}$ is delivered to the porphyrin system in a transport process catalysed by a protein termed ferrochelatase. In ferrochelatase, $Fe^{2+}$ is coordinated in a tetragonal-pyramidal fashion by three water molecules, a histidine-N, and a glutamate-O [2] (Fig. 5.3).

The $Fe^{2+}$ ion and the four nitrogen functions form a tetragonal, approximately planar system. A fifth, axial, position is occupied by the Nε of the so-called *proximal* histidine. A second, *distal*, histidine resides in the other axial position of the deoxy form of Hb. The deoxy form is also referred to as T form, where 'T' stands for 'tensed': $Fe^{2+}$ is in its high-spin electronic configuration (diameter 95 pm), in which it does not fit into the porphyrin system. Rather, this plane is watch glass-shaped, with $Fe^{2+}$ displaced at a distance of 40 pm towards the proximal His from the ideal plane formed by the four pyrrole nitrogens. In the oxy form, oxygen coordinates to $Fe^{2+}$ into the axial position opposite the proximal His; as such, it adopts a 'bent' orientation, which is additionally stabilized by hydrogen bonding interaction with the NH group of the distal histidine. Concomitantly with the intake of $O_2$, the ferrous ion switches into the low-spin configuration (diameter 75 pm), and moves, together with the proximal His, towards the now essentially flattened porphyrin plane.

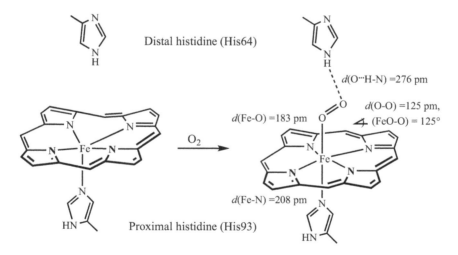

Figure 5.2 The prosthetic group of haemoglobin and myoglobin in the deoxy and oxy forms. Distances, and numbering of the histidine residues, have been adapted for oxy-Mb from [3a]. For the oxidation states of Fe and $O_2$, see the boxed part in Scheme 5.1.

---

[4] Foetal haemoglobin has a higher $O_2$ intake capacity than adult Hb. It is also present in adults at concentrations of ca. 1%. Prolonged stays at high altitudes increase this percentage—a 'doping method' occasionally employed by high-performance athletes.

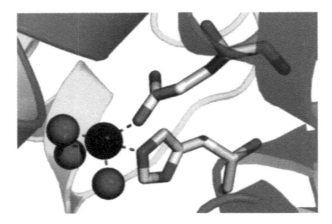

Figure 5.3 The coordination and protein environment of $Fe^{2+}$ (large black sphere) in ferrochelatase, a protein that directs the insertion of $Fe^{2+}$ into the protoporphyrin IX system. The other spheres are water molecules; the ribbon-like structures represent the protein matrix. Reprinted with permission from [2]; © American Chemical Society. *Also reproduced in full colour in Plate 4.*

The uptake of $O_2$ from breathing air into the aveoles of the lung, and the release of $O_2$ to the muscle tissues, are coupled to the uptake/release of $HCO_3^-/CO_2$. The erythrocytes releasing oxygen to tissue cells concomitantly exchange chloride for hydrogen carbonate. This process is reversed in the lungs. The overall situation is represented by Eq. (5.3). In Eq. (5.3c), the indices 'in' and 'ex' refer to intra- and extracellular, respectively, relating to the erythrocytes. Interconversion of carbonic acid and carbon dioxide, Eq. (5.3b), is catalysed by the zinc enzyme carboanhydrase, which we discuss further in section 12.2.1.

$$Hb \cdot H^+ + HCO_3^- + O_2 \rightleftharpoons Hb \cdot O_2 + H_2CO_3 \tag{5.3a}$$

$$H_2CO_3 \rightleftharpoons H_2O + CO_2 \tag{5.3b}$$

$$(HCO_3^-)_{in} + (Cl^-)_{ex} \rightleftharpoons (HCO_3^-)_{ex} + (Cl^-)_{in} \tag{5.3c}$$

The oxygen affinity of haemoglobin depends on the temperature and the pH of blood. The average standard temperature of human blood is 37 °C, the normal pH 7.4. Fever and acidaemia reduce the $O_2$ uptake capacity of Hb, the latter effect also being known as the 'Bohr effect'.[5]

When haemoglobin becomes oxidized, Eq. (5.4), methaemoglobin, MetHb, is formed. MetHb contains $Fe^{3+}$, axially coordinated to $OH^-$, and is thus no longer able to take up $O_2$. Potent oxidants for $Fe^{2+}$ in Hb and Mb are nitrous oxide NO, and ROS such as OH radicals, superoxide and peroxide. Hb can be recovered from MetHb by methaemoglobin reductase, an enzyme relying on the cofactor NADH.[6] The average content of MetHb in human blood therefore does normally not exceed 1.5%.

$$Hb(Fe^{2+}) + H_2O \rightarrow MetHb(Fe^{3+}OH) + H^+ + e^- \tag{5.4}$$

[5] Discovered by the physiologist Christian Bohr, the father of the founder of the modern atomic model, Niels Bohr.

[6] NADH is the reduced form of nicotine adenine dinucleotide, NAD; see Sidebar 9.2.

There are several—heritable—mutated forms of haemoglobin that cause severe diseases due to their drastically reduced $O_2$ binding capacity. Examples are sickle cell disease (also referred to as sickle cell anaemia) [4] and Boston disease. In the case of sickle cell disease, hydrophilic glutamate in position 6 of the β-globin chain of Hb is replaced by hydrophobic valine. This mutation reduces the $O_2$ intake capacity to about 25%, but provides some protection against malaria. Sickle cell anaemia is therefore endemic in tropical Africa in particular. Boston disease, or methaemoglobinemia, goes back to replacement of the distal histidine for tyrosinate $Tyr^-$. $Tyr^-$ firmly coordinates to the sixth position—the $O_2$ binding site—of $Fe^{2+}$ and thus blocks $O_2$ uptake. Obstruction of this sixth position at the $Fe^{2+}$ site is also achieved by carbon monoxide. CO coordinates into the free iron site of Hb and Mb very effectively: indeed, Hb·CO is 250 times more stable than Hb·$O_2$. The odourless gas CO is thus a particularly dangerous inhaled poison; see also Section 14.5.

## 5.2 Oxygen transport by haemerythrin and haemocyanin

Haemerythrin and haemocyanin are alternative oxygen transporters employed by a variety of non-vertebrates. Despite their names, they are not based on haem as the prosthetic group. Rather, they contain two iron and two copper centres, respectively, with coordination motifs similar to those also found in proteins with functions other than $O_2$ transport.

Several phyla of marine invertebrates transport oxygen with the help of haemerythrin, Hr, a protein with an overall molecular mass of typically 108 kDa, composed of eight 13.5 kDa subunits. *Monomers* of haemerythrin have been found in prokaryotic bacteria, where they also serve as $O_2$ transporters [5]. Each Hr unit contains a two-iron centre for the transport of one $O_2$. In the deoxy form, both irons are in the +II state, linked via μ-$OH^-$, μ-$Asp^-$, and μ-$Glu^-$, (**a**) in Fig. 5.4. One of the $Fe^{2+}$ is additionally coordinated to three histidines (via Nε), the other one to two histidines, leaving a free coordination site at one of the ferrous ions.

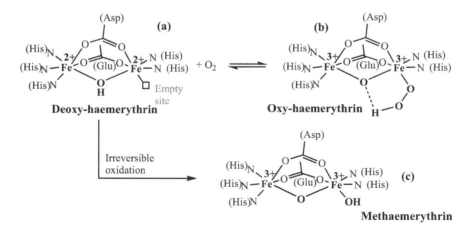

Figure 5.4 Reversible oxidation of deoxy-haemerythrin (Hr, **a**) to oxy-Hr (**b**), and the irreversible oxidation to MetHr (**c**).

Uptake of oxygen by the di-iron centre takes place by way of an oxidative addition. The $Fe^{2+}$ with the open site is oxidized to $Fe^{3+}$, while $O_2$ is reduced to superoxide $O_2^-$, coordinated to this site, and further reduced to peroxide by the second $Fe^{2+}$. Concomitantly, the proton of the bridging OH migrates to the dangling negative end of the peroxide; the hydroperoxido ligand thus formed remains in hydrogen bonding contact with the $\mu$-$O^{2-}$, (**b**) in Fig. 5.4. The two iron centres in the deoxy and the oxy forms of Hr are coupled antiferromagnetically, i.e. the spins of the two ferrous ions are in an anti-parallel alignment.[7] Irreversible oxidation leads, via semi-MetHr, to methaemerythrin MetHr, (**c**), where the sixth position at $Fe^{3+}$ is occupied by a hydroxido ligand.

The copper-containing haemocyanins, Hc, are widespread oxygen carriers in the blood of animals belonging to the phylum *mollusca*, such as snails, slugs and cephalopods, and to the phylum *arthropoda*. Examples of the latter are spiders, scorpions and crabs. The haemo-cyanins are freely suspended in the blood. The proteins are quite different for the two phyla; the active centres are, however, alike. The Hc of arthropods are typically hexamers with a molecular mass of 75 kDa per monomeric subunit; the cohesion of the subunits is mediated by $Ca^{2+}$. Further aggregation up to 24-mers can occur [6]. The functional units of the molluscs' haemocyanins have a molecular mass of 50-55 kDa. Seven to eight of these units assemble to oligomers that commonly further aggregate to form decamers and didecamers [7]. All of the subunits in these multimers are highly cooperative.

The two copper centres per monomeric Hc unit are at a distance of 460 pm in the deoxy, and 356 pm in the oxy form. In the oxygen-free form, each of the colourless $Cu^+$ (electron configuration $d^{10}$) is in a trigonal environment, (**a**) in Fig. 5.5, provided by the N$\epsilon$ of three histidines at distances $d$(Cu-N) of 190, 210 and 270 pm. Uptake of $O_2$ occurs by oxidative addition: the cuprous ions are oxidized to cupric ions ($Cu^{2+}$, $d^9$), and oxygen is reduced  to

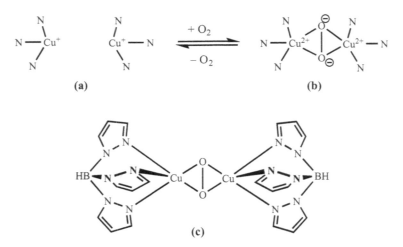

Figure 5.5 The two-copper centres in haemocyanin (**a** and **b** are the deoxy and oxy form, respectively), and a model compound, **c**, containing the ligand tris(pyrazolyl)borate(1-). The 'N' in (**a**) and (**b**) is the N$\epsilon$ of histidine.

[7]  For magnetism, see Sidebar 4.3.

peroxide $O_2^{2-}$, coordinated to the two $Cu^{2+}$ in the $\mu{:}\eta^2,\eta^2$ mode; **(b)** in Fig. 5.5. The two cupric centres are coupled antiferromagnetically, and are therefore EPR-inactive. Oxy-Hc is intensely blue due to an electronic ligand-to-metal ($O_2^{2-} \rightarrow Cu^{2+}$) charge transfer.

Appropriate ligands for model compounds are tridentate N-donors with nitrogen in a (pseudo)aromatic environment modelling histidine, such as *tris*(pyrazolyl)borate(1-), **(c)** in Fig. 5.5.

## 5.3 The respiratory chain

The respiratory chain is located in the mitochondria of the cells of eukaryotes. Here, oxygen is reduced in a four electron reduction process to water. The net reaction, Eq. (5.5), corresponds to a potential difference of $\Delta E = 1.14\,V$, or a Gibbs free energy of $-217\,kJ\,mol^{-1}$. The reduction proceeds in a cascade of electron transfer processes, with participation of iron and an iron–copper based electron transfer enzymes. The main steps are illustrated in Fig. 5.6.

$$O_2 + 4e^- + 4H^+ \rightarrow 2H_2O \qquad\qquad (5.5)$$

Electrons are delivered by NADH and succinate, which become oxidized to $NAD^+$ and fumarate, respectively, Eq. (5.6). The transfer of protons across the mitochondrial membrane generates a proton gradient, and thus couples $O_2$ reduction to the synthesis of, and energy storage by, adenosine triphosphate (ATP) from adenosine diphosphate (ADP) and inorganic phosphate.

$$NADH \rightarrow NAD^+ + 2e^- + H^+ \qquad\qquad (5.6a)$$

$$^-O_2C - CH_2 - CH_2 - CO_2^- \rightarrow {}^-O_2C - CH = CH - CO_2^- + 2e^- + 2H^+ \qquad (5.6b)$$

Four membrane-bound, interlinked protein complexes, I–IV, participate in the shuttle of electrons. These complexes are highlighted in different colour shades in Fig. 5.6.

- In complex I, nicotine-adenine dinucleotide NADH delivers two electrons to flavin mononucleotide FMN, which further shuttles the electrons in two successive one-electron steps to a [4Fe,4S] ferredoxin (Sidebar 5.1; abbreviated {FeS} in Fig. 5.6), reducing the four iron centres (formally $Fe^{+2.5}$) to $Fe^{+2.25}$. Each electron is then transferred to ubiquinone, Q, to form ubisemiquinone, QH, (first electron transfer) and ubiquinol, $QH_2$ (= ubihydroquinone; second electron transfer). The reduction of Q to $QH_2$ also requires two protons. $Q/QH_2$ floats freely in the protein matrix.

- Complex II provides additional electrons originating from organic substrates, the oxidation of succinate to fumarate in particular. The electrons are picked up by flavin-adenine dinucleotide FAD, and delivered, again via ferredoxins, to the quinone pool.

- In complex III, $QH_2$ is oxidized in a one-electron transfer process to the semiquinone ·QH by a Rieske centre. Rieske centres, Fig. 5.7, are comparable to [2Fe,2S] ferredoxins; one of the irons is, however, coordinated to two histidines. The electron is then delivered to the oxidized (ferric) form of cytochrome-c, Cyt-c. In a bifurcation process,

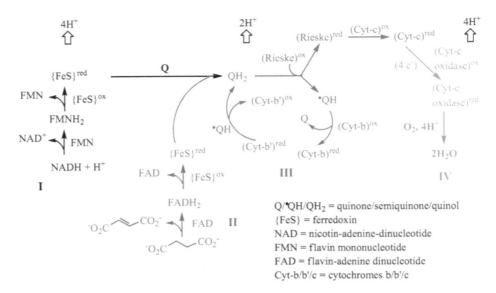

Q/·QH/QH₂ = quinone/semiquinone/quinol
{FeS} = ferredoxin
NAD = nicotin-adenine-dinucleotide
FMN = flavin mononucleotide
FAD = flavin-adenine dinucleotide
Cyt-b/b'/c = cytochromes b/b'/c

Figure 5.6 The electron and proton shuttle in the mitochondrial respiratory chain. See text for discussion of the processes taking place in the complexes I to IV. The broad arrows pointing upward indicate protons translocated across the mitochondrial membrane. For NADH, FMNH₂, FADH₂, and QH₂, see Sidebar 9.2, for {FeS} Sidebar 5.1, for the cytochromes Sidebar 5.2, and for the Rieske centre Fig. 5.7.

further oxidation of ·QH to Q is achieved via electron transfer catalysed by cytochrome b, Cyt-b, while another ·QH is reduced to $QH_2$ by Cyt-b'. Cyt-b and Cyt-b' differ in their protein environment, providing high and low redox potential, respectively.[8] For cytochromes see Sidebar 5.2.

- Four reduction equivalents are finally delivered one by one via Cyt-c to cytochrome-c oxidase, Fig. 5.8. Cyt-c oxidase contains a dinuclear copper centre ($Cu_A$), a haem-type iron centre (Cyt-a), and another haem iron (Cyt-$a_3$) coupled to a copper coordinated to three histidines ($Cu_B$). The final 4-electron reduction of $O_2$ to water takes place at this Cyt-$a_3$/$Cu_B$ complex.

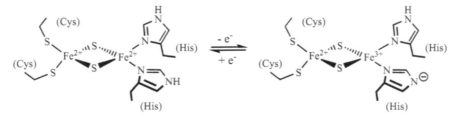

Figure 5.7 The reduced and oxidized forms of the active site of the Rieske iron proteins. The angle N-Fe-N is close to 90°; the planes of the two His are about perpendicular to each other.

[8] These cytochromes are therefore sometimes denoted Cyt-$b_H$ and Cyt-$b_L$.

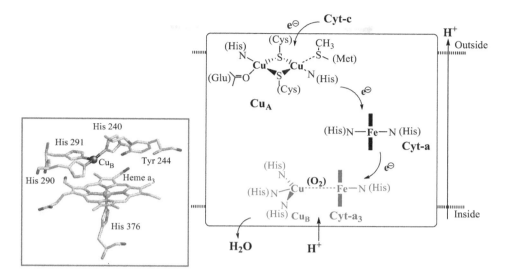

**Figure 5.8** The build-up of, and the electron transport within, cytochrome-c oxidase. Electrons are delivered to $Cu_A$ by the reduced form of Cyt-c and further shuttled, via Cyt-a, to the Cyt-$a_3$/$Cu_B$ complex, where $O_2$ is stepwise reduced to $2H_2O$. Details of Cyt-$a_3$/$Cu_B$ are shown in the inset (blue frame). $Cu_A$ is a one-electron transporter; in its reduced form, both Cu are in the oxidation state +I; in the oxidized form, the mean oxidation state is 1.5. In addition to the redox active Cu and Fe centres, Cyt-c oxidase contains a $Zn^{2+}$ and a $Mg^{2+}$ ion coordinated into the protein. The inset is reprinted with permission from [8]; © American Chemical Society. Image kindly supplied by Dr Kenneth Karlin.

The leakage of electrons to $O_2$, in particular at the complex I and complex III sites, results in the formation of the reactive oxygen species superoxide. However, organisms possess special enzymes for the 'detoxification' of superoxide: superoxide dismutases and reductases (Section 6.2). Cyt-c and Cyt-c oxidase are main targets for cyanide poisoning: $CN^-$ tightly binds to the haem-$Fe^{3+}$.

---

### Sidebar 5.1    Iron–sulfur proteins

Iron–sulfur proteins are versatile electron transporters, and therefore participate in a vast variety of (enzymatically conducted) electron transport processes in all organisms. Most commonly, the central building blocks contain one, two, three, four or 2×4 iron centres, linked—in the case of [2Fe,2S], [3Fe,4S], and [4Fe,4S]—by bridging sulfide, and connected to the protein via cysteinate(1−). For the arrangement of the clusters, see below.

Iron is usually in a tetrahedral environment; in several cases, a fifth position at one of the Fe centres can be occupied by an additional Cys or His. Other iron–sulfur proteins than those shown here exist, among these the Rieske proteins, with one of the iron ions linked to two histidines (see Fig. 5.7), and the Fe–S clusters of nitrogenase (Section 9.2).

Occasionally, an $S^{2-}$ can be exchanged for an $O^{2-}$. An example is the electron transport centre of the sulfate reducer *Desulfuvibrio vulgaris*.

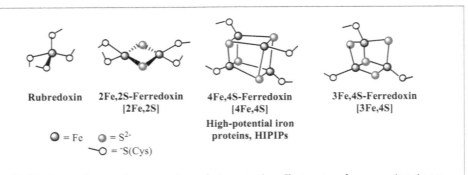

| Rubredoxin | 2Fe,2S-Ferredoxin [2Fe,2S] | 4Fe,4S-Ferredoxin [4Fe,4S] High-potential iron proteins, HIPIPs | 3Fe,4S-Ferredoxin [3Fe,4S] |

⬤ = Fe    ◐ = $S^{2-}$
    ◯ = $^-S(Cys)$

All of the iron–sulfur proteins transport one electron at a time. Electron transfer occurs via a change in oxidation state (ferrous/ferric iron). Generally, the charges are delocalized over all of the iron centres. The medium cluster charges in the reduced and oxidized states are summarized in the table below, along with the approximate ranges of redox potential typically covered by the clusters (see also [9]).

|  | Rub-redoxin | Rieske protein | [2Fe,2S] | [3Fe,4S] | [4Fe,4S] | HIPIP [4Fe,4S] |
|---|---|---|---|---|---|---|
| Cluster charge[a] | 2+/3+ | 0/1+ | 1+/2+ | 0/1+ or 1+/2+ | 1+/2+ | 2+/3+ |
| $Fe^{3+}$ count[b] | 1 | 1 | 2 | 2 to 3 | 2 | 3 |
| Range of $\Delta E$ (V) | −60 to 0 | +250 to +300 | −450 to −220 | −450 to −100 | −500 to −300 | +150 to +450 |

[a] Charge of the central iron-sulfide unit, reduced/oxidized.
[b] Formal number of ferric centres in the oxidized form of the cluster.

## Sidebar 5.2    Iron porphyrins

Porphyrins contain 22 π electrons, delocalized over the macrocycle, i.e. porphyrins are aromatic. $Fe^{2+/3+}$ coordinates to the dianionic porphyrin. The table below provides an overview of important iron porphyrins.

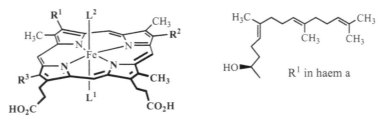

Other metal-containing porphyrins and porphyrinogenic systems include chlorophyll (with $Mg^{2+}$), cobalamin (vitamin $B_{12}$ and its derivatives, containing $Co^{+/2+/3+}$), and the factor $F_{430}$ of coenzyme-M reductase, a key enzyme, based on $Ni^{+/3+}$, in methanogenesis. For chlorophyll, see Fig. 11.2; for cobalamin and factor $F_{430}$, Fig. 10.1.

|     | Haem a | Haem b | Haem c | Cytochrome $P_{450}$ |
|-----|--------|--------|--------|---------------------|
| $L^1$ | His | His or vacant | His | Cys |
| $L^2$ | His or vacant | His or vacant | Met | $H_2O$ |
| $R^1$ | Farnesyl-OH (see above, right) | $-CH=CH_2$ | $-CH(CH_3)S-(Cys)^a$ | $-CH=CH_2$ |
| $R^2$ | $-CH=CH_2$ (vinyl) | $-CH=CH_2$ | $-CH(OH)-CH(CH_3)-S-(Cys)^a$ | $-CH=CH_2$ |
| $R^3$ | $-CHO$ (formyl) | $-CH_3$ | $-CH_3$ | $-CH_3$ |

$^a$ Connecting to the protein backbone

## Summary

In vertebrates and several representatives of other phyla, $O_2$ is taken up from inhaled air by haemoglobin, Hb, the red pigment of the erythrocytes, transported in the blood stream to tissue myoglobin, Mb, and further delivered to the mitochondria. Hb, a tetramer, and monomeric Mb contain a tetrapyrrole system (protoporphyrin IX) with high-spin $Fe^{2+}$ coordinated to the four pyrrole nitrogens and an axial histidine. Uptake of $O_2$ takes place by coordination into the 'open' axial position in a bent end-on coordination mode. The binding situation is best described by the resonance hybrid $Fe^{2+} \leftarrow {}^1O_2 \leftrightarrow Fe^{3+} \cdot \cdot O_2^-$, with iron in the low-spin state.

Many invertebrates employ alternative $O_2$ carriers, namely haemerythrin and haemocyanin. The functional unit in haemerythrin is a two-iron centre. $O_2$ is taken up by oxidative addition, i.e. both $Fe^{2+}$ are oxidized to $Fe^{3+}$, and $O_2$ is reduced to hydroperoxide $HO_2^-$. Haemocyanins are giant, free floating proteins in the blood of molluscs and arthropods. Each subunit contains two cooperative $Cu^+$ ions. Binding of oxygen to copper goes along with the reduction of $O_2$ to $O_2^{2-}$; the peroxide bridges the two $Cu^{2+}$ centres in the $\mu{:}\eta^2,\eta^2$ mode.

The oxygen taken up by Hb/Mb is finally delivered to the mitochondrial membrane, where it is reduced to water in the so-called mitochondrial respiratory chain. The electrons for this reduction are predominantly delivered by NADH and by succinate, and transported along four redox-active complexes, I to IV, with ubiquinone/ubisemiquinone/ubiquinol and iron–sulfur proteins (including a Rieske centre) playing a pivotal role in this process. The $e^-$ transport and $O_2$ reduction are coupled to the build-up of a trans-membrane proton gradient that serves the formation of ATP from ADP and inorganic phosphate.

The final reduction of $O_2$ in a four electron transfer process occurs in complex IV by cytochrome-c oxidase, Cyt-cOx, that takes up electrons one at a time from Cyt-c. The primary electron acceptor in Cyt-cOx is a dicopper centre ($Cu_A$). The electron is then transported to Cyt-a, and further to Cyt-$a_3$. Cyt-$a_3$ is in close contact with a $Cu(His)_3$ centre, $Cu_B$. The final transfer of the electrons to $O_2$ takes place at this Cyt-$a_3 \cdots Cu_B$ unit. Cyt-c and Cyt-a are porphyrin-based iron proteins, i.e. they contain the haem group.

## Suggested reading

**Ascenzi P, Bellelli A, Coletta M, et al. Multiple strategies for $O_2$ transport: from simplicity to complexity.** *IUBMB Life* 2007; 59: 600–616.

This 'critical review' provides an overview of the strategies that evolved in the kingdom of Life for oxygen transport and delivery, including more 'exotic ones' not dealt with in the present chapter.

Collman JP and Ghosh S. Recent applications of a synthetic model of cytochrome *c* oxidase: beyond functional modeling. *Inorg. Chem.* 2010; 49: 5798–5810.
A close structural and functional model of the active site of cytochrome-c oxidase is presented that provides insight into how the study of enzymes cross-fertilizes model chemistry, and vice versa.

 ## References

1. Agnez-Lima LF, Melo JTA, Silva AE, et al. DNA damage by singlet oxygen and cellular protective mechanisms. *Mutat. Res.: Rev. Mutat. Res.* 2012; 751: 15–28.

2. Hansson MD, Karlberg T, Rahardja MA, et al. Amino acid residues His183 and Glu264 in *Bacillus subtilis* ferrochelatase direct and facilitate the insertion of metal ion into protoporphyrin IX. *Biochemistry* 2007; 46: 87–94.

3. (a) Chen H, Iketa-Saiko M, Shaik S. Nature of the Fe–$O_2$ bonding in oxy-myoglobin: effect of the protein. *J. Am. Chem. Soc.* 2008; 130: 14778–14790; (b) Shaik S and Chen H. Lessons on $O_2$ and NO bonding to heme from ab initio multireference/multiconfiguration and DFT calculations. *J. Biol. Inorg. Chem.* 2011; 16: 841–855.

4. Orkin SH and Higgs DR. Sickle cell disease at 100 years. *Science* 2010; 329: 291–292.

5. Kao WC, Wang VC-C, Huang Y-C, et al. Isolation, purification and characterization of hemerythrin from *Methylococcus capsulatus* (Bath). *J. Inorg. Biochem.* 2008; 102: 1607–1614.

6. Jaenicke E, Pairet B, Hartmann H, et al. Crystallization and preliminary analysis of the 24-meric hemocyanin of the emperor scorpion (*Pandinus imperator*). *PLoS One* 2012; 7: e23548.

7. Gatsogiannis C and Markl J. Keyhole limpet hemocyanin: 9 Å cryoEM structure and molecular model of the KLH1 didecamer reveals the interfaces and intricate topology of the 160 functional units. *J. Mol. Biol.* 2008; 385: 963–983.

8. Kim E, Helton ME, Wasser IM, et al. Superoxo, μ-peroxo, and μ-oxo complexes from heme/$O_2$ and heme-Cu/$O_2$ reactivity: copper ligand influences in cytochrome *c* oxidase models. *Proc. Natl. Acad. Sci. USA* 2003; 100: 3623–3628.

9. Lill R. Function and biogenesis of iron–sulphur proteins. *Nature* 2009; 460: 831–838.

# 6 Oxidoreductases based on iron, manganese, and copper

The preceding chapter explored the transport of oxygen by carriers containing iron or copper. The transport mechanisms are of prime importance to all aerobic organisms, but also to anaerobes occasionally exposed to oxygen. For the latter, the removal and reduction of oxygen is a matter of detoxification. Organisms further process oxygen, $O_2$, and oxygen-derived species such as peroxide ($H_2O_2/HO_2^-/O_2^{2-}$) and superoxide ($O_2^{\bullet-}$) via a variety of vital processes, including the detoxification of reactive radicals (superoxide and the hydroxide radical), which are formed from undesirable side reactions.

Molecular oxygen, $O_2$, is not only employed in respiration, but also directly or indirectly in a couple of other functions essential for organisms. Thus, oxygen activates an iron- or manganese-based enzyme that in turn mediates the generation of a reactive organic radical intermediate (a tyrosyl radical) that is jointly responsible for the supply of the DNA building blocks. Other enzymes use $O_2$ to insert oxygen into an R—H bond, or to oxidize a substrate. These enzymes are termed oxygenases and oxidases, respectively. Oxygenases and oxidases employ copper or iron, or copper *and* iron in their active centres.

Reactive oxygen species, such as superoxide and peroxide, are deactivated by enzymes that catalyse disproportionation reactions (superoxide dismutases) or reductions (super-oxide reductases) or oxidations (peroxidases). These enzymes are particularly versatile with respect to the metal used: the active centres can contain iron, copper, manganese or nickel.

In this chapter, we will present selected examples of the plethora of metal-based enzymes that transform and employ oxygen directly or indirectly. Sidebar 6.1 provides an overview of the terminology for these specific redox-active enzymes. Hydrogenases (Section 13.1) have been added for completion.

---

**Sidebar 6.1**  Overview of enzymes that catalyse redox reactions in which hydrogen and oxygen are involved

Hydrogenases: $H_2 \rightleftharpoons H^+ + H^-$, followed by $H^- \rightarrow H^+ + 2e^-$

Net reaction: $H_2 \rightleftharpoons 2H^+ + 2e^-$

Can be coupled to the abstraction of H from a substrate (dehydrogenation) or the transfer of H to a substrate (hydrogenation):

substrate $- H_2 \rightleftharpoons$ substrate $+ 2H^+ + 2e^-$

Oxidoreductases: 'Oxidoreductase' is a general term, denoting the catalytic oxidation (electron abstraction from) and reduction (electron delivery to) a substrate. Versatile systems in this respect are iron–sulfur proteins and cytochromes.

Some oxidoreductases employ oxygen for the dehydrogenation of a substrate, or water for the hydrogenation of a substrate:

$$substrate - H_2 + \tfrac{1}{2}O_2 \rightleftharpoons substrate + H_2O, or$$
$$substrate - H_2 + O_2 \rightarrow substrate + H_2O_2$$

An enzyme that catalyses reductions only is a *reductase*. Correspondingly, an enzyme exclusively catalysing oxidations is an *oxidase*.

Oxygenases insert one or two oxygen atoms of $O_2$ into a substrate. Depending on the number of oxygens inserted, they are sometimes distinguished as mono- and dioxygenases:

$$Monooxygenases : substrate + \tfrac{1}{2}O_2 \rightarrow substrate = O \text{ or } substrate - OH$$

Deoxygenases catalyse the removal of oxygen from a substrate, and thus are a special case of reductases.

Peroxidases are oxygenases that employ hydrogen peroxide in the oxygenation reaction:

$$substrate + H_2O_2 \rightarrow substrate = O \text{ (or } substrate - OH) + H_2O$$

Dismutases catalyse the disproportionation (dismutation) of oxygen species containing oxygen in the oxidation state -I (peroxide) or $-\tfrac{1}{2}$ (superoxide). Accordingly, one distinguishes between catalases and superoxide dismutases:

Catalases: $H_2O_2 \rightarrow H_2O + \tfrac{1}{2}O_2$

Sub-steps : $H_2O_2 \rightarrow O_2 + 2H^+ + 2e^-$ (oxidation)
$H_2O_2 + 2H^+ + 2e^- \rightarrow 2H_2O$ (reduction)

Superoxide dismutases: $2O_2^{\bullet-} + 2H^+ \rightarrow H_2O_2 + O_2$

Sub-steps : $O_2^{\bullet-} \rightarrow O_2 + e^-$ (oxidation)
$O_2^{\bullet-} + e^- + 2H^+ \rightarrow H_2O_2$ (reduction)

## 6.1  Ribonucleotide reductases

Ribonucleotide reductases (RNRs) catalyse the reduction of nucleotides, the building blocks of ribonucleic acids (RNA), to deoxyribonucleotides and thus provide the building blocks for deoxyribonucleic acids (DNA). The catalytically active site of a RNR contains two iron ions (Scheme 6.1), or two manganese ions, or manganese plus iron. The manganese variant is expressed under iron-limited and oxidative stress conditions [1].

The reduction process is initiated by a diferric-tyrosyl radical, $Fe_2^{III}$-TyrO•, generated from the diferrous iron centre and a nearby tyrosine TyrOH[1] via an 'intermediate X' (Scheme

[1] Ribonucleotide reductases employing the tyrosyl radical (TyrO•) as a cofactor belong to the class I RNRs. In subclass Ic, with a {Mn,Fe} centre, Tyr is replaced by phenylalanine, and the role of TyrO• is adopted by $Mn^{IV}$. In class II RNRs, vitamin $B_{12}$ (adenosylcobalamin) is involved: here, the homolytic cleavage of the Co-carbon(adenosyl) bond provides the radical.

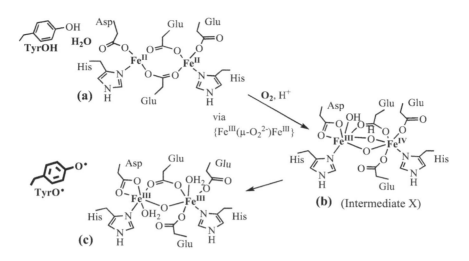

Scheme 6.1 The diiron centre of a ribonucleotide reductase in its reduced (a) and oxidized form (c). The 'intermediate X' [2], (b), with one of the iron centres in the oxidation state +IV, picks up an electron from TyrOH, thus generating a TyrO• radical. The interaction of tyrosine with the uncoordinated carboxylate oxygen of aspartate (a) is mediated by a water molecule. The distance of the Tyr-OH to the proximal iron centre is ca. 6.5 Å.

6.1). The overall four-electron process of the formation of $Fe_2^{III}$-TyrO• is described by Eq. (6.1). Two of the electrons are delivered via the oxidation—by $O_2$—of ferrous to ferric iron, one electron is provided by tyrosine, and the fourth electron is delivered 'externally' by a tryptophan from the protein surface.

$$\{Fe^{II}Fe^{II}\}+O_2+TyrOH+H^++e^- \rightarrow \{Fe^{III}\text{-}OH(\mu\text{-}O)Fe^{IV}\}+TyrOH$$

$$\rightarrow \{Fe^{III}(\mu\text{-}O)Fe^{III}\}\text{-}TyrO•+H_2O \tag{6.1}$$

The tyrosyl radical Tyr-O• conveys the abstraction of a hydrogen atom from the OH group in the 2' position of ribose [3], Eq. (6.2), via a thiyl radical CysS• (not shown in Eq. (6.2)). Concomitantly, disulfide CysS–SCys forms, the re-reduction of which is achieved by flavin mononucleotide $FMNH_2$.[2]

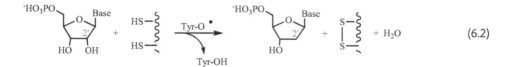

$$\tag{6.2}$$

The dimanganese RNR differs only slightly from the diiron RNR [1]. As shown in Figure 6.1a, *three* glutamates bridge the two manganese centres in the reduced ($Mn^{2+}$) form, one of which is in the $\mu\text{-}(\eta^1,\eta^2)$ coordination mode. Terminal water is coordinated to each of the manganese sites. This necessitates loss of an aqua ligand during the reaction with the oxidant in the

---

[2] For $FMN/FMNH_2$, see Sidebar 9.2.

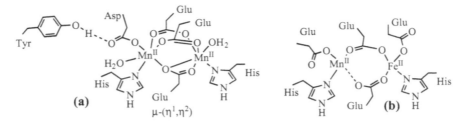

Figure 6.1 The active centres (reduced form) of the two-manganese ribonucleotide reductase (RNR) (**a**) [1], and the manganese–iron RNR (**b**) [4]. The distance of the Tyr-OH to the proximal manganese centre in (**a**) is 5.8 Å. In the {Mn,Fe} RNR (**b**), Tyr is replaced by Phe (not shown).

course of catalytic turnover. Aspartate is directly hydrogen bonded to the tyrosine. As in the case of the two-iron RNR, oxidation of the two-manganese RNR $\{Mn_2^{II}\}$ yields $\{(Mn^{III})_2\}$-TyrO·, formed via an intermediate $\{Mn^{III}, Mn^{IV}\}$ state.

In the ribonucleotide reductase of the heterometal type {Mn,Fe} (**b** in Fig. 6.1), the iron site of the {Fe,Fe} RNRs adjacent to the tyrosine (radical), Scheme 6.1, is occupied by manganese, and the aspartate is substituted for glutamate. In addition, the tyrosine present in the {Fe,Fe} and {Mn,Mn} reductases is replaced by phenylalanine in {Mn,Fe}; and a $\{Mn^{IV}, Fe^{III}\}$ radical, formed from $\{Mn^{II}, Fe^{II}\}$ upon reaction with oxygen, is used in place of the tyrosyl radical for mediating the reduction of ribose in RNA [4]. The hetero-dinuclear {Mn,Fe} cofactor is preferentially employed by extremophiles.

## 6.2 Superoxide dismutases, superoxide reductases, and peroxidases

The radical anion superoxide $O_2^-$, a reactive oxygen species formed by the diverse side reactions associated with oxygen metabolism, causes oxidative stress, for example by damaging cellular components such as DNA. Superoxide dismutases (SODs) catalyse the 'detoxification' of $O_2^-$ by its disproportionation to hydrogen peroxide and oxygen, Eq. (6.3), using cofactors based on (i) iron, (ii) manganese, (iii) iron or manganese, (iv) copper+zinc, or (v) nickel. For details of the nickel-containing SOD, see Section 10.4. Mn- and Fe-SODs are homologous with respect to their active site structure and their overall protein fold.

The disproportionation proceeds via a two-step ping-pong mechanism of alternate oxidation and reduction of superoxide, Eqs. (6.4a) and (6.4b).

$$2O_2^- + 2H^+ \rightarrow H_2O_2 + O_2 \tag{6.3}$$

$$O_2^- + \{M^{(n+1)+}\text{-}OH\} + H^+ \rightarrow O_2 + \{M^{n+}\text{-}OH_2\} \tag{6.4a}$$

$$O_2^- + \{M^{n+}\text{-}OH_2\} + H^+ \rightarrow H_2O_2 + \{M^{(n+1)+}\text{-}OH\} \tag{6.4b}$$

$(n=1: Cu; n=2: Fe, Mn, Ni)$

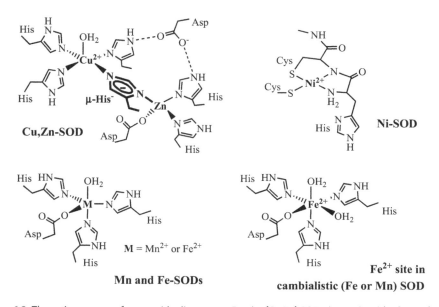

Figure 6.2 The active centres of superoxide dismutases. For the {Cu,Zn} SOD, the oxidized (Cu$^{2+}$ state) is shown. In the reduced form of the Cu/Zn SOD, the Cu–N bond to the bridging His(1–) is broken. For the other SODs, the reduced forms (Ni$^{2+}$, Fe$^{2+}$ and Mn$^{2+}$) are provided. In the oxidized (Fe$^{3+}$ and Mn$^{3+}$) versions of the Fe and Mn SODs, the axial ligand is OH$^-$. The site occupied by the aqua ligand in the ferrous form of the cambialistic SOD is the likely binding site for superoxide.

The active centres of the SODs are illustrated in Fig. 6.2. In the Cu,Zn SODs [5], Zn$^{2+}$ ensures the structural integrity of the protein but is not directly involved in electron transfer. The Cu,Zn-SODs are present in the cytoplasm (as homo-dimers) as well as in mitochondria and the extracellular space (in the form of homo-tetramers) in virtually all eukarya. Mutations of this enzyme can diminish or prevent quenching of superoxide. Mutations in SODs, or the impaired provision and/or incorporation of copper by Cu-trafficking proteins (so-called 'Cu-chaperones'), are linked to neuro-degenerative diseases such as amyotrophic lateral sclerosis.

The presence of Ni-SOD is restricted to prokarya, while the Fe- and Mn-SODs are found in both prokarya and eukarya. Iron- and manganese-based SODs are functional either with Fe *or* Mn [6], or they are active in the presence of either of Fe *and* Mn. The latter rare case, referred to as *cambialistic* SOD, has been documented for a hyperthermophilic archaeon [7].

The reaction course for the disproportionation of O$_2^-$ catalysed by the Cu,Zn SOD is shown in Scheme 6.2. The reaction proceeds according to the following successive steps:

1. The first superoxide anion is coordinated to Cu$^{2+}$. This state is stabilized by interaction between O$_2^-$ and the $=$NH$_2^+$ function of an adjacent arginine.

2. Superoxide reduces Cu$^{2+}$ to Cu$^+$; the resulting neutral O$_2$ is released. A second O$_2^-$ coordinates to Cu$^+$, and concomitantly the bridging histidine(1–) is protonated and becomes detached from the cuprous ion.

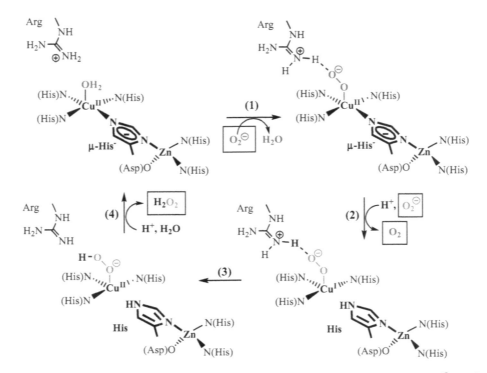

**Scheme 6.2** The disproportionation of superoxide by Cu,Zn-superoxide dismutase. The substrate $O_2^-$ and the products that are released ($O_2$ and $H_2O_2$) are boxed. For a detailed description of steps (1) to (4), see the main text.

3. $Cu^+$ reduces the coordinated superoxide to peroxide ($Cu^+$ is oxidized to $Cu^{2+}$), and the peroxide is protonated to hydroperoxide.

4. The initial state is recovered by release of $H_2O_2$, protonation of arginine, and re-establishment of the bond between the cupric ion and histidine($1-$).

For anaerobic organisms, SODs are counter-productive, since they produce oxygen. In order to cope with superoxide, however, superoxide *reductases* (SORs) instead of the dis-mutases are employed.[3] The terminal electron donor in SORs is either a monomeric, penta-coordinate iron centre ({$Fe(His)_4Cys$}, (**a**) in Fig. 6.3) [8], or a dinuclear iron centre belonging to the rubrerythrins, (**b**) in Fig. 6.3 [9].

In the ferrous state of the mononuclear centre, Figure 6.3a, four histidine residues are coordinated in the equatorial plane of a tetragonal pyramid, the axial site of which is occu-pied by cysteinate. The second axial position serves to bind superoxide, and facilitates the electron transfer to superoxide, Eq. (6.5a). Re-reduction of the ferric to the ferrous state, Eq. (6.5b), is accomplished by an external electron donor [8].

$$\{Fe^{2+}(His)_4Cys\}O_2^- + 2H^+ \rightarrow \{Fe^{3+}(His)_4Cys\} + H_2O_2 \tag{6.5a}$$

---

[3] SORs are not restricted to anaerobic bacteria, but have also been found in microaerophiles, i.e. organisms thriving in environments with a low level of oxygen.

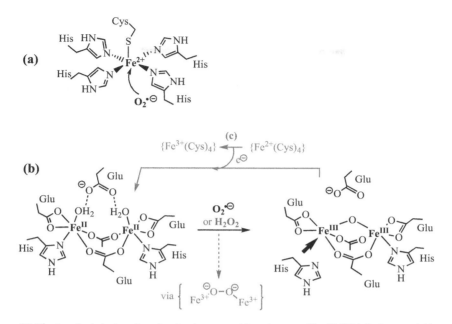

Figure 6.3 The terminal electron transfer sites in superoxide reductases. The {Fe(His)$_4$Cys} centre (a) has an open axial position available for coordination and reduction of superoxide. The rubrerythrin site (b) is also present in some peroxidases, for which a peroxido intermediate is proposed [9]. The electron relay from {Fe$^{2+}$(Cys)$_4$} (c) to the oxidized form of rubrerythrin is shown in blue; electrons are delivered one by one. The iron centre that receives the electron is indicated by a bold arrow. For {Fe(Cys)$_4$}=rubredoxin, see Sidebar 5.1.

$$\{Fe^{3+}(His)_4Cys\} + e^- \rightarrow \{Fe^{2+}(His)_4Cys\} \tag{6.5b}$$

In many cases, the electron involved in the reduction of Fe$^{3+}$ to Fe$^{2+}$ is transferred across the protein by a distant rubredoxin-type {Fe(Cys)$_4$} centre (we discuss rubredoxin in Sidebar 5.1). The reduction equivalents (the electrons) are delivered to {Fe(Cys)$_4$} by NAD(P)H, and then relayed to the SOR, (c) in Fig. 6.3.

The mechanism of electron transfer by rubrerythrins in SORs resembles that of iron-based *peroxidases* with rubrerythrins constituting the active centre. Peroxidases target (and thus eliminate) hydrogen peroxide, generated with the help of SODs and SORs, for the oxidation of organic substrates, Eq. (6.6).

$$RH + H_2O_2 \rightarrow ROH + H_2O \tag{6.6}$$

## 6.3 Oxygenases and oxidases

Oxygenases and oxidases are ubiquitous in all organisms. Monooxygenases catalyse the incorporation of one atom of O$_2$ into an organic substrate, while dioxygenases incorporate both; oxidases catalyse the oxidative dehydrogenation of a substrate, accompanied by the reduction of $O_2$ to $H_2O$ (via $H_2O_2$). Oxygenases and oxidases contain copper or iron at their catalytic site(s).

In the following, we will first provide a general overview of selected oxygenation and oxidation reactions, and then provide a more detailed insight into mechanistic aspects.

Examples of Cu-based enzymes are galactose oxidase (an oxidase) and tyrosinase (an oxygenase and oxidase). Galactose oxidase [10] catalyses the two-electron reduction of primary alcohols to aldehydes, Eq. (6.7a). The reduction equivalents delivered in this reaction are then employed for the reduction of $O_2$ to $H_2O_2$, Eq. (6.7b).

$$RCH_2OH \rightarrow RCHO + 2H^+ + 2e^- \tag{6.7a}$$

$$O_2 + 2H^+ + 2e^- \rightarrow H_2O_2 (\rightarrow H_2O + \tfrac{1}{2}O_2) \tag{6.7b}$$

Tyrosinases catalyse the successive hydroxylation of phenols to o-hydroquinones (oxygenase activity) and the oxidation of hydroquinones to quinones (oxidase activity). An example is the oxidative hydroxylation of tyrosine to Dopa, the precursor for the neurotransmitter dopamine,[4] Eq. (6.8), followed by the oxidation of Dopa to the respective quinone, Eq. (6.9); this species is then further oxidized to indole quinone. The formation of the brown skin pigment melanin, and the browning of fruit and vegetables are examples of such activity [11].

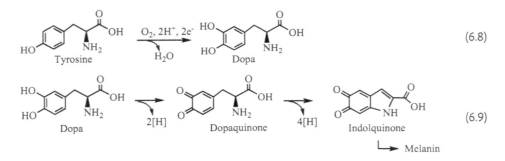

$$\tag{6.8}$$

$$\tag{6.9}$$

The oxidation of Dopa to the o-quinone in Eq. (6.9) is also catalysed by catechol oxidase. Catechol oxidases generally convert catechols oxidatively into quinones. The same enzyme also catalyses the oxidation of ascorbate (vitamin C) to dehydro-ascorbate, and is termed 'ascorbate oxidase' in this context [12], Eq. (6.10). The acceptor for the reduction equivalents [H] in Eq. (6.10) can be $O_2$, providing vitamin C with antioxidant properties.

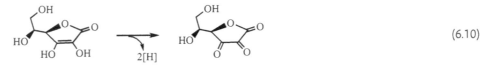

$$\tag{6.10}$$

Many oxidations and oxygenations are catalysed by members of the $P_{450}$ enzyme superfamily [13] with a haem iron as the prosthetic group.[5] A typical reaction is the oxygenation

---

[4] Dopa is L-3,4-dihydroxyphenylalanine; in dopamine, the carboxylic acid function is replaced by an amino group.

[5] See Sidebar 5.2. The denotation $P_{450}$ refers to the absorption maximum (at 450 nm) of the carbon monoxide adduct of the ferrous form of the cytochrome.

of an organic substrate, RH, to an alcohol, ROH, Eq. (6.11), with concomitant oxidation of NADPH.

$$RH + O_2 + NADPH + H^+ \rightarrow ROH + H_2O + NADP^+ \tag{6.11}$$

Another group of iron-based enzymes relies on Rieske centres.[6] The oxidative demethylation of the herbicide 'dicamba', Eq. (6.12), is an example of a reaction that features an enzyme with a terminal Rieske oxygenase [14]. Along with the Rieske type diiron prosthetic group, the active site of this enzyme incorporates an {Fe(His)$_2$Glu} centre. During the course of this reaction, the methyl group of dicamba (=2-methoxy-3,6-dichlorobenzoic acid) is oxygenated to formaldehyde.

$$\tag{6.12}$$

Copper and iron centres can also cooperate in enzymatic oxidation reactions, exemplified by cytochrome-c oxidase, Cyt-c. Cyt-c oxidase couples the oxidation of cytochrome-c with the reduction of O$_2$. This process, which terminates the respiratory chain, is outlined in Fig. 5.8 of Section 5.3.

The mechanisms of function of selected enzymes that we briefly sketch above with respect to the reactions they catalyse, will now be considered in more detail, focussing on structural and mechanistic aspects.

Tyrosinases contain two copper centres (type 3), with both copper ions in a trigonal-pyramidal coordination environment provided by three histidines. In the di-cuprous form, the distance between the two Cu$^+$ is 3.4 Å. The proposed mechanism for the oxygenation of tyrosine proceeds according to the following reaction steps (Scheme 6.3).

1. Oxygen oxidizes the two cuprous centres and is oxidatively added—in the form of peroxide—in between the two Cu$^{2+}$ ions. The coordination mode of the peroxido ligand is symmetrically side-on bridging, hence $\mu{:}\eta^2,\eta^2$.

2. Tyrosinate coordinates to one of the cupric sites, and O$_2^{2-}$ is protonated to HO$_2^-$.

3. One of the oxygens of the hydroperoxido ligand is transferred into the *ortho* position of TyrO$^-$, forming a catecholate. Simultaneously, a His is protonated and becomes detached from Cu.

4. (a) Protonation of the catecholate provides catechol, restores the coordination of His, and leaves a $\mu$-oxido bridged {Cu$^{2+}$} centre.

   (b) In a side reaction, catecholate directly delivers two electrons to the two cupric centres, thus forming an *o*-quinone and an aqua-bridged di-cuprous centre.

5. Proceeding from (4a), the starting situation is recovered by a two-electron reduction of {Cu$^{2+}$}$_2$ to {Cu$^+$}$_2$, and protolytic removal of the oxido bridge.

---

[6] See Fig. 5.7 in the previous chapter.

## Sidebar 6.2    Copper proteins: classification

Copper is an essential element for all living organisms. A human being weighing 70 kg contains ca. 150 mg of Cu bound to proteins. Both the under- and oversupply of copper can cause severe dysfunctions of physiological processes, briefly outlined in Section 14.2.2. In copper-based enzymes, copper's activity sees it switching between the cupric ($Cu^{2+}$) and the cuprous ($Cu^+$) oxidation states.

With respect to the geometry and ligand sphere of copper centres in copper proteins, and the related spectroscopic characteristics, one discriminates between the following main classes of copper proteins: types 1, 2, and 3, $Cu_A$, $Cu_B$, and $Cu_Z$. The typical coordination arrangements of types 1, 2, and 3, $Cu_A$ and $Cu_B$ are shown full colour in Plate 5; for $Cu_Z$ see Fig. 9.5 in Section 9.3.

**Type 1**, also referred to as 'blue copper proteins': Copper is in a trigonal-planar or flat trigonal-pyramidal coordination environment of a sulfur (cysteinate) and two nitrogen ligands, commonly the Nε of two histidines. A ligand-to-metal ($CysS^- \rightarrow Cu^{2+}$) charge transfer at 600 nm (extinction coefficient 3000 $M^{-1}cm^{-1}$) is responsible for the blue colour.

EPR characteristics: Four signal components by coupling of the single electron of $Cu^{2+}$ ($d^9$) with the nuclear spin $I = 3/2$ of the isotope $^{65}Cu$ (70% abundance). The nuclear hyperfine quadrupole coupling constant, $A_{\parallel} = 5 \times 10^{-4} cm^{-1}$, is particularly small.

Type 1 Cu proteins are mostly involved in $e^-$ transport. Examples of single type 1 copper centres are plastocyanin and azurin. In azurin, there are two weakly interacting axial ligands (methionine and a backbone carbonyl) in addition to the equatorial $His_2Cys$ set.

**Type 2**: Copper is typically in a square planar (or pyramidal) coordination environment of three histidines plus an additional O- or N-functional amino acid residue and, in several cases, an axial O-donor. Optical behaviour and EPR spectroscopic hyperfine coupling constant ($A_{\parallel} = 18 \times 10^{-4} cm^{-1}$) are 'normal'.

Type 2 copper centres are functional units in oxidases, reductases, oxygenases, and dismutases; examples are nitrite reductase, galactose oxidase, and Cu,Zn superoxide dismutase.

**Type 3**: Two closely spaced copper centres, in a trigonal coordination environment provided by three histidine residues. In the oxidized form, the two cupric sites are bridged by $OH^-$, $O^{2-}$, or $O_2^{2-}$, μ-{O}. The oxidized forms of the type 3 centres are blue due to a μ-{O} $\rightarrow Cu^{2+}$ charge transfer. The two $Cu^{2+}$ centres are strongly coupled antiferromagnetically, and hence are EPR-silent.

Type 3 centres function as transporter for oxygen (haemocyanins) and promote oxygen activation (catecholoxidase, tyrosinase).

Combinations of type 1, type 2, and type 3 copper proteins are sometimes referred to as type 4. Examples are ceruloplasmin and ascorbate oxidase. Ceruloplasmin contains six Cu centres (three type 1, two type 3, and one type 2); it functions as a copper storage protein and ferrous oxidase. Ascorbate oxidase, a four-copper centre, accommodates one representative each of types 1, 2, and 3.

There are additional copper sites that do not fit into this scheme: In **$Cu_A$ centres**, two copper ions are coordinated by two His, one Met, a protein backbone carbonyl, and two bridging $Cys^-$. In **$Cu_B$ centres**, copper is coordinated to three His in a trigonal-pyramidal arrangement; $Cu_B$ centres are thus related to type 3. Examples of $Cu_A$ and $Cu_B$ are the primary and final electron transfer units of cytochrome-c oxidase. Nitrous oxide reductase contains a **$Cu_Z$ centre**. Here, four copper ions are coordinated by seven histidine residues, and additionally bridged by sulfide.

Ascorbate oxidase, the catalyst for the oxidation of ascorbate to dehydroascorbate, Eq. (6.10), is an example of a multi-copper enzyme, containing the mononuclear types 1 and 2, and the dinuclear type 3 copper centre; see Fig. 6.4. $O_2$ is reduced to $H_2O$ at the dinuclear site.

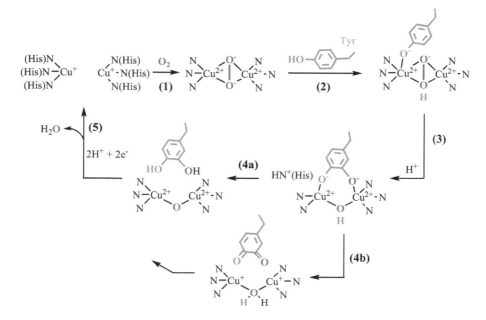

Scheme 6.3 The proposed mechanism for the oxidation of phenols (such as tyrosine) to catechols (such as Dopa) and *o*-quinones, as catalysed by tyrosinase. See text for details.

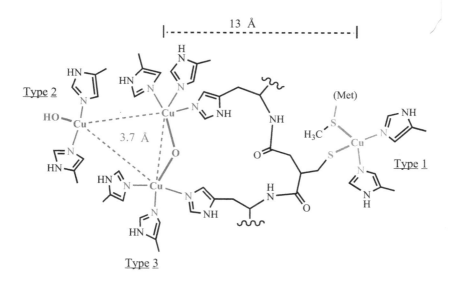

Figure 6.4 The four-copper centre of ascorbate oxidase contains three distinct copper sites (types 1, 2, and 3; see also Sidebar 6.2 for the classification of copper enzymes).

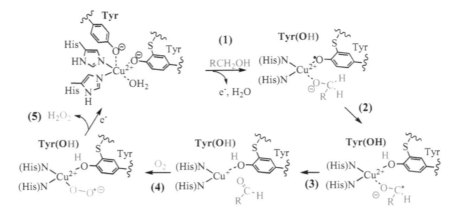

Scheme 6.4 The oxidation of primary alcohols to aldehydes at the catalytic centre of galactose oxidase. Dashed lines are weak bonds. For steps (1) to (5) see the text.

The enzyme galactose oxidase (GO) is responsible for the oxidation of galactose to galactohexodialdose[7] and, more generally, for the oxidation of primary alcoholic functions to aldehydes, as depicted in Eq. (6.7a). GOs are type 2 copper enzymes. The reaction course, assembled in Scheme 6.4, comprises the following reaction steps:

1. Exchange of $H_2O$ for the alcoholate $RCH_2O^-$, coupled to the protonation of the axial tyrosine and its concomitant de-coordination, and further coupled to the removal of an $e^-$ from the equatorial Tyr (linked to the protein matrix via sulfide provided by cysteinate) $TyrO^- \rightarrow TyrO^{\cdot} + e^-$.

2. Transfer of $H^{\cdot}$ from the coordinated alcoholate to the Tyr radical: $RCH_2O^- + TyrO^{\cdot} \rightarrow RC^{\cdot}HO + TyrO^-$.

3. Reduction of $Cu^{2+}$ to $Cu^+$ through $RCHO^{\cdot}$ and detachment of the aldehyde.

4. Coordination of oxygen $O_2$ to the free coordination site, and simultaneous oxidation of $Cu^+$ to $Cu^{2+}$/reduction of $O_2$ to superoxide.

5. Reconstitution of the initial situation by the reduction plus protonation of superoxide to hydrogen peroxide.

Iron oxygenases can be haem-iron or non-haem iron enzymes. As noted, the haem-iron oxygenases act on cytochrome $P_{450}$. The proposed reaction path is shown in Scheme 6.5:

1. The organic substrate RH is bonded through hydrophobic interaction into the protein pocket in the proximity of the ferric centre (but does not directly coordinate to Fe); concomitantly, the axial aqua ligand is released.

2. Ferric iron is reduced to ferrous iron.

3. Oxygen coordinates to iron via oxidative addition, forming superoxide $(Fe^{2+} + O_2 \rightarrow Fe^{3+} - O_2^-)$.

[7] Galactohexodialdose is the $2e^-$ oxidation product of galactose, a hexa-aldose. The two aldehyde functions are at C1 and C6.

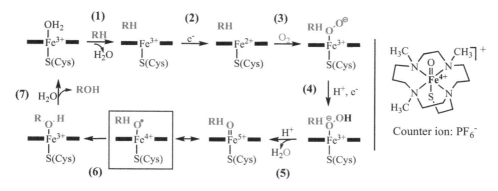

Scheme 6.5 The reaction course for the oxidation of hydrocarbons catalysed by $P_{450}$. The compound at the right, with the pentadentate $N_4S$ ligand derived from trimethyl-cyclam [15], models the boxed intermediate with $Fe^{IV}$. See text for steps (1) to (7).

4. Further reduction of the superoxido ligand by external electron delivery generates the hydroperoxido moiety $Fe^{3+}$-$O_2H^-$.

5. Hydroperoxide is then reduced to water in a two-electron transfer from $Fe^{3+}$ to $HO_2^-$. The oxido-$Fe^{5+}$ intermediate thus formed is resonance-stabilized with an $Fe^{4+}$-oxyl radical.

6. RH now adds heterolytically to the oxyl intermediate, forming an alcohol coordinated to the ferric centre.

7. Exchange of the alcohol ROH for water restores the initial situation.

The intermediate high-valence states of iron have been modelled with various ligand systems [15], one of which is shown in Scheme 6.5.

In non-haem iron oxygenases, the iron at the catalytic site is typically coordinated by two histidines and an acid residue (glutamate or aspartate). Within this latter group, the

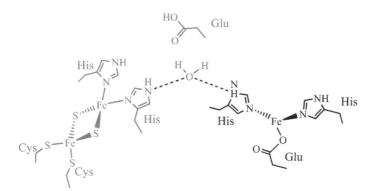

Figure 6.5 The active site of the the dicamba O-demethylase/monooxygenase according to [14a]. The Rieske centre is in blue, the {Fe(His)$_2$(Glu)} site in black. Only part of the hydrogen bonding network (in grey) is shown. For the reaction catalysed by this enzyme, see Eq. (6.12) on p. 73.

so-called 'Rieske oxygenases' contain, in addition to the $\{Fe(His)_2(Glu/Asp)\}$ site, a Rieske centre, $\{Cys_2Fe(\mu\text{-}S)_2Fe(His)_2\}$, Fig. 6.5. Reduction equivalents are delivered to these centres by NAD(P)H, flavins, or ferredoxins. The reaction sequence, with NADH as the primary electron donor, is depicted in Eq. (6.13). The activation of $O_2$ in the final step of this reaction sequence likely occurs via a ferric hydroperoxido intermediate, $\{Fe^{3+}OOH\}$.

$$NADH \rightarrow NAD^+ + H^+ + 2e^-, \text{ or } [FeS]^{red} \rightarrow [FeS]^{ox} + e^- \qquad (6.13a)$$

$$[Rieske]^{ox} + e^- \rightarrow [Rieske]^{red}; \; [Rieske]^{red} \rightarrow [Rieske]^{ox} + e^- \qquad (6.13b)$$

$$\{Fe^{3+}(His)_2Glu\} + e^- \rightarrow \{Fe^{2+}(His)_2Glu\} \qquad (6.13c)$$

$$\tfrac{1}{2}O_2 + \text{substrate} + \{Fe^{2+}(His)_2Glu\}(2\times) \rightarrow \text{substrate-OH} + \{Fe^{2+}(His)_2Glu\} \qquad (6.13d)$$

 ## Summary

The reduction of ribonucleotides to deoxyribonucleotides is catalysed by ribonucleotide reductase RNR, an enzyme that commonly contains a two-iron centre, but may also rely, in some organisms, on manganese or Mn plus Fe. The di-iron RNR uses $O_2$ to generate a diferric-tyrosyl radical. This tyrosyl radical initiates the abstraction, via the intermediate formation of a thiyl radical, of the 2'-OH group of ribose, thus converting ribose to deoxyribose.

In many processes of oxygen consumption and recirculation, superoxide $O_2^{\cdot-}$ is an intermediate or by-product. For the removal of this potentially toxic reactive radical, superoxide dismutases (SODs) and superoxide reductases (SORs) are available, containing the redox-active metals Fe, Mn, Ni, or Cu. SODs catalyse the dismutation of $O_2^{\cdot-}$ to $O_2$ and $H_2O_2$. An example is copper–zinc SOD with a type 2 copper centre; $Zn^{2+}$ takes over a structural role. SORs guide the reduction of $O_2^{\cdot-}$ to $H_2O_2$. Representatives of this group are enzymes with the cofactor $\{Fe(His)_4Cys\}$, or with rubrerythrin as the terminal electron donor. Rubrerythrins contain two iron centres bridged by two glutamates and additionally coordinated to histidine and glutamate.

Oxygenases catalyse the transfer of oxo groups (delivered by $O_2$) to a substrate, while oxidases promote the dehydrogenation of substrates. Both groups of enzymes rely on copper or iron. Examples of enzymes with Cu-based cofactors are galactose oxidase and tyrosinase. Galactose oxidase, with type 2 Cu in the active centre, catalyses the $2e^-$ oxidation of galactose to galactohexadialose or, more generally, the oxidation of alcohols to aldehydes. Tyrosinases are type 3 copper proteins that catalyse the oxygenation of tyrosine to Dopa, and the oxidation of Dopa to dopaquinone, indolquinone, and the brown skin pigment melanin. Ascorbate oxidases contain one of each Cu centres, type 1, type 2 and type 3; they oxidize ascorbate (vitamin C) to dehydroascorbate.

Iron-based oxygenases can contain haem or non-haem centres. Examples of the former are enzymes belonging to the cytochrome-$P_{450}$ family; an example of the latter is the dicamba demethylase/monooxygenase. Enzymes containing the cofactor $P_{450}$ insert oxygen into the C–H bond of a substrate—thus creating an alcoholic function—by passing through an $\{Fe^{4+/5+}(O)\}$ intermediate state. The active site of the dicamba oxygenase contains a Rieske plus an $\{Fe(His)_2(Glu)\}$ centre. This enzyme oxygenates a methyl ether function at the substrate; the methyl group is released as formaldehyde.

# Suggested reading

Högbom M. Metal use in ribonucleotide reductase R2, di-iron, di-manganese and heterodinuclear – an intricate bioinorganic workaround to use different metals for the same reaction. *Metallomics* 2011; 3: 110–120.
The build-up and function of active centres and protein surroundings of the ribonucleases is addressed on a comparative level, and is presented in relation to diiron centres of other $O_2$ activating enzymes such as monooxygenases.

Crichton RR and Declercq J-P. X-ray structures of ferritins and related proteins. *Biochim. Biophys. Acta* 2010; 1800: 706–718.
The proteins accommodating the two-iron centre of superoxide dismutases and reductases are members of a large superfamily of proteins, the ferritin superfamily. Structures of this superfamily are introduced comparatively.

Silavi R, Divsalar A, Saboury AA. A short review on the structure–function relationship of artificial catecholase/tyrosinase and nuclease activities of Cu-complexes. *J. Biomol. Struct. Dynamics* 2012; 30: 752–772.
The potential of model compounds of tyrosinases and catecholases in medicinal and environmental issues is examined.

# References

1. Boal AK, Cotruvo JA, Jr, Stubbe J, et al. Structural Basis for activation of class Ib ribonucleotide reductase. *Science* 2010; 329: 1526–1530.

2. (a) Mitić N, Clay MD, Saleh L, et al. Spectroscopic and electronic structure studies of intermediate X in ribonucleotide reductase R2 and two variants: a description of the Fe$^{IV}$–oxo bond in the Fe$^{III}$–O–Fe$^{IV}$ dimer. *J. Am. Chem. Soc.* 2007; 129: 9049–9065; (b) Shanmugam M, Doan PE, Lees NS, et al. Identification of protonated oxygenic ligands of ribonucleotide reductase intermediate X. *J. Am. Chem. Soc.* 2009; 131: 3370–3376.

3. Nordlund P and Reichard P. Ribonucleotide reductases. *Annu. Rev. Biochem.* 2006; 75: 681–706.

4. (a) Andersson CS, Öhrström M, Popović-Bijelić A, et al. The manganese ion of the heterodinuclear Mn/Fe cofactor in *Chlamydia trachomatis* ribonucleotide R2c is located at metal position 1. *J. Am. Chem. Soc.* 2012; 134: 123–125; (b) Roos K and Siegbahn PEM. Oxygen cleavage with manganese and iron in ribonucleotide reductase from *Chlamydia trachomatis*. *J. Biol. Inorg. Chem.* 2011; 16: 553–565.

5. (a) Antonyuk SV, Strange RW, Marklund SL, et al. The structure of human extracellular copper–zinc superoxide dismutase at 1.7 Å resolution: insight into heparin and collagen binding. *J. Mol. Biol.* 2009; 388: 310–326; (b) Mera-Adasme R, Mendizábal F, Gonzales M, et al. Computational studies of the metal binding site of the wild-type and the H46R mutant of the copper,zinc superoxide dismutase. *Inorg. Chem.* 2012; 51: 5561–5568.

6. Jackson TA, Gutman CT, Maliekal J, et al. Geometric and electronic structures of manganese-substituted iron superoxide dismutase. *Inorg. Chem.* 2013; 52: 3356–3367.

7. Nakamura T, Torikai K, Uegaki K, et al. Crystal structure of the cambialistic superoxide dismutase from *Aeropyrum pernix* K1 – insights into the enzyme mechanism and stability. *FEBS J.* 2011; 278: 598–609.

8. (a) Bonnot F, Duval S, Lombard M, et al. Intermolecular electron transfer in two-iron superoxide reductase: a putative role for the desulfredoxin center as an electron donor to the iron active site. *J. Biol. Inorg. Chem.* 2011; 16: 889–898; (b) Kurtz DM, Jr. Avoiding high-valent iron intermediates: superoxide reductase and rubrerythrin. *J. Inorg. Biochem.* 2005; 100: 679–693.

9. Dillard BD, Demick JM, Adams MWW, et al. A cryo-crystallographic time course

for peroxide reduction by rubrerythrin from *Pyrococcus furiosus. J. Biol. Inorg. Chem.* 2011; 16: 949–959.

10. Lee Y-K, Whittaker MM, Whittaker JW. The electronic structure of the Cys-Tyr• free radical in galactose oxidase determined by EPR spectroscopy. *Biochemistry* 2008; 47: 6637–6649.

11. (a) Olivares C and Solano F. New insights into the active site structure and catalytic mechanism of tyrosinase and its related proteins. *Pigment Cell Melanoma Res.* 2009; 22: 750–760; (b) Fairhead M and Thoeny-Meyer L. Bacterial tyrosinases: old enzymes with new relevance to biotechnology. *New Biotechnol.* 2012; 29: 183–191.

12. Quintanar L, Stoj C, Taylor AB, et al. Shall we dance? How a multicopper oxidase chooses its electron transfer partner. *Acc. Chem. Res.* 2007; 40: 445–452.

13. Fasan R. Tuning P450 enzymes as oxidation catalysts. *ACS Catal.* 2012; 21: 647–666.

14. (a) Dumitru R, Jiang WZ, Weeks DP, et al. Crystal structure of dicamba monooxygenase: a Rieske nonheme oxygenase that catalyzes oxidative demethylation. *J. Mol. Biol.* 2009; 392: 498–510; (b) D'Ordine RL, Rydel TJ, Storek MJ, et al. Dicamba monooxygenase: structural insights into a dynamic Rieske oxygenase that catalyzes an exocyclic monooxygenation. *J. Mol. Biol.* 2009; 392: 481–497.

15. Bukowski MR, Koehntop KD, Stubna A, et al. A thiolate-ligated non-heme oxoiron(IV) complex relevant to cytochrome P450. *Science* 2005; 310: 1000–1002.

# 7 Oxo-transfer proteins based on molybdenum, tungsten, and vanadium

The early transition metals vanadium, molybdenum, and tungsten are used in industrial contexts, commonly in steel production or, in the form of their oxides, to catalyse oxidation reactions, exploiting the ease of the change in oxidation states: +V and +IV in the case of vanadium; +VI, +V, and +IV in the case of molybdenum and tungsten. In addition, catalysts based on molybdenum sulfide are versatile desulfurization agents and thus are employed in the desulfurization of processed crude oil.

Nature also employs these metals as active centres in enzymes in a variety of oxidation and reduction reactions. For example, the conversion of dinitrogen to ammonium ions as catalysed by nitrogenase (containing a molybdenum–iron–sulfur cofactor)—a reduction—is described in Chapter 9.

Molybdenum in particular is indispensable for all organisms. In human beings, four molybdenum enzymes with redox functions have been identified. Examples of vital reactions catalysed by these enzymes are (i) the degradation of the purine metabolite xanthine (thus preventing renal failure), (ii) the detoxification of sulfite through its oxidation to sulfate, and (iii) the oxidation and thus detoxification of aldehydes (such as acetaldehyde, a metabolite of alcohol). In all of these reactions, Mo cycles through the +IV, +V, and +VI oxidation states.

The daily requirement of the essential trace element molybdenum for humans is about 150 to 500 µg; this corresponds to roughly 10 g of molybdenum within a life span of 75 years. Though one of the less abundant elements, Mo is comparatively omnipresent and, due to the water solubility of molybdate $MoO_4^{2-}$, easily available. However, acidification of the soil as a consequence of acid rain, and copper contamination of soils, disrupts the molybdenum balance. In acid media, polyoxidomolybdates form; the presence of copper gives rise to the formation of insoluble Cu,Mo cubane clusters, leading to severe problems for grazing cattle.

In addition to the overview in the present chapter, we provide brief accounts of molybdopyranopterins in Sections 9.3 (nitrate reductase) and 10.2 (formylmethanofuran dehydrogenase).

In some enzymes of prokarya—and thermophilic and hyperthermophilic archaea in particular—molybdenum can be replaced by tungsten; unlike molybdenum, however, tungsten does not play a global role in life.

Vanadium is also unable to take over functions in enzymatic processes in vertebrates and other phyla of more developed organisms. Vanadium can, however, replace molybdenum in some bacterial nitrogenases. Vanadium is also used in the active centre of vanadate-dependent haloperoxidases (VHPOs) in algae, fungi, lichen, and some *Streptomyces* bacteria.

Interestingly, and in contrast to Mo and W, vanadium does *not* operate in these VHPOs via a change in oxidation state. Rather, vanadate(V) functions as a Lewis acid centre.

## 7.1 Molybdo- and tungsto-pyranopterins

The cofactor ligand system that incorporates molybdenum is commonly referred to as pyranopterin or molybdopterin. In the literature, the term 'molybdopterin' is occasionally used for both the Mo-free and the Mo-loaded pterin ligand. In order to avoid confusion, we will employ the term 'pyranopterin' for the metal-free ligand system, and 'molybdo-pyrano-pterin' (or Mo-pyranopterin) for the ligand coordinated to Mo; we will refer to the corres-ponding tungsten variant as tungsto-pyranopterin. For the ligand pyranopterin, see Fig. 7.1.

In the Mo/W-pyranopterins, molybdenum and tungsten cycle between the oxidation states +IV, +V, and +VI, thus enabling two-electron transfer processes. These electrons are usually relayed sequentially to a second and third site, commonly a ferredoxin or haem-type protein and/or FAD.[1] The general redox reaction catalysed by the enzymes is shown in Eq. (7.1) [1], where X and XO can be organic or inorganic substrates.

$$X + H_2O \rightleftharpoons XO + 2e^- + 2H^+ \tag{7.1}$$

The reaction may run from left to right (catalysed by oxidases/dehydrogenases), from right to left (reductases), or in either direction (oxidoreductases). However, not all of the pro-cesses fit into this general scheme. Exceptions include the transmutations catalysed by for-mate dehydrogenase, carbon monoxide dehydrogenase, and acetylene hydratase, AH. The reaction promoted by AH—the hydration of acetylene—is the only process catalysed by the pyranopterin-based proteins where the net reaction is a true non-redox process.

One usually distinguishes between four families[2] of Mo/W-pyranopterins, which differ according to the coordination sphere of molybdenum or tungsten, as shown in Fig. 7.1. The tungsten enzymes, predominantly present in thermophilic archaea, commonly are members of the aldehyde:ferredoxin oxidoreductase and, in some cases, the DMSO reductase family. Four *mammalian* molybdoenzymes have so far been characterized: sulfite oxidase, xanthine oxidase, nitrate reductase, and the 'amidoxime reducing component'. The latter catalyses the reductive deoxygenation of N-hydroxylated compounds [2].

We will now consider an assortment of representatives of the four Mo/W-pyranopterin families, including the net reactions catalysed by the enzymes, and mechanistic aspects for selected reactions.

### 7.1.1 The xanthine oxidase family

This family is also referred to as 'xanthine dehydrogenase' and 'aldehyde oxidase' family.

Aldehyde oxidases: Aldehyde oxidases catalyse the cleavage of a C—H bond with con-comitant formation of a C—O bond, and thus the oxidation of aldehydes to the respective carboxylic acids, Eq. (7.2).

---

[1] For FAD, see Sidebar 9.2; for ferredoxins, Sidebar 5.1.
[2] Along with the four common families (xanthine dehydroxynase family, sulfite oxidase family, DMSO reductase family, aldehyde:ferredoxin oxidoreductase family), two tungsten-based enzymes (considered here within the DMSO reductase family) are sometimes treated as extra families, namely formate dehydrogenase and acetylene hydratase [1].

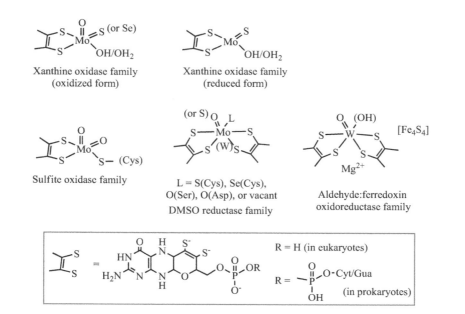

Figure 7.1 Representatives of the active sites of the four families of oxidases/reductases depending on molybdenum and tungsten. For the xanthine-oxidase family, the oxidized (Mo$^{VI}$) and reduced (Mo$^{IV}$) forms are provided. Only the fully oxidized forms are shown for the three other families. Formate dehydrogenase, a member of the DMSO reductase family, can also contain W instead of Mo. The inset illustrates the pyranopterin ligand.

$$RCHO + H_2O \rightarrow RCO_2H + 2H^+ + 2e^- \tag{7.2}$$

During catalytic turnover, Mo$^{VI}$ is reduced to Mo$^{IV}$, and the electrons are further delivered sequentially—mediated by the pyranopterin—to two non-identical [2Fe,2S] ferredoxins and FAD, as depicted in Fig. 7.2, and finally to NAD$^+$. Aldehyde oxidoreductases catalyse both the reaction represented by Eq. (7.2) *and* the back reaction. The same type of reaction is catalysed by the archaean aldehyde:ferredoxin oxidoreductase, which contains the tungsto-pyranopterin cofactor and a [4Fe,FS] ferredoxin as primary electron acceptor. We discuss the mechanistic aspects of aldehyde oxidases in the context of this tungsto enzyme in Section 7.1.4.

Xanthine dehydrogenase catalyses the oxidation of hypoxanthine to xanthine, and of xanthine to uric acid; for the latter step see Eq. (7.3). Hypoxanthine derives from guanine; the enzyme is thus involved in purine metabolism.

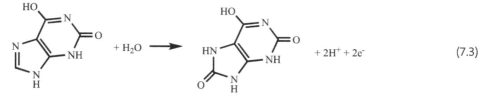

The main steps of the mechanism are provided in Scheme 7.1. In step (**1**), the hydroxido ligand at molybdenum is activated by an active site glutamate, allowing for the transfer of

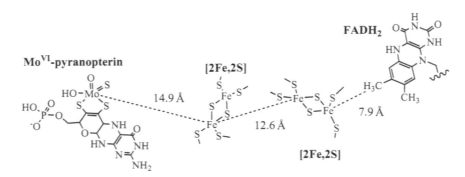

Figure 7.2 The alignment of metal cofactors involved in the relay of electrons from the molybdo-pyranopterin cofactor to flavin-adenine dinucleotide, FAD, in the xanthine oxidase family. Redrawn on the basis of Romão MJ. *Dalton Trans.* 2009; 4053–4068 (see Suggested reading).

an oxo group to the substrate with simultaneous hydride transfer from the substrate to the sulfido ligand, step (**2**). Molybdenum is thereby reduced from the +VI to the +IV state. The intermediate +V state is generated in step (**3**) by release of a reduction equivalent, [H] ≡ $H^+ + e^-$. Release of a second [H], and of the oxidation product uric acid by replacement of urate for a hydroxido ligand, step (**4**), restores the initial situation.

Carbon monoxide dehydrogenase converts carbon monoxide to carbon dioxide, Eq. (7.4). As in the case of aldehyde oxidases, the electrons are channelled from the molybdo-protein to an iron–sulfur protein (containing two [2Fe,2S] ferredoxins) and the FAD of a flavoprotein.

$$CO + H_2O \rightarrow CO_2 + 2H^+ + 2e^- \tag{7.4}$$

Carbon monoxide dehydrogenase is remarkable insofar as it contains a dinuclear molybdenum-copper centre [3]. As shown in Scheme 7.2, (**1**), the cuprous ion is in a slightly bent linear geometry, linked to molybdenum via a μ-sulfido ligand and to the protein via cysteinate. During turnover, CO is inserted between the copper–sulfide linkage, forming an intermediate thiocarbonate, $\mu{:}\eta^2,\eta^1{-}(CSO_2^{2-})$, that bridges $Mo^{VI}$ and $Cu^I$; step (**2**). In step (**3**), $CO_2$ is released with simultaneous reduction of $Mo^{VI}$ to $Mo^{IV}$. The initial situation is restored by the transfer of two electrons to a ferredoxin, step (**4**).

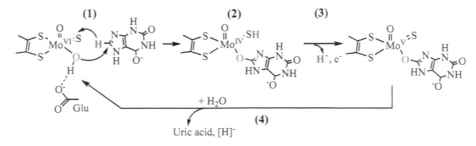

Scheme 7.1 The proposed catalytic cycle for the dehydrogenation/oxidation of xanthine to uric acid as catalysed by xanthine dehydrogenase.

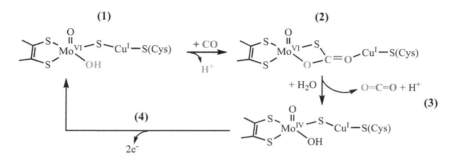

Scheme 7.2 Oxidation of CO to $CO_2$ at the Mo-Cu active centre of CO dehydrogenase. See text for steps (1) to (4).

### 7.1.2 The sulfite oxidase family

The oxidation of sulfite to sulfate, Eq. (7.5)—the final step in oxidative degradation of cysteine and methionine in higher organisms (plants and animals)—is catalysed by sulfite oxidases. For eukarya, sulfite oxidases are important for sulfite detoxification. *Bacterial* sulfite dehydrogenases have a key function in the global geochemical cycle of sulfur (Chapter 8).

$$HSO_3^- + H_2O \rightarrow SO_4^{2-} + 3H^+ + 2e^- \qquad (7.5)$$

Sulfite oxidases do not employ ferredoxins; rather the final electron recipient is a cytochrome *b* or *c* type haem iron. According to the proposed mechanism (Scheme 7.3), the initial step (1) is the attack of sulfite at the equatorial oxido group of the molybdenum centre, generating a transient sulfato-$Mo^{IV}$ complex. Sulfate is then hydrolytically released (2), and the starting situation is restored by the delivery of two electrons, one by one, to a nearby haem iron centre (3).

Nitrate reductases (NR) enable the reduction of nitrate to nitrite; Eq. (7.6). The NR belonging to the sulfite oxidase family is an assimilatory[3] nitrate reductase present in eukarya such as plants, algae and fungi. Prokaryotic nitrate reductases belong to the DMSO reductase

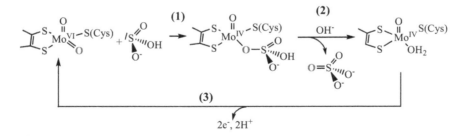

Scheme 7.3 The oxidation of hydrogensulfite to sulfate as mediated by sulfite oxidases.

[3] In this context, 'assimilatory' refers to the fact that the product (here: nitrite) remains in the organism, where it is further metabolized to ammonium ions. In 'dissimilatory' nitrate reduction, nitrate is also reduced to nitrite, and further to $N_2$ in several enzymatic steps, a process which is coupled to the formation of ATP.

family; they can be assimilatory or dissimilatory. Generally, nitrate reductases are important enzymes in the overall nitrogen cycle; for details, see Section 9.3. The final electron acceptor in the eukaryotic NRs are haem $b$ + FAD, in the prokaryotic NRs [4Fe,4S]/[3Fe,4S] ferredoxins + haem $b/c$.

$$NO_3^- + 2H^+ + 2e^- \rightarrow NO_2^- + H_2O \tag{7.6}$$

### 7.1.3 The dimethyl sulfoxide (DMSO) reductase family

The DMSO reductase family differs from the other families insofar as it is larger and more diverse with respect to the variability of the ligand sphere of the molybdenum or tungsten centre, and also with respect to the spectrum of reactions catalysed by its members. Subtle but important differences in the second coordination sphere at the catalytic site contribute to this diversity. The amino acid directly coordinated to the metal can be serine in DMSO reductase, cysteinate in acetylene hydratase, selenocysteinate[4] in formate dehydrogenase, aspartate in membrane-bound assimilatory nitrate reductase, or vacant as in arsenite oxidase.

DMSO reductase [4] catalyses the reduction of dimethyl sulfoxide to dimethyl sulfide, DMS, Eq. (7.7), a gas with the characteristic odour of cooked cabbage. DMS is an important gas in the global sulfur cycle (Chapter 8). In aquatic systems, microbial activity can convert $(CH_3)_2S$ oxidatively to DMSO,[5] a potential cryoprotectant.

$$(CH_3)_2SO + 2H^+ + 2e^- \rightarrow (CH_3)_2S + H_2O \tag{7.7}$$

Formate dehydrogenase can contain tungsten or molybdenum, linked to the protein via selenocysteinate, SeCys. Eq. (7.8) represents the overall reaction.

$$HCO_2^- \rightarrow CO_2 + H^+ + 2e^- \tag{7.8}$$

According to the proposed mechanism, formate binds directly to molybdenum by replacing SeCys, (1) in Scheme 7.4. The anionic SeCys is stabilized by a nearby cationic arginine, Arg$^+$. Transfer of hydride from the coordinated formate to the sulfido ligand, and electronic rearrangement of Mo$^{VI}$=S(H$^-$) to Mo$^{IV}$-SH with simultaneous release of $CO_2$ yields the reduced centre; step (2). In the final step, (3), the original state is restored by transfer of the electrons to a four-iron ferredoxin, [4Fe,4S].

Tungsten-based acetylene hydratase represents a rare case within the superfamily of Mo/W enzymes because it does not catalyse a redox reaction. Rather, the net reaction is the addition of the H and OH of water to the triple-bonded acetylene (ethyne) carbons, forming an intermediate vinyl alcohol (ethenol) that rapidly rearranges to acetaldehyde (ethanal) as the final product; Eq. (7.9), [5].

$$HC \equiv CH + H_2O \rightarrow \{H_2C = CHOH\} \rightarrow H_3C - CHO \tag{7.9}$$

---

[4] Not to be mistaken for selanylcysteine, with a HSe-S- terminus.
[5] An example for microbial conversion is the oxidation of DMS to DMSO by anaerobic photosynthetic purple bacteria.

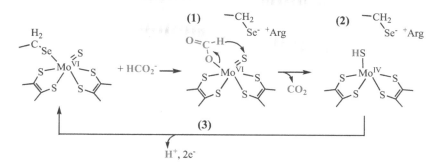

**Scheme 7.4** Proposed mechanism for the dehydrogenation of formate.

However, the activity of the acetylene hydratase requires activation by a strong reductant, likely indicating that the *active* form contains tungsten in the oxidation state +IV.

### 7.1.4 The aldehyde:ferredoxin oxidoreductase family

Aldehyde oxidoreductases catalyse the oxidation of aldehydes to carbonic acids (and vice versa) according to Eq. (7.2) by oxidatively inserting the oxo group of a water molecule into the C–H bond of the aldehyde function. The proposed mechanism [1b] is shown in Scheme 7.5 for the archaean tungsten-based aldehyde:ferredoxin oxidoreductase. In a first step (**1**), a nearby water molecule is activated by hydrogen bonding interaction with a glutamate in the vicinity of the active centre. Simultaneously, the aldehydic carbonyl group is activated by hydrogen bonding to a tyrosine. In step (**2**), the activated aldehyde is attacked electrophilically by the $W^{VI}$ centre. A hydride is then transferred to the oxido ligand, step (**3**), accompanied by reduction of $W^{VI}$ to $W^{IV}$. Re-oxidation of $W^{IV}$ to $W^{VI}$, step (**4**), occurs electron by electron, via $e^-$ transfer to a four-iron ferredoxin. The paramagnetic ($d^1$) tungsten(+V) state formed as an intermediate has been established by EPR spectroscopy [6].

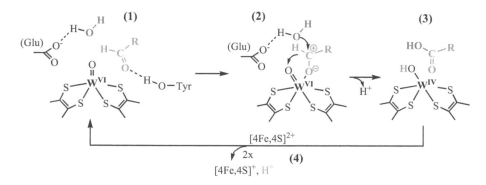

**Scheme 7.5** Proposed mechanism for the oxidation of aldehydes to carbonic acids, shown here for an archaean (tungsten-based) aldehyde:ferredoxin oxidoreductase. For the dithiolene ligand, see Fig. 7.1. Steps (**1**) to (**4**) are explained in the text.

Interestingly, the molybdenum substituted enzyme is inactive: as suggested by density functional calculations, the formation of {Mo$^{VI}$=O} according to step (4) is *endo*thermic by 59 kJ mol$^{-1}$ [7].

## 7.2 Vanadate-dependent haloperoxidases

Vanadium can substitute for molybdenum in the FeMo/V-cofactor of nitrogenase (Section 9.1) in some bacteria and cyanobacteria—in accordance with the similar chemistry of the two elements, reflecting their diagonal relationship.[6] In contrast, replacement of Mo by V in the molybdo-pyranopterin cofactor inactivates these enzymes. Nonetheless, several groups of organisms employ vanadium in oxo-transfer reactions from $H_2O_2$ to substrates such as halides (Cl$^-$, Br$^-$, I$^-$), pseudohalides (N$_3^-$, SCN$^-$), and organic sulfides. The oxidation of halides Hal$^-$ results in the formation of a two electron oxidation product Hal$^+$, such as HOBr, Br$_2$, and Br$_3^-$ in the case of Hal = Br. This reaction is exemplified by the formation of hypobromous acid in Eq. (7.10). This oxo-transfer, catalysed by vanadate-dependent haloperoxidases (VHPOs), corresponds to a two-electron oxidation of the substrate. In the absence of a substrate, singlet oxygen $^1O_2$ forms, Eq. (7.11); this rapidly decays to triplet-$O_2$.

$$Br^- + H_2O_2 + H^+ \rightarrow HOBr + H_2O \tag{7.10}$$

$$2H_2O_2 \rightarrow {}^1O_2 + 2H_2O \tag{7.11}$$

Prochiral organic sulfides are oxidized enantioselectively to sulfoxides, Eq. (7.12), a reaction type with an important potential for *in vivo* as well as *in vitro* organic syntheses.

$$RSR' + H_2O_2 \rightarrow R(R')S = O + H_2O \tag{7.12}$$

VHPOs are expressed by many marine macro algae, by fungi, lichen and some *Streptomyces* bacteria. The Hal$^+$ intermediates are potent halogenations reagents, Eq. (7.13); VHPOs thus supply a plethora of halogenated organic compounds—including antibiotics [8]—into the habitats of organisms. Due to their potential as efficient oxidants, Hal$^+$ such as present in hypohalous acids also act as anti-fouling agents. Fungi, which contain a chloroperoxidase and thus produce HOCl, exploit the particularly efficient oxidative potential of hypochlorite to degrade the lignocellulose cell walls of, and thus to penetrate into, their 'guests'.

$$RH + HOBr \rightarrow RBr + H_2O \tag{7.13}$$

In the active centre of VHPOs, vanadate $H_2VO_4^-$, is coordinated to the Nε of a histidine of the protein matrix [9]. Vanadium is in a trigonal bipyramidal coordination environment,[7] and experiences ion-pair and hydrogen bonding interactions with several active site amino acids (Fig. 7.3).

X-ray structure analyses of the algal bromoperoxidases indicate that the substrate bromide is not in direct contact with the vanadium centre. Rather, the interaction with vanadate

[6] Pairs of diagonally adjacent elements in the periodic table, such as vanadium (4$^{th}$ period, group 5) and molybdenum (5$^{th}$ period, group 6), can have similar atomic or ionic radii, and have related chemistry as a result.

[7] Interestingly, the same motif is present in some vanadate-inhibited phosphatases, a fact of relevance for the antidiabetic potential of vanadate (Section 14.3.4).

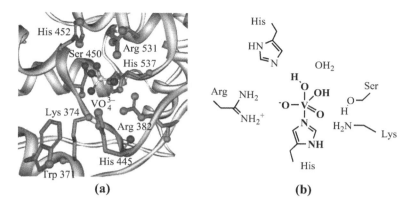

Figure 7.3 (a) The vanadate binding pocket of vanadate-dependent bromoperoxidase from the marine macro-alga *Ascophyllum nodosum*. Reprinted from [9b], p. 29. Copyright (2013), with permission from Elsevier. Part (a) kindly supplied by Dr Jens Hartung. (b) The coordination and protein environment of vanadium in VHPOs. *Panel (a) also reproduced in full colour in Plate 6.*

is mediated via an active site amino acid such as serine or arginine. The proposed overall reaction mechanism for the oxidation of halide to hypohalous acid by VHPOs is shown in Scheme 7.6. Accordingly, vanadate does not change between different oxidation states (such as +V and +IV) during turnover, but retains its +V state throughout the catalytic process [10]: In step (**1**), the oxido group of $H_2VO_4^-$ is exchanged for peroxide. The peroxido ligand is then protonated to hydroperoxide, (**2**), and the hydroperoxido ligand is attacked nucleophilically by bromide. Vanadium is in a distorted tetragonal-pyramidal environment in both the peroxido and the hydroperoxido intermediates. In step (**3**), hypobromous acid is released, and the original catalytic centre is restored.

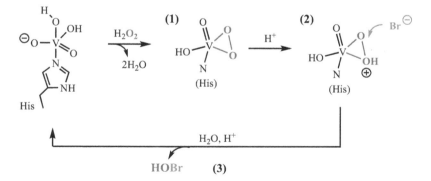

Scheme 7.6 The process of catalytic bromide oxidation at the active centre of vanadate-dependent bromoperoxidase.

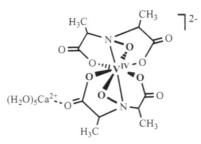

Figure 7.4 The molecular structure of amavadin from the fly agaric (*Amanita muscaria*). The overall charge of the {Ca(H$_2$O)$_5$}{V(hida)$_2$} system is zero; hida = N-oxyiminodiacetate(3−).

Several species of the mushroom genus *Amanita* contain a low molecular mass vanadium compound, dubbed amavadin [11]. An example is the fly agaric[8] (*A. muscaria*). In amavadin (Fig. 7.4), non-oxido vanadium(IV) is in an octa-coordinated environment of two trianionic ligands derived from N-hydroxyiminodiacetic acid, H$_3$hida. The two negative charges of the anionic complex [V(hida)$_2$]$^{2-}$ are equilibrated by Ca$^{2+}$. The role of amavadin is elusive. It is tempting, however, to assume that amavadin is a relic of an ancient but now redundant component of a redox-active enzyme. In any case, amavadin exhibits catalase activity, Eq. (7.11), and peroxidase activity, Eq. (7.14).

$$C_6H_{12} + H_2O_2 + Br^- + H^+ \rightarrow C_6H_{11}Br + 2\,H_2O \qquad (7.14)$$

## 7.3 Model chemistry

Why do Mo and W enzymes employ the pterin ligand, coordinated to the metal via its unusual dithiolene moiety? Dithiolene(2−), or ene-dithiolate(2−), is a 'non-innocent' ligand. In other words, the ligand is able to shift a ligand electron towards the metal centre, thus initiating a quasi one-electron reduction of the metal. This is illustrated in Scheme 7.7 by mesomeric formulae for the molybdenum-dithiolene moiety.

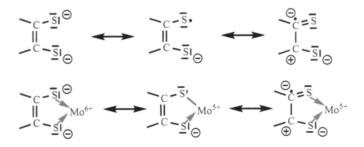

Scheme 7.7 Three mesomeric formulae of the Mo-dithiolene moiety. The blue arrow indicates the S$^-$ → Mo$^{VI}$ donor bond (the bonding electron pair is exclusively provided by the thiolate); the blue dash represents the S–Mo$^V$ covalent bond (where the electron pair is equally provided by S and Mo).

---

[8] *Amanita muscaria* is also known as toad stool.

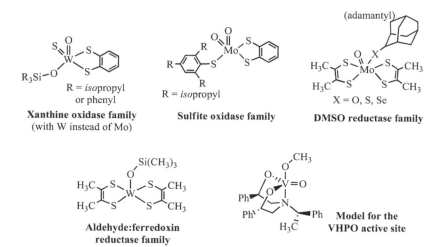

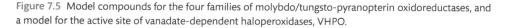

**Figure 7.5** Model compounds for the four families of molybdo/tungsto-pyranopterin oxidoreductases, and a model for the active site of vanadate-dependent haloperoxidases, VHPO.

The non-innocent character of the dithiolene moiety of the pyranopterin ligand facilitates the off- and on-transport of electrons in the catalytic cycle, but also allows for versatile electron transfer between model complexes and substrates. Models of the molybdo- and tungsto-pyranopterins are thus commonly based on complexes with rather simple non-biogenic dithiolene ligands. Selected examples [1a,12,13] that mimic the immediate coordination environment of the members of the four Mo/W-pyranopterin families are collated in Fig. 7.5.

A typical model for an oxo-transfer reaction is the oxidation of phosphines[9] $PR_3$ to phosphine oxides $R_3P=O$ with DMSO. The overall reaction as catalysed by the model complexes LMo=O (where L is the dithiolene ligand), Eq. (7.15), involves the transfer of the oxo group from the oxido-molybdenum(VI) moiety to the phosphine, Eq. (7.15a), and the deoxygenation of DMSO by return transport of the oxo group to the molybdenum(IV) moiety, Eq. (7.15b).

$$(CH_3)_2S=O+PR_3 \rightarrow (CH_3)_2S+O=PR_3 \tag{7.15}$$

$$LMo^{VI}O+PR_3 \rightarrow LMo^{IV}+O=PR_3 \tag{7.15a}$$

$$LMo^{IV}+(CH_3)_2S=O \rightarrow LMo^{VI}O+(CH_3)_2S \tag{7.15b}$$

Fig. 7.5 also contains a model for the active site in vanadate-dependent haloperoxidases. The complex models the trigonal-bipyramidal geometry and the $O_4N$ donor set of vanadium in VHPOs, as well as the enantioselective oxidation of sulfides to sulfoxides [14]; see Eq. (7.12) in Section 7.2.

---

[9] The term phosphane for $PR_3$, where R is H or any organic residue binding through C, is more correct (and is in accordance with IUPAC recommendations for nomenclature)—but is not in general use.

Summary

The elements molybdenum, tungsten, and vanadium are involved in a plethora of catalytically conducted two-electron transfer reactions, commonly accompanied by the attachment/insertion of $O^{2-}$ to/into a substrate (oxygenation) and/or the detachment of $O^{2-}$ from a substrate (deoxygenation). Hence, most of these active centres function as redox catalysts. An exception is acetylene hydratase; the net reaction catalysed by this enzyme is the hydration of ethyne to ethanal.

In the widespread Mo and the less common W enzymes, the metal centres are coordinated to one or two dithiolene groupings of pterin ligands. During turnover, the metal centres change between the oxidation states +VI and +IV, running through an intermediate +V state. The vanadium enzymes— vanadate-dependent haloperoxidases, VHPOs—contain vanadate $H_2VO_4^-$ linked via a histidine to the protein matrix. Vanadium does not change its oxidation state during turnover.

Molybdenum and tungsten enzymes can be allocated to four groups, depending on the ligands that are present in addition to the pterin(s). In the oxidized form, the members of the xanthine oxidase family contain the ligands $O^{2-}$, $OH/H_2O$, and $S^{2-}$ or $Se^{2-}$ coordinated to Mo, along with a single pterin. Members of the sulfite oxidase family, again exclusively based on Mo, carry two oxido and one cysteinate ligand, plus a pterin. In the DMSO reductase family, two pterins coordinate to the metal ion, in most cases Mo, and the coordination sphere is supplemented by $O^{2-}/S^{2-}$ and an amino acid function: cysteinate, selenocysteinate, serinate, or aspartate. Formate dehydrogenase is an example for a tungsten-based centre in this family. Finally, the aldehyde:ferredoxin oxidoreductase family, present in thermophilic archaea, contains tungsten ligated to two pterins and $O^{2-}$. In all cases, the electrons delivered by the substrate to the metal centres are relayed, mediated by the pterin, to external electron acceptors, viz. ferredoxins, cytochromes and FAD.

VHPOs can be present in algae, lichen, fungi and *Streptomyces* bacteria. They catalyse the two-electron oxidation, by $H_2O_2$, of halides $X^-$ to an $X^+$ species (for example hypohalous acid). The $X^+$ species can then halogenate secondary substrates, organic compounds in particular. Substrates alternative to halides include pseudohalides and sulfides.

The dithiolene moiety of the pyranopterins is non-innocent and thus facilitates electron transfer processes between the substrate, the metal centre and the pterin. Model complexes of Mo/W-pyranopterins containing basic dithiolene ligands in many cases effectively mimic the original biological processes.

## Suggested reading

Romão MJ. **Molybdenum and tungsten enzymes: a crystallographic and mechanistic overview.** *Dalton Trans.* 2009; 4053–4068.

Provides an excellent and coherent overview of mechanistic and structural aspects of the three main families of molybdo- and tungsto-pyranopterins with many illustrative figures.

(a) Megalon A, Fedor JG, Walburger A, et al. **Molybdenum enzymes and their maturation.** *Coord. Chem. Rev.* 2011; 255: 1159–1178; (b) Hille R. **The molybdenum oxitransferases and related enzymes.** *Dalton Trans.* 2013; 42: 3029–3040.

These reports are recent reviews on systematic and functional aspects of molybdo-pyranopterin enzymes.

Rehder D. **The future of/for vanadium.** *Dalton Trans.* 2013; 142: 11749–11761.

The article refers to the biology of vanadium in relation to vanadium-based systems in industrial applications.

# References

1. (a) Schulzke C. Molybdenum and tungsten oxidoreductase models. *Eur. J. Inorg. Chem.* 2011; 1189–1199. (b) Bevers LE, Hagedoorn P-L, Hagen WR. The bioinorganic chemistry of tungsten. *Coord. Chem. Rev.* 2009; 253: 269–290.

2. Havemeyer A, Lang J, Clement B. The fourth mammalian molybdenum enzyme mARC: current state of research. *Drug Metabol. Rev.* 2011; 43: 524–539.

3. Dobbek H, Gremer L, Kiefersauer R, et al. Catalysis at a dinuclear [CuSMo(=O)OH] cluster in a CO dehydrogenase resolved at 1.1-Å resolution. *Proc. Natl. Acad. Sci. USA* 2002; 99: 15971–15976.

4. Hanson GR and Lane I. Dimethyl sulfoxide (DMSO) reductase, a member of the DMSO reductase family of molybdenum enzymes. *Biol. Magn. Reson.* 2010; 29: 169–199.

5. Seiffert GB, Ullmann GM, Messerschmidt A, et al. Structure of the non-redox-active tungsten/[4Fe:4S] enzyme acetylene hydratase. *Proc. Natl. Acad. Sci. USA* 2007; 104: 3073–3077.

6. Veloso-Bahamonde R, Ramirez-Tagle R, Arratia-Perez R. DFT modeling of the tungsten(V) cofactor of the hyperthermophilic *Pyrococcus furiosus* tungsto-bispterin enzyme via the calculated EPR parameters. *Chem. Phys. Lett.* 2010; 491: 214–217.

7. Liao R-Z. Why is the molybdenum-substituted tungsten-dependent formaldehyde ferredoxin oxidoreductase not active? A quantum chemical study. *J. Biol. Inorg. Chem.* 2013; 18: 175–181.

8. Kaysser L, Bernhardt P, Nam S-J, et al. Merochlorins A–D, cyclic meroterpenoid antibiotics biosynthesized in divergent pathways with vanadium-dependent chloroperoxidases. *J. Am. Chem. Soc.* 2012; 134: 11988–11991.

9. (a) Littlechild J, Rodriguez EG, Isupov M. Vanadium-containing bromoperoxidase – insights into the enzymatic mechanism using X-ray crystallography. *J. Inorg. Biochem.* 2009; 103: 617–621; (b) Wischang D, Radlow M, Schulz H, et al. Molecular cloning, structure, and reactivity of the second bromoperoxidase from *Ascophyllum nodosum*. *Bioorg. Chem.* 2012; 44: 25–34.

10. Coletti A, Galloni P, Sartorel A, et al. Salophen and salen oxo vanadium complexes as catalysts of sulfides oxidation with $H_2O_2$: mechanistic insights. *Catal. Today* 2012; 192: 44–55.

11. da Silva JAL, Fraústo da Silva JJR, Pombeiro AJL. Amavadin, a vanadium natural complex: Its role and applications. *Coord. Chem. Rev.* 2013; 257: 2388–2400.

12. Holm RH, Solomon EI, Majumdar A, et al. Comparative molecular chemistry of molybdenum and tungsten and its relation to hydroxylase and oxotransferase enzymes. *Coord. Chem. Rev.* 2011; 255: 993–1015.

13. Enemark JH, Cooney JJA, Wang J-J, et al. Biomimetic inorganic chemistry. *Chem. Rev.* 2004; 104: 1175–1200.

14. Wu P, Santoni G, Fröba M, et al. Modelling the sulfoxygenation activity of vanadate-dependent peroxidases. *Chem. Biodivers.* 2008; 5: 1913–1926.

# 8 The sulfur cycle

We commonly associate 'sulfur' with volcanic activity such as the *solfataras*—the hot volcanic exhalations consisting of water vapour and hydrogen sulfide ($H_2S$) which, on contact with air, are oxidized to elemental sulfur and sulfur dioxide. We also associate sulfur with the smell of rotten eggs (smacking of $H_2S$), the stinging smell of $SO_2$ accompanying fireworks, or, more pleasantly, the scent of truffle (thanks to dimethyl sulfide ($(CH_3)_2S$). Positive associations with sulfur also go along with mineralogical finds. Examples are 'fool's gold', otherwise known as pyrite, $FeS_2$, or gypsum, as exemplified by the giant crystals of $CaSO_4 \cdot 2H_2O$ in caverns of the Naica mine in Chihuahua, Mexico.

For the medieval alchemist, sulfur and mercury represented combinations of Empedocles' four elements.[1] Accordingly, fire (hot and dry) and air (hot and moist) meld to form sulfur, while mercury was considered an association of water (cold and fluid) and earth (cold and dry). The formation of the various metals and minerals was attributed to combinations of mercury and sulfur in different purities and proportions, known as 'sulfur–mercury theory'; this theory goes back to the Persian/Arab natural philosopher and alchemist Jābir ibn Hayyān (Latinized: Geber), who lived from 721 to 815 AD. The theory was revived in the 13th/14th century by the works of an alchemist referred to as pseudo-Geber. (The modern chemist is, of course, aware of the fact that mercury and sulfur combine to form cinnabar, HgS.) In the early 16th century, sulfur became associated with 'flammability' in terms of 'spirit' and 'mind', a notion that goes back to Paracelsus and persisted for about two-and-a-half centuries.

Some of the mystery and spirit related to sulfur still prevails. Dimethyl sulfide, a main end-product of sulfur metabolism in marine bacteria and released from the oceans into the troposphere, has long been considered almost the sole cause of cloud formation and, hence, rainfall. But is this really the case? And how do organisms in an oxic aquatic environment, with sulfate (and hence $S^{+VI}$) as the predominant sulfur source, manage to activate sulfate and *reduce* it down to the oxidation state –II needed for the integration into essential amino acids? And how is it that other organisms, living in anoxic settings, thrive on the *oxidation* of sulfur present in the form of $H_2S$ and sulfides?

Both geochemical and biogenic sulfur cycling, including their tight interconnections, will be addressed in this chapter.

---

[1] Empedocles' (490–430 BCE) four elements were later popularized and supplemented by a fifth element by Aristotle (384–322 BCE). The fifth element, the aether, was introduced as the divine substance that makes up planets and stars.

## 8.1 Environmental sulfur cycling

Several of the main steps of the sulfur cycle are documented in Scheme 8.1. Volatile forms of sulfur, i.e. sulfur dioxide ($SO_2$) and hydrogen sulfide ($H_2S$), are released from the lithosphere into the atmosphere via volcanic exhalations, the decay of organic materials, and both natural and anthropogenic combustion. Once in the atmosphere, $SO_2$ and $H_2S$ are rapidly oxidized to sulfate/sulfuric acid, Eq. (8.1), essentially catalysed by transition metal ions present in mineral dust particles [1]. $H_2SO_4$ is washed out of the atmosphere by fog and rain, and re-deposited to the surface. In moist and aquatic surface areas the strong acid $H_2SO_4$ dissociates to form sulfate $SO_4^{2-}$. A substantial part of this sulfate is taken up by marine organisms, biogenically reduced to hydrogen sulfide ($H_2S$), Eq. (8.2), and incorporated into organic sulfur compounds, as detailed in Section 8.2.

$$H_2S + 4H_2O \rightarrow H_2SO_4 + 8H^+ + 8e^- \tag{8.1a}$$

$$SO_2 + 2H_2O \rightarrow H_2SO_4 + 2H^+ + 2e^- \tag{8.1b}$$

$$SO_4^{2-} + 9H^+ + 8e^- \rightarrow HS^- + 4H_2O \tag{8.2}$$

Marine organisms, coral reef dwelling algae in particular, synthesize the amino acid methionine, employing $H_2S$ as a substrate. Methionine is then further metabolized to dimethylsulfoniopropionate. Marine microbes feeding on this organic sulfur compound as both a carbon and (to some extent) sulfur source release substantial amounts of sulfur in the form of the volatiles methanethiol[2] ($CH_3SH$) [2a] and dimethyl sulfide (($CH_3)_2S$) [2b].

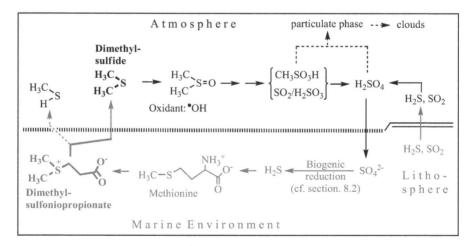

Scheme 8.1 Cycling of sulfur between the atmosphere and the marine hydrosphere. The cycling involves redox processes and both biogenic and abiotic methylation and demethylation of sulfur. Two of the key compounds are dimethylsulfoniopropionate and its metabolite dimethyl sulfide. For additional details, see the text.

---

[2] $CH_3SH$ is also known as methyl mercaptan. For the potential significance of $CH_3SH$ and iron sulfides for the development of primordial life, cf. Sidebar 2.1 in Chapter 2.

While the highly reactive $CH_3SH$ is subject to rapid turnover, or assimilated by bacteria into sulfur-based amino acids, the comparatively stable $(CH_3)_2S$ is emitted from the ocean surface waters and transported into the troposphere, where it contributes to the generation of aerosol particles and thus cloud formation.[3] $(CH_3)_2S$ is further oxidized, mainly by OH radicals, via dimethyl sulfoxide $(CH_3)_2SO$ (Eq. (8.3)) to methanesulfonic acid $(CH_3SO_3H)$ and sulfurous acid $(H_2SO_3)$, and finally to sulfuric acid $H_2SO_4$. These compounds go into the particulate phase and into cloud droplets, and eventually become re-deposited with the rain.

$$(CH_3)_2S + 2OH \rightarrow (CH_3)_2SO + H_2O \qquad (8.3)$$

We will discuss several of the key steps in the biogenic conversion of inorganic sulfur compounds into organic sulfur species relevant for catabolic processes in the following section. Sidebar 8.1 provides an overview of key inorganic and organic sulfur compounds that are life-sustaining and/or play a substantial role in sulfur cycling.

---

### Sidebar 8.1    A selection of sulfur compounds involved in Life

The oxidation state of sulfur can range from +VI (in its most oxidized form) to –II (the most reduced form). The main inorganic sulfur compounds in the lithosphere are sulfates such as gypsum $CaSO_4 \cdot 2H_2O$, elemental sulfur $S_8$, and sulfides. Examples for sulfides are pyrite $FeS_2$ (with $Fe^{II}$ and $S^{-I}$) and troilite $FeS_x$ ($S^{-II}$; $x$ is close to 1). Some magnetobacteria employ greigite $Fe^{II}Fe^{III}_2S_4$ (Section 4.2) for orientation in the Earth's magnetic field, and sulfur bacteria can store metastable $S_6$ in their cellular structures. The main sulfur source in the oceans is sulfate(VI) ($c = 28$ mM); sulfate is also a main constituent in blood serum, with a mean concentration of 0.3 mM. In anoxic surroundings, e.g. in the proximity of deep sea black smokers, $H_2S$ and metal sulfides prevail. Inorganic sulfur compounds in the atmosphere include hydrogen sulfide ($H_2S$), sulfur dioxide ($SO_2$), sulfur trioxide ($SO_3$), and sulfuric acid ($H_2SO_4$).

The chart below provides major inorganic sulfur compounds in their physiologically relevant protonation states, including the mean oxidation number of sulfur (in bold).

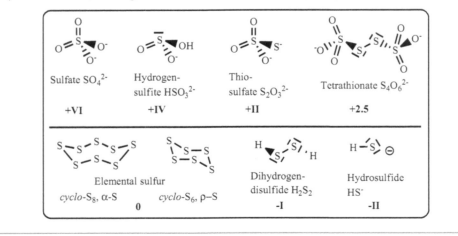

| Sulfate $SO_4^{2-}$ | Hydrogen-sulfite $HSO_3^{2-}$ | Thio-sulfate $S_2O_3^{2-}$ | Tetrathionate $S_4O_6^{2-}$ |
| --- | --- | --- | --- |
| **+VI** | **+IV** | **+II** | **+2.5** |

| Elemental sulfur *cyclo*-$S_8$, α-S    *cyclo*-$S_6$, ρ–S | | Dihydrogen-disulfide $H_2S_2$ | Hydrosulfide $HS^-$ |
| --- | --- | --- | --- |
| **0** | | **-I** | **-II** |

---

In organic sulfur compounds, common oxidation states of sulfur are +VI (sulfates), +IV (sulfonates), +II (sulfones), 0 (sulfoxides), –I (disulfides), and –II (sulfides, thiocyanate). Chondroitin sulfate (Fig. 8.1) is an example of a physiologically active sulfate, as are the biogenic sulfonation reagents adenosinemonophosphate-sulfate (APS) and phospho-APS (PAPS) shown in Scheme 8.2. Iron-sulfur clusters such as the cubane $[Fe_4S_4Cys_4]$ (Sidebar. 5.1), contain inorganic and organic sulfide. Clusters of this general type are widespread cofactors in redox-active enzymes. Other examples for sulfur-containing cofactors in enzymatic reactions include the dithiolene moiety in molybdopterins, represented here by sulfite oxidase (Section 7.1), methyl-coenzyme-M in methanogenesis (Section 10.2), acetyl-coenzyme-A ('active acetic acid', a broadly employed transmitter of $C_2$ fragments), and taurine, which is essential for cardiovascular function.

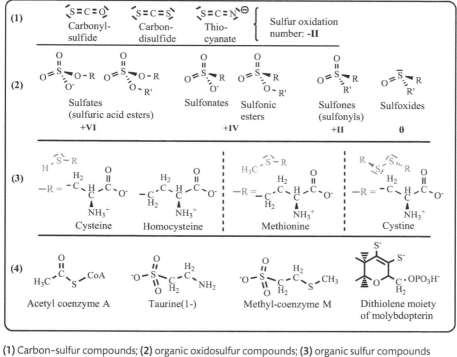

(1) Carbon–sulfur compounds; (2) organic oxidosulfur compounds; (3) organic sulfur compounds derived from hydrosulfides; (4) examples of functional organic sulfur compounds.

## 8.2  Biogenic metabolism of sulfur

With a concentration of 0.3 mM, sulfate is the fifth-most abundant solute in blood serum, out-classed only by hydrogencarbonate (25 mM), glycine (2.3 mM), lactate (1.5 mM), and hydrogen-phosphate (1.2 mM). Sulfate ($SO_4^{2-}$) enters the body with nutrients, is resorbed in the intestines, and desorbed and eventually re-resorbed mainly via the kidneys. Plasma sulfate levels are maintained through the 'renal clearance mechanisms' [4a], i.e. by filtering and reabsorbing sulfate in the kidneys. Due to its comparatively high charge, the strongly hydrophilic sulfate cannot cross cell walls directly. Rather, the transport of $SO_4^{2-}$ by membrane-bound sulfate transporters is commonly coupled to the transport of $Na^+$ or $H^+$—a process referred to as co-transport.

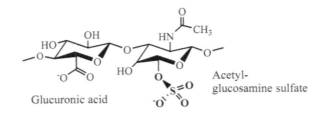

Figure 8.1 Formula unit of chondroitin sulfate, a main component of cartilage.

Along with the dietary supply of sulfate, the oxidation of organic sulfur compounds (in particular cysteine and methionine) in the diet can contribute to sulfate homeostasis. Sulfate is required for proper cell growth and development; it is involved in various activation and detoxification processes, and in the build-up of structural components of membranes and tissues [4b]. An example is chondroitin sulfate (Fig. 8.1), an oligosaccharide that functions as a main constituent in structural tissue components such as cartilage. Sulfate is reduced to sulfite in liver cells, a reaction catalysed by sulfite oxidase/reductase (Section 7.1), a molybdenum-dependent enzyme responsible for the redox interconversion of sulfite and sulfate, Eq. (8.4). The physiological oxidation of the reductant sulfite is a detoxification process.

$$SO_4^{2-} + 2e^- + 3H^+ \rightleftharpoons HSO_3^- + H_2O \qquad (8.4)$$

More generally, sulfate is the main—and for prototrophic[4] microbial organisms the sole—starting species for the synthesis and supply of organic sulfur compounds, in particular the amino acids cysteine/cystine and methionine. Here, sulfur is in its lowest oxidation states: –II in cysteine and methionine, and –I in cystine. Sulfate-reducing bacteria are also found in the human colon. Examples are bacteria belonging to the genera *Desulfovibrio*, *Desulfomonas*, and *Desulfobacter*. The net reaction carried out by the bacteria is the reduction of sulfate by hydrogen to hydrogen sulfide, Eq. (8.5).

$$SO_4^{2-} + 4H_2 + H^+ \rightarrow HS^- + 4H_2O \qquad (8.5)$$

The first step in microbial sulfate activation is the formation of adenosine-5′ phosphosulfate (APS) and, depending on the microbe, 3′-phosphoadenosine-5′-phosphosulfate PAPS [5], Scheme 8.2; this step is catalysed by enzymes termed sulfotransferases [6]. The activated sulfate is then either transferred to functional groups of proteins and polysaccharides (a process called sulfonation[5]), or reduced to sulfite. These reduction processes are catalysed by thioredoxins, small proteins with a dithiol/disulfide active site. The respective reductase for APS also contains an $[Fe_4S_4]$ cluster cofactor.

Scheme 8.3 provides an overview of the various steps of the (microbial) reduction of sulfate to sulfide, and re-oxidation of sulfide to sulfate. Conversion of activated sulfate to sulfide can

---

[4] *Prototrophic* microorganisms synthesize their cellular constituents from *inorganic* sources.

[5] Also known as sulfation or sulfatization. *Sulfonation* refers to the mechanistic aspect (transfer of a sulfone fragment), *sulfation* (and sulfatization) to the resulting product: a sulfate (sulfuric acid ester). See also Sidebar 8.1.

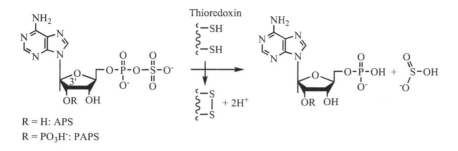

**Scheme 8.2** The reduction of activated sulfate (with $S^{+VI}$) to hydrogensulfite ($S^{+IV}$) by thioredoxin. Sulfate is activated by adenosinephosphate, ATP. APS = adenosine-5'-phosphosulfate, PAPS = 3'-phospho-ATP.

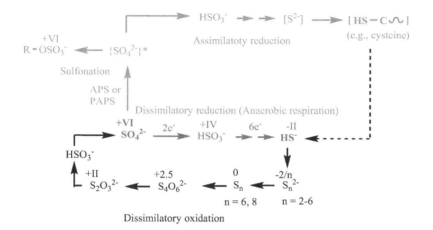

**Scheme 8.3** The main steps in biogenically relevant sulfur cycling. $\{SO_4^{2-}\}$* refers to activated sulfate (for APS and PAPS see Scheme 8.2), [$S^{2-}$] to low-valent sulfur in an organic matrix. R is a protein or polysaccharide that binds to the sulfate via a phenolic, alcoholic, or hydroxylamine function; see Fig. 8.1 for an example. Sulfur oxidation states are indicated on top of the formulae.

occur by dissimilatory reduction and, alternatively or additionally, by assimilatory reduction. *Assimilatory* reduction is directed towards the incorporation of the reduced—sulfidic—sulfur into organic compounds. Cysteine, homocysteine, and methionine are the most prominent examples. In the case of *dissimilatory* reduction—a form of anaerobic respiration (oxidation without oxygen)—sulfate is reductively converted to inorganic sulfide, essentially HS⁻. The reverse process, dissimilatory *oxidation*, leads back to sulfate.

*Thioalkalivibrio*, a haloalkaliphilic bacterium thriving in soda lakes, is an example of a microorganism that oxidizes hydrogen sulfide HS⁻ via oligosulfides $S_n^{2-}$ ($n = 3-8$) and elemental sulfur to tetrathionate $S_4O_6^{2-}$ (along with some tri- and pentathionate), and further to sulfite and sulfate [7]. This bacterium can even resort to carbon disulfide[6] and thiocyanate as

---

[6] For the conversion of $CS_2$ into $H_2S$ and $CO_2$ by the zinc-dependent $CS_2$ hydrolase from an acidothermophilic archaeon, see Section 12.2.2.

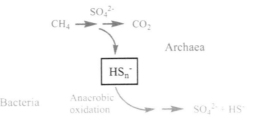

Scheme 8.4 Methane oxidation/sulfate reduction, carried out by archaea (dark blue), coupled to the bacterial disproportionation of sulfur to sulfate and sulfide (light blue). In polysulfide (boxed), sulfur is in the mean oxidation state $-(II/n)$.

a sulfur source. In the latter case, the primary products, generated by hydrolysis, are cyanate and $H_2S$. $H_2S$ is then oxidized to sulfate, and cyanate converted to ammonia and carbon dioxide, Eq. (8.6). Electron acceptors in the oxidation process are nitrate, nitrite, and $N_2O$, with $N_2$ as the final product of reduction.

$$NCS^- + H_2O \rightarrow H_2S + NCO^- (\rightarrow \rightarrow SO_4^{2-} + NH_3 + CO_2) \tag{8.6}$$

Another issue of considerable interest, the cooperation between methane oxidizing (methanotrophic) archaea and sulfate reducing bacteria in oceanic sediments, will be resumed in Section 10.3. Scheme 8.4 represents a special pathway, where sulfate is re-delivered by bacterial disproportionation of polysulfide, the reduction product formed by archaean reduction of sulfate [8].

 Summary

The majority of sulfur compounds released into the atmosphere by biological and non-biological processes ends up as sulfuric acid, which is washed down by rain into the oceans. Once transported into the hydrosphere, sulfuric acid is reductively metabolized by the aquatic microflora. A metabolite of central importance is dimethylsulfoniopropionate (DSP). DSP is further broken down to $CH_3SH$ and $(CH_3)_2S$. The latter is considered a 'culprit'—after aerial oxidation to dimethyl sulfoxide and sulfuric acid—for the condensation of water vapour into water droplets and thus rain.

Sulfate is, next to hydrogencarbonate and hydrogenphosphate, the most abundant inorganic anion in blood serum. Its plasma concentration, 0.3 mM, is essentially maintained by desorption and resorption in the kidneys. Sulfate is used as such in activation and detoxification processes, or esterified with carbohydrates in supporting tissues, for example cartilage. Biologically relevant sulfur compounds with sulfur in medium oxidation states are sulfonates $RSO_3^-$ and sulfones $R_2SO_2$.

Microbial sulfate reduction is initiated by the activation of sulfate through its linkage to adenosine monophosphate, i.e. the formation of adenosine-5′-phosphosulfate (APS). Activated sulfate ($S^{+VI}$) in APS is reduced to $S^{-II}$ either via assimilatory reduction to inorganic sulfide, or via dissimilatory reduction to sulfide as an integral constituent of organics such as cysteine and methionine. Bacteria living in anoxic habitats are capable of dissimilatory oxidation of sulfide. An example is the haloalkaliphile *Thioalkalivibrio*.

## Suggested reading

**Thomas D and Surdin-Kerjan Y. Metabolism of sulfur amino acids in *Saccharomyces cerevisiae*. *Microbiol. Mol. Biol. Rev.* 1997; 61: 503–532.**
A review that covers sulfate assimilation pathways towards the synthesis of cysteine and methionine, including transcriptional regulation of the respective pathways.

## References

1. Harris E, Sinha B, van Pinxteren D, et al. Enhanced role of transition metal ion catalysis during in-cloud oxidation of $SO_2$. *Science* 2013; 340: 727–730.

2. (a) Reisch CR, Stoudemayer MJ, Varaljay VA, et al. Novel pathway for assimilation of dimethylsulphoniopropionate widespread in marine bacteria. *Nature* 2011; 473: 208–211; (b) Vila-Costa M, Simó R, Harada H, et al. Dimethylsulfoniopropionate uptake by marine phytoplankton. *Science* 2006; 314: 652–654.

3. Quinn PK and Bates TS. The case against climate regulation via oceanic phytoplankton sulphur emission. *Nature* 2011; 480: 51–56.

4. (a) Markovich D and Aronson PS. Specificity and regulation of renal sulfate transporters. *Annu. Rev. Physiol.* 2007; 69: 361–375; (b) Markovich D. Physiological roles and regulations of mammalian sulfate transporters. *Physiol. Rev.* 2001; 81: 1499–1533.

5. Bhave DP, Hong JA, Keller RI, et al. Iron–sulfur cluster engineering provides insight into the evolution of substrate specificity among sulfonucleotide reductases. *ACS Chem. Biol.* 2011; 7: 306–315.

6. Gamage N, Barnett A, Hempel N, et al. Human sulfotransferases and their role in chemical metabolism. *Toxicol. Sci.* 2006; 90: 5–22.

7. Sorokin DY and Kuenen JG. Haloalkaliphilic sulfur-oxidizing bacteria in soda lakes. *FEMS Microbiol. Rev.* 2004; 29: 685–702.

8. Milucka J, Ferdelman TG, Polerecky L, et al. Zero-valent sulphur is a key intermediate in marine methane oxidation. *Nature* 2012; 491: 541–546.

# 9 Nitrogenase and nitrogen cycle enzymes

Nitrogen, in a form accessible for living organisms, is a limiting factor for the development, thriving, and reproduction of all extant life, including the nutritional supply for a rapidly rising human population. Consequently, there is an ever increasing need to secure bio-available nitrogen, essentially in the form of ammonium salts, as a main constituent for the production of artificial fertilizers. Presently, industrially furnished nitrogen compounds for agricultural needs are about equal to the bio- and geochemically conducted conversion of otherwise inaccessible nitrogen, in particular inert molecular nitrogen present in the atmosphere and dissolved in aquatic systems. To some extent, the global demand for artificial fertilizers disrupts the natural nitrogen cycle.

Nature has available an efficient, though complex, approach to 'fix' molecular nitrogen, i.e. to convert $N_2$ reductively to ammonium ions for direct use in life processes. This interconversion, termed nitrogen fixation, is one of the focal points of this chapter. Biological nitrogen fixation is the natural counterpart of industrial nitrogen fixation by the Haber–Bosch process. As has recently been demonstrated, the enzymes promoting biological $N_2$ fixation also catalyse the conversion of carbon monoxide into hydrocarbons, and thus mimic the Fischer–Tropsch process.

Another main subject to be addressed in this chapter is the procedural methodology by which nature cycles nitrogen between ammonia (with nitrogen in its lowest oxidation state), $N_2$, and nitrate (with nitrogen in its highest oxidation state). The understanding of the versatility and differentiation by which this biochemical processing of nitrogen occurs has been a vital issue in bioinorganic chemistry in the past decades, and will be of prime interest also in the oncoming years.

One of the key molecules in nitrification (the conversion of ammonia into nitrate) and denitrification (the transformation of nitrate into $N_2$) is nitric oxide (NO), a gas commonly considered to be highly toxic. However, in terms of what had been stated by Paracelsus about 500 years ago, i.e. 'only the dose makes a thing not a poison' (*Dosis sola venenum facit*), NO also acts as a multifunctional messenger, and induces luminescence in luminescent organisms. Both aspects of NO, the toxic and the beneficial ones, will be addressed in the last section of this chapter.

## 9.1 Overview and native nitrogenase

A compilation of basic organic and inorganic nitrogen compounds is provided in Sidebar 9.1. The majority of nitrogen on our planet, about $2 \times 10^{17}$ t, is deposited in mineralized form in rocky and sedimentary materials of Earth's mantle and crust, mainly in the form of nitrate ($N^{+V}$) and ammonium ($N^{-III}$). Nitrides (with the anion $N^{3-}$) are extremely rare. An example is osbornite TiN, which has also been found in meteorites and cometary dust particles. Our atmosphere contains $4 \times 10^{15}$ t of nitrogen almost exclusively in the form of $N_2$, the oceans $10^{12}$ t (nitrate, nitrite, ammonium, and dissolved $N_2$). Soil organisms accommodate $3 \times 10^{11}$ t, and animals and plants $10^{10}$ t of nitrogen in its reduced form 'fixed' into organic materials such as amines and nucleobases.

Living organisms can also resort to using nitrogen-containing bedrock as potential nitrogen sources. Mica schist in particular can contain appreciable amounts of nitrogen in the form of interlayer $NH_4^+$, which becomes released as the rock weathers [1]. The main access, however, to utilizable nitrogen requires the reduction of $N_2$ (aerial or dissolved in water), a process termed 'nitrogen fixation'.

---

### Sidebar 9.1   Nitrogen compounds

Nitrogen–hydrogen compounds include $NH_3$ (ammonia; synthesis from $H_2$ and $N_2$ according to the Haber–Bosch process) and ammonium ions ($NH_4^+$), $N_2H_4$ (hydrazine), $HN_3$ (hydrazoic acid; hydrogen azide). Salts derived from $HN_3$, azides, e.g. $NaN_3$, are used as fungicides and bactericides in bio-assays). Nitrides such as $Na_3N$ formally derive from ammonia. Ammonia is an efficient complexing agent for many metal ions. An example is the intensely blue tetraammine complex formed with $Cu^{2+}$, trans-$[Cu(H_2O)_2(NH_3)_4]^{2+}$. The ammonium ion is a Brønsted acid: aqueous solutions of ammonium salts consequently are acidic.

Nitrogen–oxygen compounds include $N_2O$ (dinitrogen monoxide, nitrous oxide, 'laughing gas'), which is a particularly efficient greenhouse gas, NO (nitrogen monoxide, nitric oxide; synthesis by combustion of ammonia according to the Ostwald process), red-brown $NO_2$ (nitrogen dioxide) in a temperature-dependent equilibrium with $N_2O_4$ [with water, $NO_2$ forms nitrous acid ($HNO_2$)+nitric acid ($HNO_3$)], and $N_2O_5$ (dinitrogen pentoxide). In environmental contexts, nitrogen oxides are often subsumed as $NO_x$. Salts derived from $HNO_3$ are termed nitrates, those derived from $HNO_2$ nitrites. Nitrites $NO_2^-$ can further be oxidized to peroxonitrites $ONOO^-$. Hyponitrous acid 'HNO' (actually HON=NOH) plays a role as an intermediate in denitrification. Hydroxylamine ($NH_2OH$) contains N in the oxidation state –I.

Organic nitrogen compounds include amines (primary: $NH_2R$; secondary: $NHR_2$; tertiary: $NR_3$), heterocyclic nitrogen compounds (for a selection see below, upper row), amides (1a) and peptides (1b), amino acids (2), hydroxamic acids (3), nitro compounds $R-NO_2$, nitrosamines (4), diazo compounds (5), Schiff bases (6), and nitrosothiols RS-NO.

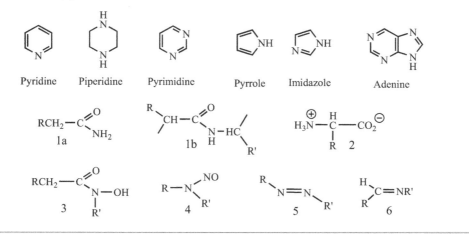

| Pyridine | Piperidine | Pyrimidine | Pyrrole | Imidazole | Adenine |

Other nitrogen compounds are hydrocyanic acid (HCN) and cyanide (CN⁻), cyanate (NCO⁻), and thiocyanate (NCS⁻), all of which can be generated in the frame of metabolic processes. Cyanide in particular is toxic, but can also occur as a ligand in enzymes (an example is iron-only hydrogenase). Amides of carbonic acid include: carbamic acid, $O=C(OH)NH_2$, and esters thereof (carbamates), $O=C(OR)NH_2$; urea $O=C(NH_2)_2$.

More generally, nitrogen fixation is the biogenic as well as the non-biogenic transformation of elemental $N_2$ into nitrogen compounds, a conversion which must overcome the bonding energy of 949 kJ mol⁻¹ between the two triply bonded nitrogen atoms. In principle, reduction of dinitrogen to ammonia can occur abiotically, e.g. by ferrous iron. The *biogenic* fixation, resulting in the formation of ammonium ions $NH_4^+$, is carried out by free living nitrogen-fixing bacteria (*Azotobacter*) and cyanobacteria ('blue-green algae', *Anabaena*), some archaea, and by symbiotic fungi (*Rhizobium*) mainly associated with leguminous plants.

Biogenic fixation accounts for about half of the overall nitrogen supply. Non-biogenic non-anthropogenic fixation, which can be promoted by electric discharge (lightning) in the troposphere, and by short-wave UV and the solar wind (basically high velocity protons) and cosmic radiation (protons and γ rays) in the stratosphere, Eq. (9.1), accounts for 10%. The remaining 40% of worldwide nitrogen conversion occur via the Haber–Bosch process, the combustion of fossil fuels such as coal and crude oil, and products based on crude oil (petrol, gasoline, diesel), processes by which organic nitrogen compounds present in the fossils are oxidized to $NO_x$. Aerial oxidation of $N_2$ to $NO_x$ also occurs on the surface of dust particles in the presence of catalytically active ingredients (such as $VO_x$), and in combustion engines.

$$N_2 \rightarrow 2N; N+O_2 \rightarrow NO+O; NO+xO \rightarrow NO_{x+1} \tag{9.1}$$

The conditions required for industrial and biogenic nitrogen fixation are listed in Table 9.1. Both processes employ an iron-based catalyst. However, while the Haber–Bosch process requires high pressure and temperature, biogenic $N_2$ fixation proceeds at ambient conditions and with considerably better yields and higher efficiency with respect to utilization by plants: About 80% of the nitrogen in *industrially* produced fertilizers is lost to the environment.

The major biological pathways linked to the nitrogen cycle are outlined in Scheme 9.1. The electrons necessary for the reduction of $N_2$ to the ammonium cation are provided by

Table 9.1 Conditions for industrial and biogenic nitrogen fixation

| Haber–Bosch | Biogenic |
| --- | --- |
| $N_2 + 3H_2 \rightleftharpoons 2NH_3$ | $N_2 + 10H^+ + 8e^- \rightarrow 2NH_4^+ + H_2$ (for Fe/Mo-nitrogenase) |
| Temperature: 500 °C | Temperature: ca. 20 °C (thermophiles: up to 92 °C) |
| Pressure: 200–450×10⁵ Pa | Pressure: 10⁵ Pa |
| Catalyst: Fe (+ $Al_2O_3 + K_2O + ...$) | Catalyst: Nitrogenase (Fe-, Fe/Mo-, or Fe/V-S cluster) |
| % Conversion: 17% | % Conversion: 75% (for Mo-nitrogenase) |
| Annual production ca. 2×10⁸ t | Annual production ca. 10⁸ t |

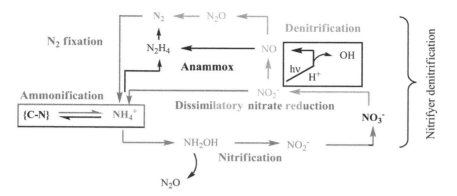

**Scheme 9.1** Biogenic pathways within the nitrogen cycle (see also [2]). 'Anammox' is short for *anaerobic ammonium oxidation*, according to the net reaction $NH_4^+ + NO_2^- \rightarrow N_2 + 2H_2O$. Dissimilatory nitrate reduction ($NO_3^- \rightarrow NO_2^- \rightarrow NH_4^+$) is a main path providing $NH_4^+$ for anammox in aquatic oxygen minimum zones [3]. A substantial amount of reactive nitrogen in the oceans is removed by heterotrophic denitrification ($NO_3^- \rightarrow NO_2^- \rightarrow\rightarrow N_2$). {C-N} symbolizes organic nitrogen compounds such as amino acids and nucleotides. While nitrifyer-denitrification is related to bacterial activity, the formation of the greenhouse gas $N_2O$ by reduction of $NH_4^+$ via $NH_2OH$ (bottom left) is linked to archaea [4]. Apart from the biogenic formation of NO (light blue trace), soil nitrite can also be the origin of atmospheric NO, formed photolytically from $HNO_2$ ($\equiv HONO$) released into the atmosphere [5] (boxed area).

respiratory oxidation of organic carbon to $CO_2$, as symbolized by Eq. (9.2), where {$CH_2O$} represents a molecule such as glucose. The reducing equivalents are supplied via iron proteins, gated by the energy releasing hydrolysis of $Mg^{2+}$-activated adenosine (Ad) triphosphate (ATP) to adenosine diphosphate (ADP) and inorganic phosphate, Eq. (9.3). Some of the protons involved in the overall reaction are reduced to $H_2$; the enzyme enabling the process of nitrogen fixation, nitrogenase, thus also attains some hydrogenase activity. The overall reaction going on in the reaction centre of the molybdenum nitrogenase is formulated in Eq. (9.4.), where $P_i$ represents inorganic phosphate $H_2PO_4^-$.

$$\{CH_2O\} + \tfrac{1}{2}O_2 \rightarrow CO_2 + 2H^+ + 2e^- \tag{9.2}$$

$$\text{Ad-triphosphate(Mg)} + H_2O \rightarrow \text{Ad-diphosphate(Mg)} + H_2PO_4^- \tag{9.3}$$

$$N_2 + 10H^+ + 8e^- + 16ATP \rightarrow 2NH_4^+ + H_2 + 16ADP + 16P_i \tag{9.4}$$

The enzyme molybdenum nitrogenase consists of two components (Fig. 9.1): the homo-dimeric iron protein, and the $\alpha_2\beta_2$ tetrameric iron–molybdenum protein [6a]. The 64 kDa iron protein contains a single [4Fe–4S] cluster (see also Sidebar 5.1 for iron–sulfur clusters) at the interface of the two protein subunits, and two ATP linked to the protein. Electrons (commonly delivered by NADH) are shuttled to the Fe–Mo protein in a process which is driven by the hydrolysis of ATP.

The iron–molybdenum protein contains two $M$ and two $P$ clusters arranged in such a way that the tetramer attains $C_2$ symmetry. The $P$ cluster is a double cubane with an $Fe_8S_7$ core, consisting of two subclusters, [4Fe–4S] and [4Fe–3S], bridged by cysteinate [7], in an

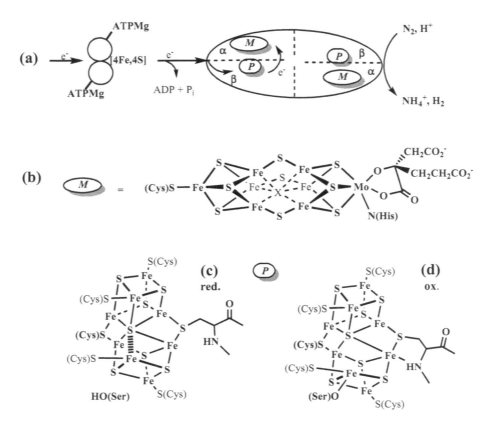

Figure 9.1 Symbolic representation of the organization of the molybdenum-nitrogenase (**a**), the catalytic centre {Fe$_7$MoS$_9$} (the *M* cluster, **b**, see also Fig. 9.2) [6], and the *P* cluster in the oxidized (ox., **c**) and reduced form (red., **d**), respectively [7]. The hatched line in the reduced *P* cluster (**c**) typifies a comparatively long bond. X in the *M* cluster stands for electron density corresponding to a light scatterer such as N, O or, most likely, C. The six central iron ions are in a trigonal-prismatic arrangement. *P* and *M* cluster are separated by 20 Å; the distance between two *M* clusters is 70 Å.

oxidation-state dependent overall arrangement; Fig. 9.1, bottom. The central core of the *M* cluster, where the final reduction of N$_2$ occurs, is also a double cubane of composition Fe$_7$MS$_9$, where M is Fe (iron-only nitrogenase), V (vanadium-nitrogenase) or—in most cases—Mo (molybdenum-nitrogenase; shown in Fig. 9.1b and in Fig. 9.2; see also the book cover).[1] The complete iron–molybdenum cofactor, or FeMo-co for short, is a 200 kDa protein. Molybdenum is in an octahedral coordination environment, coordinated to three triply bridging sulfides, the Nδ of a histidine, and the vicinal carboxylate plus hydroxo groups of homocitrate. The cluster is further linked to the protein matrix through a cysteinate attached to the iron centre opposite Mo. The way in which the activation and reduction of N$_2$ takes place is presently in the process of being enlightened.

---

[1] The currently most common nitrogenase, Mo-nitrogenase, appeared only 1.5–2.2 Ga ago, and hence substantially later than the V- and Fe-only nitrogenases.

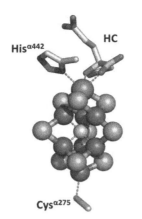

**Figure 9.2** Structure of the FeMo-co; HC=homocitrate. Reprinted with permission from [8a]. Copyright (2013) American Chemical Society. Kindly supplied by Dr Markus Ribbe. *Also reproduced in full colour in plate 7.*

Electron density has been detected in the central cavity of the $Fe_7MoS_9$ cage that corresponds to a light atom such as C, N, or O, with C providing the best fits with data from X-ray emission spectroscopy [6b]. Additional current data based on $^{13}C$ and $^{14}C$ labelling experiments [8a,b] also point to $X = \mu_6$-carbide ($C^{4-}$), with the methyl group of adenosyl-methionine as the source [8c]; see also Section 13.1.[2] The present conception for the reduction path has been derived from model chemistry, and will be discussed in this context in Section 9.2.

In addition to $N_2$, other unsaturated molecules can likewise be substrates for nitrogenases. Examples are ethyne which, in the presence of $D_2O$, is reductively protonated to the $Z$-isomer of dideuteroethene, Eq. (9.5); and isonitrile which is converted to a mix of primary amine, methane, and ethene, Eq. (9.6). In vanadium nitrogenase (with an $\alpha_2\beta_2\delta_2$ substructure), expressed by *Azotobacter*, Mo in the $M$ cluster is replaced by V in the absence of molybdenum or at low temperatures [9a,b]. Vanadium nitrogenase is even able to catalyse the reduction of carbon monoxide [9c], Eq. (9.7), and thus links the Haber–Bosch and Fischer–Tropsch processes. Similarly, CO can also be substrate for the molybdenum nitrogenase. In this case, however, no methane is formed.

$$H-C\equiv C-H + H_2 + D_2O \longrightarrow \underset{D}{\overset{H}{>}}C=C\underset{D}{\overset{H}{<}} \tag{9.5}$$

$$H_3C-N\equiv C + H_2 \longrightarrow \longrightarrow CH_3NH_2, CH_4, H_2C=CH_2 \tag{9.6}$$

$$\begin{array}{l} CO \\ \Big\downarrow H^+ + e^- \\ \longrightarrow CH_4, CH_2=CH_2, CH_2=CH-CH_3, CH_2=CH-CH_2-CH_3, n\text{-}C_4H_{10} \end{array} \tag{9.7}$$

While nitrogenases commonly fall back on NADH as the source of reduction equivalents (electrons), $N_2$ fixation can also be coupled to CO-dehydrogenase (COD) and superoxide

---

[2] For adenosyl methionine see also Section 13.2.4.

oxidoreductase (SOR) [10]. As depicted in the sequence represented by Eq. (9.8), COD (containing the molybdopterin cofactor, Section 7.1) can oxidize CO to $CO_2$ with concomitant formation of $H_2O_2$ and the superoxide radical anion $O_2^{\cdot-}$, which is then oxidized to $O_2$ by a Mn-based SOR (Section 6.2). The electrons that are delivered are channelled into the $N_2$ reduction pathway.

$$CO + H_2O + \tfrac{3}{2}O_2 \rightarrow CO_2 + \tfrac{1}{2}H_2O_2 + O_2^- + H^+$$
$$O_2^- \rightarrow O_2 + e^-$$
$$N_2 + 6e^- + 8H^+ \rightarrow 2NH_4^+ \qquad (9.8)$$

## 9.2 Nitrogenase models and model reactions

The following main steps in $N_2$ reduction catalysed by the $\{Fe_7MoS_9\}$ cluster have been elucidated [11]; see also Scheme 9.2:

(i)   Hydrogen atoms are assembled sequentially via the $\mu_3$-S close to Mo ('$\mu_3$-S*'), and located at the $\mu_2$-S and $\mu_3$-S*, and iron atoms adjacent to the bridging sulfides;

(ii)  $N_2$ is bound, $\eta^1$ and/or, more likely, $\eta^2$, to Fe*, and H is transferred to $N_2$ to yield $N_2H$, $NNH_2$ or HNNH, and finally $N_2H_4$;

(iii) The N–N bond is broken to yield Fe–$NH_2$ and/or Fe–$NH_3$.

The role of the central atom $\mu_6$-X (X=N or, more likely, C) is such that it allows for the elongation of *one* of the six X–Fe bonds, viz. the bond to Fe*, thus affecting the preferential coordination of $N_2$ to Fe*.

Ideally, an enzyme model is both a good structural *and* a good functional model, i.e. it features the structure of the active centre of the enzyme, and it copies the function of the enzyme under environmental/physiological circumstances (such as temperature, pressure, pH, salinity, oxic/anoxic conditions) where it is optimally active. In practice, and specifically in the case of nitrogenase, this dual requirement is still far from being fulfilled, and consequently we are still obliged to cope with the world's immense need of 'fixed' nitrogen (mainly

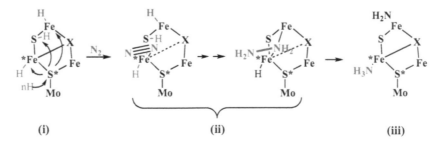

**Scheme 9.2** Several of the intermediates of $N_2$ reduction at the *M* cluster, based on ref [11]. X is probably $C^{4-}$ [8]; for the steps (i), (ii) and (iii) see text.

in the form of nitrates and ammonium salts) by relying on the harsh conditions intrinsic to the Haber–Bosch synthesis.

While the FeMo-co (the *M* cluster) itself can be synthesized *in vitro*—in systems containing cofactor proteins—from ferrous iron, sulfide, molybdate(VI), and homocitrate in the presence of adenosyl-methionine and $Mg^{2+}$-ATP [12], structural models for this cluster are restricted to those without any functional activity. There are, however, promising approaches for functional model chemistry that have provided some insight into the reaction sequence and reaction intermediates of biological nitrogen fixation, a process which requires 6 electrons and eight protons (not counting the hydrogenase side reaction, Eq. (9.4) in Section 9.1) for the conversion of dinitrogen to ammonium ions.

The first *functional* nitrogenase models were based on low-valent phosphane-stabilized dinitrogen complexes of molybdenum ($Mo^0$) and vanadium ($V^{-I}$). These simple complexes, prepared from THF-stabilized chlorido complexes under reductive conditions in the presence of phosphanes ($PR_3$) and dinitrogen, Eq. (9.9), yield ammonia and/or hydrazine when treated with protic reagents, Eq. (9.10).

$$Mo/VCl_3(thf)_3 + Na + N_2 + PR_3 \rightarrow\rightarrow [Mo(N_2)_2(PR_3)_4]/[V(N_2)_2(PR_3)_4]^- \tag{9.9}$$

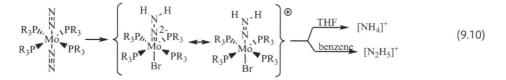

(9.10)

In the case of molybdenum, the intermediate hydrazide(2–)/diazene complex has been characterized. Based on the identification of several other intermittently formed species [13,14], reaction intermediates such as those shown in Scheme 9.3 have been proposed and partially identified—species that are generated by successive addition of electrons and protons to the coordinated and thus activated dinitrogen. The activation of $N_2$ on coordination is documented by a shift of the $\nu(N{\equiv}N)$ from 2331 cm$^{-1}$ for free dinitrogen to lower wavenumbers, e.g. 2033 and 1980 cm$^{-1}$, for the symmetric and antisymmetric stretching vibrations, respectively, in *cis*-[$Mo(N_2)_2(dppe)_2$], dppe$=Ph_2P(CH_2)_2PPh_2$.

A more recent development towards a functional model is the molybdenum complex {Mo}$N_2$ shown in Fig. 9.3, with the molybdenum centre in a sterically demanding environment [14]. The steric crowding prevents dimerization (formation of a {Mo}$_2\mu$-$N_2$ species), but

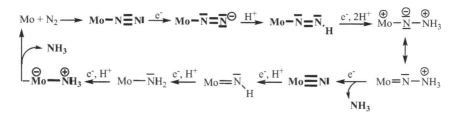

Scheme 9.3 Selection of proposed and characterized (in bold) intermediates in $N_2$ reduction.

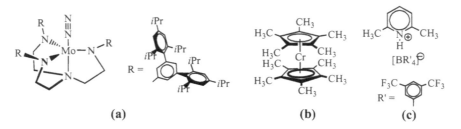

Figure 9.3 The three components employed for the catalytic dinitrogen reduction in the 'Yandulov–Schrock cycle' [14]. (a) The catalyst with the substrate $N_2$ coordinated to $Mo^{III}$; R=3,5-*bis*[(tri-isopropyl)phenyl] phenyl. (b) The reducing agent pentamethyl-chromatocene. (c) The proton source lutidinium boranate.

still allows for the entry of $N_2$, reduction equivalents and protons. The reduction equivalents are provided by pentamethyl-chromocene, the protons by 2,6-lutidinium boranate. Both of these co-reactants are also displayed in Fig. 9.3. Intermediates which have been identified in the course of this reaction are provided in bold in Scheme 9.3.

One should keep in mind that these merely functional models are based on the assumption that the *heterometal* (Mo or V) is the centre of activity for $N_2$ reduction, an assumption which appears to be reasonable given the fact that the iron-only nitrogenase is less effective than the Mo- and V-nitrogenases. It should be noted, however, that the availability of the Mo/V site presupposes a free coordination site at the hetero metal in the cofactor, a situation which might be effected by protonation of one of the coordinating homocitrate functions. On the other hand, there are six geometrically distorted iron centres available: four-coordinate iron in a trigonal pyramidal environment, with the iron exposed; Fig. 9.1 b.

The nitrido $Fe^V$ complex shown in Fig. 9.4, stabilized by a tris(imidazolyl)-boranate ligand, releases high yields of $NH_3$ when treated with water as a proton source and cobaltocene $\eta^5$-$(C_5H_5)_2Co$ ($Cp_2Co$) as the reductant [15]. This reaction may be considered to model a potential iron site as the site of $N_2$ reduction in the nitrogenase cofactor, and also provides insights into the initial step in the Haber–Bosch process: the splitting of molecular $N_2$ at the surface of the activated iron catalyst and concomitant production of iron-bound surface nitride species, followed by the stepwise hydrogenation of $N^{3-}$ and release of ammonia.

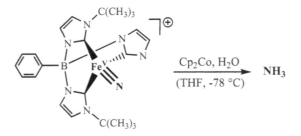

Figure 9.4 Conversion of $Fe^V$ bound nitride ($N^{3-}$) to ammonia by reductive protonation. The reductant is cobaltocene $Cp_2Co$, the proton source $H_2O$ [15].

## 9.3 Denitrification

As depicted in Scheme 9.1, ammonium ions can also be provided by nitrate reduction via nitrite. The reductases responsible for the reduction of nitrite to ammonium ions are either solely based on haems, or on a combination of haem and [4Fe–4S] proteins. Nitrite $NO_2^-$ is at the cross-over point towards ammonia on the one hand, and the reductive degradation to dinitrogen on the other hand. The sequence $NO_3^- \rightarrow NO_2^- \rightarrow NO \rightarrow N_2O \rightarrow N_2$,[3] Eq. (9.11), goes by the name of *denitrification* or, to be more exact, heterotrophic denitrification, where heterotrophic relates to the fact that organic carbon (instead of $CO_2$) is employed as the source for growth. The reverse process, the oxidation of $NH_4^+$ to $NO_3^-$, is termed *nitrification* and runs through hydroxylamine $NH_2OH$ and nitrite $NO_2^-$, Eq. (9.12). The overall process, combining the reaction sequences denoted by Eqs. (9.11) and (9.12), is sometimes also referred to as nitrifyer-denitrification.

$$NO_3^- + 2H^+ + 2e^- \rightarrow NO_2^- + H_2O \qquad (9.11a)$$

$$NO_2^- + 2H^+ \rightarrow NO^+ + H_2O; \; NO^+ + e^- \rightarrow NO \qquad (9.11b)$$

$$2NO + 2e^- \rightarrow {}^-ONNO^-; \; {}^-ONNO^- + 2H^+ \rightarrow N_2O + H_2O \qquad (9.11c)$$

$$N_2O + 2H^+ + 2e^- \rightarrow N_2 + H_2O \qquad (9.11d)$$

$$NH_4^+ + H_2O \rightarrow NH_2OH + 3H^+ + 2e^- \qquad (9.12a)$$

$$NH_2OH + H_2O \rightarrow NO_2^- + 5H^+ + 4e^- \qquad (9.12b)$$

$$NO_2^- + H_2O \rightarrow NO_3^- + 2H^+ + 2e^- \qquad (9.12c)$$

Reaction (9.11a), the start of denitrification, is catalysed by an enzyme containing the molybdopterin cofactor (Section 7.1). A proposed catalytic process is sketched in Scheme 9.4. During the first step in this process, nitrate is coordinated to the molybdenum centre in its reduced form ($Mo^{IV}$). Two electrons are then transferred to nitrogen with concomitant transfer of an oxo group to Mo (now in its oxidized form, $Mo^{VI}$), followed by the release of nitrite.

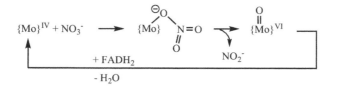
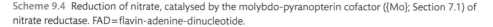

Scheme 9.4 Reduction of nitrate, catalysed by the molybdo-pyranopterin cofactor ({Mo}; Section 7.1) of nitrate reductase. FAD=flavin-adenine-dinucleotide.

---

[3] For the denitrification path $NO_2^- \rightarrow NO \rightarrow N_2$ (and hence bypassing $N_2O$), coupled to the oxidation of methane, see Eq. (2.8) and ref. [8] in Chapter 2.

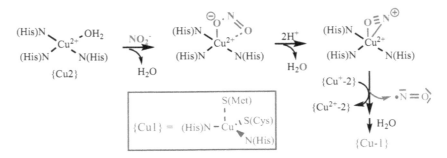

Scheme 9.5 Proposed mechanism for the reduction of nitrite to nitric oxide catalysed by nitrite reductase.

The catalytic centre is finally restored by two-electron reduction of oxidomolybdenum(VI) by the reduced form of flavin-adenine-dinucleotide, $FADH_2$ (see Sidebar 9.2).

The next step, the one-electron reduction of nitrite to nitric oxide NO, Eq. (9.11b), is catalysed by nitrite reductase [16a,b] (Scheme 9.5). The reaction centre of this enzyme contains a type 1 and a type 2 copper (Sidebar 6.2), {Cu-1} and {Cu-2}. {Cu-1} conveys $e^-$ transfer, while {Cu-2} is responsible for substrate binding. Nitrite reductases based on haem-type cofactors (dubbed nitrophorines) are also known [16c]; the reaction sequence, Eq. (9.13), resembles that for the copper-based enzyme. Eventually, nitrite reductase can also operate with a molybdo-pyranopterin cofactor (Section 7.1).

$$\{Fe^{2+}\}+NO_2^- \rightarrow \{Fe^{2+}-NO_2^-\} \tag{9.13a}$$

$$\{Fe^{2+}\text{-}NO_2^-\}+2H^+ \rightarrow \{Fe^{2+}\text{-}NO^+\}+H_2O \tag{9.13b}$$

$$\{Fe^{2+}\text{-}NO^+\} \rightarrow \{Fe^{3+}NO\} \rightarrow \{Fe^{3+}\}+NO \tag{9.13c}$$

Further reduction of NO to nitrous oxide, $N_2O$ (more widely known as laughing gas), is catalysed by nitric oxide reductase, the catalytic site of which consists of a haem-type iron (haem-$b_3$: one axial histidine; Sidebar 5.2) and a non-haem iron, the so-called $Fe_B$, at a distance $d(Fe...Fe)$ of 3.9 Å, as illustrated in Fig. 9.5a).

Two additional, more distant haems are involved in electron transfer to the haem-$b_3$/$Fe_B$ reaction centre. The iron in $Fe_B$ is coordinated, in a slightly distorted trigonal-pyramidal geometry, to the N$\varepsilon$[4] of three histidines and, in the apex, one glutamate ligand. As suggested in Eq. (9.11c), hyponitrite is likely formed as a transient species by N–N bond formation between two coordinated, and thus activated and subsequently reduced NO molecules, followed by protonation of hyponitrite and $N_2O$ formation. There are structural similarities between the *substrate* binding sites of nitric oxide reductase and cytochrome-c oxidase, suggesting a common ancestor for the two enzyme families.

The final step of denitrification, Eq. (9.11d), again a two-electron reduction, is catalysed by nitrous oxide reductase. $N_2O$ is thermodynamically unstable; the Gibbs free energy for the net reaction represented by Eq. (9.11d) is −340 kJ mol⁻¹. The reductive decomposition

---

[4] The imidazole nitrogen adjacent to the side-chain linking to the protein backbone is referred to as N$\delta$, while the nitrogen function opposite this linkage is N$\varepsilon$.

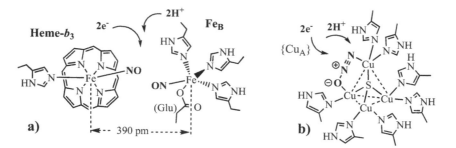

Figure 9.5 (a) The catalytically active site of nitric oxide reductase, based on [17], and assuming that the $Fe^{2+}$ centres of haem-$b_3$ (substituents at the porphyrin rings omitted) and of $Fe_B$ both coordinate nitric oxide. NO is subsequently reduced to $NO^-$, two of which couple to form $^-ONNO^-$ and, by concomitant delivery of protons, form $N_2O$ and water; Eq. (9.11c). (b) The active site of nitrous oxide reductase [18], with the substrate bound to the two copper sites of the four-copper ($Cu_Z$) centre which, in the resting state, accommodate an aqua (or a hydroxido) ligand.

is, however, hampered kinetically, making $N_2O$ a comparatively stable gas: the mean dwell-time in the atmosphere amounts to 114 years. This stability gives rise to environmental concern: $N_2O$ is a greenhouse gas, and about 290 times as effective as $CO_2$ in this respect.

The enzyme nitrous oxide reductase contains a dinuclear $Cu_A$ centre (a similar $Cu_A$ is also present in cytochrome-c oxidase, section 5.3), which is responsible for electron transport to a tetranuclear copper centre, $Cu_Z$, functioning as the catalytic site. The four copper ions of $Cu_Z$ (Fig. 9.5b) [18]) are in a slightly distorted tetrahedral arrangement, linked by a fourfold bridging ($\mu_4$) inorganic sulfide, and ligated to seven imidazoles of histidine residues. Two of the copper centres are additionally coordinated by $H_2O/OH^-$. In addition, there is a chloride close by.

It has been proposed that, in the first step of the reaction catalysed by nitrous oxide reductase, $N_2O$ docks to the two-copper site of the fully reduced ($\{Cu_4^+\}$)$Cu_Z$ by replacing the $H_2O/OH^-$ ligand. $N_2O$ then undergoes two-electron reduction, resulting in the release of $N_2$ and formation of an oxido-$\{Cu_2^{2+}, Cu_2^+\}$ centre, which receives two electrons via $Cu_A$. Channelling of two protons to the active site then restores the active aqua-$\{Cu_4^+\}$ centre.

*Nitrification*, i.e. the oxidation of ammonia to nitrate, Eq. (9.12), starts with the oxidation, by oxygen, of ammonia to hydroxylamine. As indicated by the overall reaction, Eq. (9.14), the net electron balance is consumption of two electrons: 2 $e^-$ are released as $NH_3$ is oxidized to $NH_2OH$, but 4 $e^-$ are required for the reduction of an oxygen molecule to two $O^{2-}$, one of which becomes inserted into $NH_2OH$, while the other one ends up as water. The enzyme catalysing this reaction, ammonia monooxygenase (or, more precisely, ammonium monooxygenase), has not yet been structurally characterized. EPR spectroscopic evidence suggests the presence of type 2 copper sites as well as haem and non-haem iron, the latter possibly related to particulate methane monooxygenase [19], an enzyme containing a dinuclear, oxido-bridged $\{Fe_2\}$ centre.

$$NH_4^+ + O_2 + H^+ + 2e^- \rightarrow NH_2OH + H_2O \qquad (9.14)$$

In the second step of the oxidation sequence, the oxidation of $NH_2OH$ to nitrite, four electrons are released, Eq. (9.12b). The enzyme involved in this oxidation, hydroxylamine oxidoreductase, contains 24 haems at its active centre [20]. The key haem at the active site, $P_{460}$, belongs to the cytochrome-*c* family (Cyt-*c*, Section 5.3). The attribution '460' reflects its absorption maximum at 460 nm. The iron centre of $P_{460}$ is linked to an axial histidine; the sixth (second axial) position of $P_{460}$ is available for substrate binding and processing.

### Sidebar 9.2   Common redox-active cofactors in physiological redox processes (selection)

The oxidized forms of the functional units are highlighted in blue, the reduced in grey.

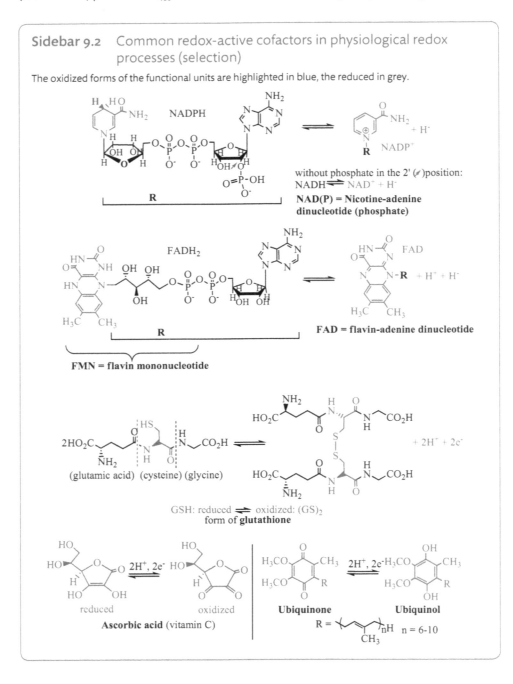

## 9.4 Nitric oxide

The radical nitrogen monoxide (nitric oxide; Scheme 9.6) forms biogenically as an intermediate product in denitrification (cf. the previous section) and, in higher organisms, from arginine (see below). Non-biogenically, NO forms in the troposphere and stratosphere from $N_2$ and $O_2$ by electric discharge (lightning, Eq. (9.15)), bombardment with cosmic rays (primarily protons) or high-energy UV light; it also forms anthropogenically in the wake of the combustion of fuels. In the presence of oxygen, NO is readily oxidized to nitrogen dioxide, Eq. (9.16). Under the influence of UV light, i.e. on sunny days, tropospheric $NO_2$ is split into NO and oxygen atoms, which oxidize alkanes (from exhaust gases) to alkyl hydroperoxides, and react with molecular oxygen to form ozone, Eq. (9.17),[5] a phenomenon referred to as 'summer smog' in urban areas. In the stratosphere, NO catalyses ozone depletion, Eq. (9.18), causing seasonal ozone holes in Antarctic and, more recently, also in Arctic regions.

$$N_2 \rightarrow 2N; N+O_2 \rightarrow NO+O \tag{9.15}$$

$$2NO+O_2 \rightarrow 2NO_2 \tag{9.16}$$

$$NO_2 + h\nu \rightarrow NO+O \tag{9.17a}$$

$$O + C_2H_6 + NO_2^{\bullet} \rightarrow C_2H_5O_2H + NO^{\bullet} \tag{9.17b}$$

$$O + O_2 \rightarrow O_3 \tag{9.17c}$$

$$NO + O_3 \rightarrow NO_2 + O_2 \tag{9.18a}$$

$$NO_2 + O \rightarrow NO + O_2 \tag{9.18b}$$

NO is toxic to higher organisms at micromolar concentrations due to damages going along with its radical character, and because it binds to the Cu and Fe centres of functional proteins that contain these metal ions in their active centre. For example, haemoglobin binds NO ten-thousand-fold more effectively than $O_2$; consequently, NO can seriously impede the ability of haemoglobin to operate as an oxygen transporter. On coordination, NO is reduced to $NO^-$ which is isoelectronic with $O_2$ and binds, as $O_2$, in the bent end-on mode.

More generally, NO interacts with the iron centres of various haem-type proteins, forming $\{FeNO\}^6$ and $\{FeNO\}^7$ adducts. In this so-called Enemark–Feltham notation for nitrosyl–iron complexes, the upper index indicates the sum of electrons delivered by the valence shell of the iron centre (5, 6, or 7 for $Fe^{3+}$, $Fe^{2+}$, or $Fe^+$) and NO (1 for the neutral radical NO, 2 for the anion $NO^-$). Additional NO toxicity arises from the fact that NO nitrosylates amines, via the

Scheme 9.6 The two mesomeric forms of nitric oxide.

---

[5] Atmospheric hydroperoxides form also according to the following reaction sequence: $RH + \cdot OH$ (formed from $H_2O_2$) $\rightarrow R\cdot + H_2O$; $R\cdot + O_2 \rightarrow ROO\cdot$; $ROO\cdot + HOO\cdot$ (generated from $O_2$ and H atoms) $\rightarrow ROOH + O_2$.

intermediate formation of nitrous acid, to form carcinogenic nitrosamines, as illustrated in Eq. (9.19). Nitrous acid is also formed, under physiological conditions, by reduction of nitrate present in, e.g. leafy vegetables.

The toxicity of NO is further due to its ability to degrade iron–sulfur clusters in FeS proteins, and the ease of its oxidation, by $O_2$ and superoxide $O_2^-$, to $NO_2$ and peroxonitrite $ONOO^-$, Eq. (9.20a), both powerful oxidants which cause tissue and DNA damage. The concentration of free NO in blood and tissue is controlled by oxy-Hb/Mb (Hb=haemoglobin, Mb=myoglobin), which oxidize NO to nitrate, Eq. (9.20b).

$$NO + H_2O \rightarrow HNO_2 + e^- + H^+ \tag{9.19a}$$

$$R_2NH + HNO_2 \rightarrow R_2N\text{-}NO + H_2O \tag{9.19b}$$

$$NO + O_2^- \rightarrow ONO_2^- \tag{9.20a}$$

$$NO + Hb \cdot O_2 \rightarrow \rightarrow NO_3^- \tag{9.20b}$$

However, nitric oxide is also bio-synthesized, as it acts (in sub-nanomolar concentrations) as an important multi-functional messenger and neurotransmitter, as well as in cardiovascular signalling. NO targets metal centres in haem-type proteins, and augments cellular levels of cyclic guanosine monophosphate cGMP. cGMP is a second messenger, responsible for signal transduction, and thus also helps to relax the smooth muscle tissue of, e.g. blood vessels, causing vasodilation (improved blood flow) and hence counteracting high blood pressure. The former use of sodium nitroprusside $Na_2[Fe(NO)(CN)_5]$ in the treatment of hypertension relates to the release of NO from this coordination compound.

Increased NO levels initiating increased cGMP levels are also responsible for the tumescence of the erectile vascular tissue and hence for penis erection; the function of Viagra™ stems from the 'protection' of cGMP from degradation by phosphoesterases (enzymes which catalyse the hydrolytic splitting of the phosphoester bond). Therapeutic effects of NO, including its antibacterial potential, will be dealt with in more detail in Section 14.5.

Ascorbate and deoxy-Hb/Mb can reduce nitrite to nitric oxide, a reaction which constitutes a pathway for the in vivo production of NO. Biosynthesis of NO is also effected by oxidation of the =NH (or $=NH_2^+$) function of arginine, Figure 9.6. The oxidation product, along with NO, is citrulline. This process is catalysed by the enzyme nitric oxide synthase (or NO-synthase, NOS), which is activated by the calcium modulating protein calmodulin (Section 3.5). NO-synthase contains a haem and redox-active biopterin at the active site [21,22]; cf. Figure 9.6. The haem belongs to the cytochrome $P_{450}$ family (Section 5.3), i.e. one of its axial positions is occupied by the sulfide of cysteinate.

Three functionally different NOSs are known: (1) nNOS in neurons initiates signal transduction and thus takes part in mnemonic functions, (2) iNOS in macrophages induces the liberation of NO in case of infections and thus acts as a killing agent for infectious germs of the immune system, and (3) eNOS in the endothelial tissue cells controls the tonicity of the vascular muscles and thus the blood pressure.

NO is also used by the glow worm (lightning bug, firefly) to switch on its glow organs [23]. The glow worm's luminescence has been traced back to the oxidation, by $O_2$, of luciferyl-AMP (AMP=adenosine-monophosphate) via peroxoluciferin to oxoluciferin (Fig. 9.7). To

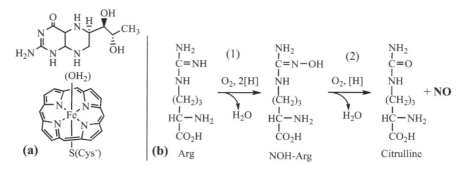

**Figure 9.6** Nitric oxide synthase. (**a**) The redox-active cofactor biopterin in its reduced (tetrahydro) form, and the catalytic haem centre (substituents at the porphyrin omitted). (**b**) The two successive mono-oxigenase reactions, by which arginine (Arg) is oxidized to NOH-Arg and further to citrulline. [H] represents reduction equivalents such as delivered by NADPH (cf. Sidebar 9.2). In step (1), $4e^-$ ($2e^-$ by [H] and $2e^-$ by Arg) are delivered to reduce one $O_2$; in step (2), [H] delivers $1e^-$ and NOH-Arg $3e^-$ to reduce the second $O_2$ and release NO.

start the $O_2$ consuming formation of peroxoluciferin, the glow worm triggers NO synthesis. The NO blocks mitochondrial cytochromes (symbolized by a flash in Fig. 9.7) by coordination of NO to Fe, and consequently screens oxygen from consumption in respiration. The $O_2$ thus becomes available for initiating luminescence. Other organisms capable of bioluminescence also employ this mechanism. An example is *Nocticula scintillans*, a dinoflagellate responsible for marine phosphorescence.

Nitric oxide thus is not just a toxic molecule, but also—at low concentrations—beneficial and even essential for biochemical processes and response. Consequently, NO is synthesized and provided naturally in organisms. Similar considerations apply to other molecules and ions commonly considered toxic. Examples are CO, and $H_2S$, to be addressed in Section 14.5.

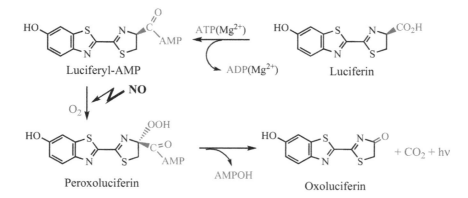

**Figure 9.7** NO-induced luminescence. The flash arrow indicates indirect action of NO (blockade of $O_2$ binding sites in cytochromes).

##  Summary

The main nitrogen species present in water and air is $N_2$, a particularly inert molecule which needs to be converted to ammonium ions (nitrogen fixation) or oxidic nitrogen compounds to become biologically available. Nitrogen fixation by bacteria and cyanobacteria is a particularly important natural process with a roughly equivalent amount of ammonia being industrially produced by the Haber–Bosch process. Some ammonia can also be provided by weathering of bedrock such as mica schist. Ultimately, crustal geological sources of nitrogen are mostly biogenic in nature.

The overall nitrogen cycle encompasses nitrogen fixation ($N_2 \rightarrow NH_4^+$), nitrification ($NH_4^+ \rightarrow NO_2^-$ and $NO_3^-$) dissimilatory nitrate reduction ($NO_3^- \rightarrow NO_2^- \rightarrow NH_4^+$), anammox ($NH_4^+ + NO_2^- \rightarrow N_2$) and (nitrifyer-)denitrification ($NO_3^- \rightarrow N_2$). Nitrogen fixation is catalysed by nitrogenases which accommodate, in their reaction centre, the $M$ cluster $\{Fe_7MS_9\}$, M = Mo, V, or Fe, the most prominent of which is the FeMo-co (iron-molybdenum cofactor). Along with $N_2$, alkynes, isonitriles and CO can be substrates for nitrogenases.

Functional models include dinitrogen complexes of $Mo^0$ and $V^{-I}$, and the Yandulov-Schrock systems, in which molybdenum is coordinated to a sterically hindered tripodal tetradentate amine. Model studies have provided insight into intermediates in $N_2$-fixation, including Mo–N=NH and Mo≡N. Denitrification occurs in four steps: (1) $NO_3^- \rightarrow NO_2^-$ (catalysed by nitrate reductase, which contains the molybdopterin cofactor); (2) $NO_2^- \rightarrow NO$ (nitrite reductase: 2 copper centres, viz. $\{Cu-1\}$ and $\{Cu-2\}$); (3) $NO \rightarrow N_2O$ (nitric oxide reductase: haem-Fe and $\{Fe(His)_3Glu\}$); (4) $N_2O \rightarrow N_2$ (nitrous oxide reductase: $\{Cu_4(\mu_4\text{-}S)\}$). Nitrification is a two-step process: (1) $NH_4^+ \rightarrow NH_2OH$ (ammonia monooxygenase: Cu-2, haem and non-haem Fe); (2) $NH_2OH \rightarrow NO_2^-$ ($P_{460}$). Haem-NO, formed as an intermediate in denitrification and, in eukarya, by the oxidation of arginine (catalysed by NO-synthase), is toxic to more developed organisms because it binds strongly to iron (e.g. in haemoglobin). Atmospheric sources of NO include its formation from $O_2$ and $N_2$ by lightning, and photolysis of $HNO_2$. NO is beneficial in physiological concentrations because it counteracts hypertension, kills infectious germs, and takes part in mnemonic functions. Many organisms capable of luminescence (such as the glow worm) synthesize NO to block off the access of $O_2$ to haemoglobin, and thus exploit $O_2$ in the oxidation of luciferin, a process by which energy is liberated in the form of visible light.

## Suggested reading

**Canfield DE, Glazer AN, Falkowsky PG. The evolution and future of Earth's nitrogen cycle.** *Science* 2010; 330: 192–196.

Emphasizes the disruption of the natural nitrogen cycle—eutrophication of the water bodies and atmospheric input of the greenhouse gas $N_2O$—by industrially produced nitrogen-based fertilizers.

**Einsle O, Tezcan FA, Andrade SLA, et al. Nitrogenase MoFe-protein at 1.16 Å resolution: a central ligand in the FeMo-cofactor.** *Science* 2002; 297: 1696–1700.

The first high-resolution X-ray study of the nitrogenase is described. The article thus marks the 'atmosphere of departure' into extensive studies of the functioning of natural $N_2$ fixation over ten years ago.

**Voss M and Montoya JP. Oceans apart.** *Nature* 2009; 461: 49–50.

The authors provide a briefing on the loss of $N_2$ from the oceans by microbial metabolism in the oxygen minimum zones. The $N_2$ production can be traced back to either nitrate reduction, or to coupling of nitrite reduction to the oxidation of ammonia.

**Lehnert N and Scheidt WR. Preface for the inorganic chemistry forum: the coordination chemistry of nitric oxide and its significance for metabolism, signaling, and toxicity in biology.** *Inorg. Chem.* 2010; 49: 6223–6225.

A preface to a series of articles dealing with the toxic and beneficial effects of nitric oxide.

**Lee SC and Holm RH. The clusters of nitrogenase: synthetic methodology in the construction of weak-field clusters.** *Chem. Rev.* **2004; 104: 1135–1158.**
Biologically relevant {Fe(Mo)S} clusters are addressed 'as synthetic challenges in bioinorganic chemistry', and the pursuit of synthetic analogues is delineated.

# References

1. Morford SL, Houlton BZ, Dahlgren RA. Increased forest ecosystem carbon and nitrogen storage from nitrogen rich bedrock. *Nature* 2011; 477: 78–81.

2. (a) Canfield DE, Glazer AN, Falkowsky PG. The evolution and future of Earth's nitrogen cycle. *Science* 2010; 330: 192–196; (b) Lam P, Lavik G, Jensen MM, et al. Revising the nitrogen cycle in the Peruvian oxygen minimum zone. *Proc. Natl. Acad. Sci. USA* 2009; 106: 4752–4757.

3. (a) Kartal B, Maalcke WJ, de Almeida NM, et al. Molecular mechanism of anaerobic ammonium oxidation. *Nature* 2011; 479: 127–130; (b) Prokopenko MG, Hirst MB, DeBrabandere L, et al. Nitrogen losses in anoxic marine sediments driven by *Thioploca*-anammox bacterial consortia. *Nature* 2013; 500: 184–198.

4. Santoro AS, Buchwald C, McIlvin MR, et al. Isotopic signature of $N_2O$ produced by marine ammonia-oxidizing archaea. *Science* 2011; 333: 282–285.

5. Su H, Cheng Y, Oswald R, et al. Soil nitrite as a source of atmospheric HONO and OH radicals. *Science* 2011; 333: 1516–1587.

6. (a) Einsle O, Tezcan FA, Andrade SLA, et al. Nitrogenase MoFe-protein at 1.16 Å resolution: a central ligand in the FeMo-cofactor. *Science* 2002; 297: 1696–1700; (b) Lancaster KM, Roemelt M, Ettenhuber P, et al. X-ray emission spectroscopy evidences a central carbon in the nitrogenase iron-molybdenum cofactor. *Science* 2011; 334: 974–977.

7. (a) Peters JW, Stowell MHB, Soltis SM, et al. Redox-dependent structural changes in the nitrogenase P-cluster. *Biochemistry* 1997; 36: 1181–1187; (b) Drennan CL and Peters JW. Surprising cofactors in metalloenzymes. *Curr. Opin. Struct. Biol.* 2003; 13: 220–226.

8. (a) Wiig JA, Lee CC, Hu Y, et al. Tracing the interstitial carbide of the nitrogenase cofactor during substrate turnover. *J. Am Chem. Soc.* 2013; 135: 4982–4983; (b) Boal AK and Rosenzweig AC. A radical route for nitrogenase carbide insertion. *Science* 2012; 337: 1617–1618; (c) Wiig JA, Hu Y, Lee CC, et al. Radical SAM-dependent carbon insertion into the nitrogenase M-cluster. *Science* 2012; 337: 1672–1675.

9. (a) Fay AW, Blank MA, Lee CC, et al. Characterization of isolated nitrogenase FeVco. *J. Am. Chem. Soc.* 2010; 132: 12612–12618; (b) Rehder D. Vanadium nitrogenase. *J. Inorg. Biochem.* 2000; 80: 133–136; (c) Hu Y, Lee CC, Ribbe MW. Extending the carbon chain: hydrocarbon formation catalyzed by vanadium/molybdenum nitrogenases. *Science* 2011; 333: 753–755.

10. Ribbe M, Gadkari D, Meyers O. $N_2$ fixation by *Streptomyces thermoautotrophicus* involves a molybdenum-dinitrogenase and a manganese superoxide oxidoreductase that couple $N_2$ reduction to the oxidation of superoxide produced from $O_2$ by a molybdenum-CO dehydrogenase. *J. Biol. Chem.* 1997; 272: 26627–26633.

11. Dance I. Ramifications of C-centering rather than N-centering of the active site FeMo-co of the enzyme nitrogenase. *Dalton Trans.*, 2012; 41: 4859–4865.

12. Curatti L, Hernandez JA, Igarashi RY, et al. In vitro synthesis of the iron–molybdenum cofactor of nitrogenase from iron, sulfur, molybdenum, and homocitrate using purified proteins. *Proc. Natl. Acad. Sci. USA* 2007; 104: 17626–17631.

13. Lehnert N and Tuczek F. The reduction pathway of end-on coordinated dinitrogen II. Electronic structure and reactivity of Mo/W-$N_2$, -NNH, and -$NNH_2$ complexes. *Inorg. Chem.* 1999; 38: 1671–1682.

14. (a) Yandulov DV and Schrock RR. Catalytic reduction of dinitrogen to ammonia at a single molybdenum center. *Science* 2003; 301: 76–78; (b) Schrock RR. Catalytic reduction of dinitrogen to ammonia by molybdenum: theory versus experiment. *Angew. Chem. Int. Ed.* 2008; 47: 5512–5522.

15. Scepaniak JJ, Vogel CS, Khusniyarov MM, et al. Synthesis, structure, and reactivity of an iron(V) nitride. *Science* 2011; 331: 1049–1052.

16. (a) Li H-T, Chang T, Chang W-C, et al. Crystal structure of C-terminal desundecapeptide nitrite reductase from *Achromobacter cycloclastes. Biochem. Biophys. Res. Commun.* 2005; 338: 1935–1942; (b) Dell'Acqua S, Pauleta SR, Moura I, et al. The tetranuclear copper active site of nitrous oxide reductase: the CuZ center. *J. Biol. Inorg. Chem.* 2011; 16: 183–194; (c) He C, Ogata H, Knipp M. Formation of the complex of nitrite with ferriheme b $\beta$-barrel protein nitrophorin 4 and nitrophorin 7. *Biochemistry* 2010; 49: 5841–5851.

17. Hino T, Matsumoto Y, Nagano S, et al. Structural basis of biological $N_2O$ generation by bacterial nitric oxide reductase. *Science* 2010; 330: 1666–1670.

18. (a) Haltia T, Brown K, Tegoni M, et al. Crystal structure of nitrous oxide reductase from *Paracoccus denitrificans* at 1.6 Å resolution. *Biochem. J.* 2003; 369: 77–88; (b) Dell'Acqua S, Pauleta SR, Moura I, et al. The tetranuclear copper active site of nitrous reductase: the CuZ center. *J. Biol. Inorg. Chem.* 2011; 16: 183–194.

19. Gilch S, Meyer O, Schmidt I. Electron paramagnetic studies of the copper and iron containing soluble ammonia monooxygenase from *Nitrosomonas europaea. Biometals* 2010; 23: 613–622.

20. Prince RC and George GN. The remarkable complexity of hydroxylamine oxidoreductase. *Nat. Struct. Biol.* 1997; 4: 247–250.

21. Li H, Igarashi J, Jamal J, et al. Structural studies of costitutive nitric oxide synthases with diatomic ligands bound. *J. Biol. Inorg. Chem.* 2006; 11: 753–768.

22. Delker SL, Ji H, Li H, et al. Unexpected binding modes of nitric oxide synthase inhibitors effective in the prevention of cerebral palsy phenotype in an animal model. *J. Am. Chem. Soc.* 2010; 132: 5437–5442.

23. Trimmer BA, Aprille JR, Dudzinski DM, et al. Nitric oxide and the control of firefly flashing. *Science* 2001; 292: 2486–2488.

# 10 The methane cycle and nickel enzymes

Methane, $CH_4$, plays a key role in the global carbon cycle; it is therefore important to be able to reveal its sources and sinks. As the main component of natural gas, methane is an increasingly important energy source in industrial production, in heating households and powering alternative fuel cars. But methane is also increasingly contributing to the greenhouse effect. Although present only in trace amounts in our atmosphere (the mixing ratio is 1.8 ppm by volume), its contribution to the greenhouse effect is about half that of carbon dioxide $CO_2$ (around 400 ppm = 0.04%). This tells us that $CH_4$ is about 25 times more effective than $CO_2$ in absorbing thermal infrared radiation, emitted from our planet's surface and re-radiated into the atmosphere.

Since pre-industrial times (about 250 years ago), the amount of atmospheric methane has tripled, and—despite of a retention time of only 9–15 years—is continuously increasing. The main primary anthropogenic sources of $CH_4$ are coal and oil extraction, hydraulic fracturing ('fracking') of deeply buried shale and sandstone, leaking gas pipes, and the burning of biomass. Secondary sources include the release of methane encapsulated in methane hydrates from lakes and mud in permafrost areas and deep sea reservoirs as a result of global warming. Methane hydrates are believed to store more energy than all other fuels combined.

Natural sources of methane production can be of biogenic and non-biogenic origin. The conversion of cellulose to methane by microbes in the stomach of ruminants is an example of the biogenic production of $CH_4$, serpentinization for the abiotic formation of methane. 'Serpentinization' refers to geochemical processes by which $CO_2$ is reduced to methane by ferrous minerals.

The central enzyme in the production of methane by methanogens, coenzyme M reductase, contains a nickel-based cofactor. But microorganisms do not only *generate*, but also *consume* methane: Most of the methane produced in sea floor sediments abiotically—for example, by serpentinization—and biogenically, by methanogens, is in fact channelled into aerobic or anaerobic methane oxidation, carried out by prokaryotes, so-called methanotrophs that use methane as their carbon source. Here, nickel enzymes are again involved, making nickel central to the global carbon, oxygen, and nitrogen cycles.

Nickel-based enzymes are also involved in numerous other processes, including:

- hydrogenation and dehydrogenation reactions;
- the synthesis of acetyl-coenzyme-A and the conversion of carbon monoxide plus water to carbon dioxide plus reduction equivalents;
- the dismutation of superoxide;
- the hydrolysis of urea.

The last section of this chapter provides some insight into these nickel enzymes. The versatile use of nickel in enzymatic reactions reflects this metal's flexible coordination chemistry, and its ability to cycle through the oxidation states +I, +II, and +III in a potential range that spans ca. 1.5 V.

## 10.1  Introduction

Natural biogenic methane production and release into the atmosphere is achieved by anaerobic methanogenic archaea associated with termites and ruminants (where they promote fermentation and thus digestion of cellulose in the guts). These archaea also dwell in wetland soil, rice fields and agricultural waste. In tropical areas, trees take over a 'chimney function' by moving methane, microbially produced in wet low-level zones, to their leaves and releasing $CH_4$ into the atmosphere. Gram-negative bacteria, such as *Escherichia coli*, also produce $CH_4$, resorting to activated methylphosphonate $(CH_3-PO_2(OH)OR^{2-})$[1] as a methane source [1]. Several saprophytic fungi have been shown to aerobically produce methane (i.e. under oxic conditions), resorting to the thiomethyl function of the amino acid methionine as a precursor [2]. The main sources for reductive methane production by methanogenic archaea are $CO_2$, CO and acetate; the reducing agent here is commonly $H_2$.

Most of the methane in deep *underground* reservoirs was formed non-biogenically from bulk organics by heat and pressure—in a way similar to coal and oil, and often in the same places. Another source of abiotic methane production goes back to a process referred to as serpentinization: In this process, protons (stemming from water) and $CO_2$ are reduced to $H_2$ plus $CH_4$. The reduction equivalents are delivered by ferrous iron in minerals such as olivine $(Mg_xFe_{2-x})SiO_4$. In this process, olivine is converted to magnetite $Fe^{II}Fe^{III}{}_2O_4$ and serpentine, a magnesium silicate. The overall process is represented in the (non-balanced) Eq. (10.1a). The hydrogen released in this redox conversion can further react with $HCO_3^-$ or $CO_2$ to form methane, Eq. (10.1b), in the presence of a catalyst such as pentlandite [3], an iron–nickel sulfide of composition $(Fe,Ni)S_{0.9}$.

$$(Mg,Fe)_2SiO_4 + H_2O + CO_2 \rightarrow Fe_3O_4 + Mg_3Si_2O_5(OH)_4 + H_2 + CH_4 \qquad (10.1a)$$

$$4H_2 + CO_2 \rightarrow CH_4 + 2H_2O \qquad (10.1b)$$

The main proportion of methane released into the atmosphere is destroyed through the reaction with OH radicals to form methyl radicals and water. The methyl radicals can react with another OH radical to make formaldehyde.

## 10.2  Methanogenesis

Methanogenesis—the conversion of inorganic and organic carbon compounds to $CH_4$—is a complex process that involves functional centres in enzymes, containing iron, molybdenum,

---

[1]  R represents ribose-5-phosphate. The hydrogen afforded for the formation of $CH_4$ by cleavage of the methyl–phosphorus bond is delivered by methyl group of 5'-deoxyadenosine.

cobalt, and nickel, with a porphyrin-related nickel centre in the final step of reductive $CH_4$ release.

Biomass such as cellulose is broken down by bacteria, protozoa and fungi primarily to acetate, lactate, propionate, butyrate and ethanol. Lactate, propionate, butyrate, and ethanol are further fermented by bacteria to formate ($HCO_2^-$), $CO_2$, and $H_2$ which, along with acetate ($CH_3CO_2^-$), are substrates in methanogenesis [4]. Acetic acid is converted to $CO_2$ and $CH_4$ by so-called type-1 methanogenic bacteria, Eq. (10.2a). Formate and $CO_2$, together with $H_2$, act as substrates for $CH_4$ production by type-2 methanogens; cf. Eq. (10.2b) for the net reaction starting from $CO_2$. Formate can provide both $CO_2$ and reduction equivalents, Eq. (10.3). [H] in Eqs. (10.2b) and (10.3a) stands for $H^+ + e^-$ or $\frac{1}{2}H_2$.

$$CH_3CO_2H \rightarrow CH_4 + CO_2 \tag{10.2a}$$

$$CO_2 + 8[H] \rightarrow CH_4 + 2H_2O \tag{10.2b}$$

$$HCO_2^- \rightarrow CO_2 + [H] + e^- \tag{10.3a}$$

$$HCO_2^- + H_2O \rightarrow HCO_3^- + H_2 \tag{10.3b}$$

Most of the $H_2$ in methanogenesis is delivered by bacteria that produce hydrogen; the enzymatic generation of $H_2$ from formate as represented by Eq. (10.3b) is coupled to the synthesis of ATP from ADP and phosphate [5], and hence to an energy storage process. Approximately one third of biologically generated methane derives from $CO_2$, while most of the remaining $CH_4$ originates from the conversion of the methyl group of acetate [6].

The main steps of the overall process of biogenic methanogenesis starting from $CO_2$ are summarized in Scheme 10.1; the cofactors of the enzymes participating in the various steps are displayed in Figure 10.1.

The initiating step in methanogenesis, step (a) in Scheme 10.1, is the two-electron reduction of $CO_2$ and the concomitant transfer and attachment of a formyl group to methanofuran, resulting in the formation of formylmethanofuran. This process is catalysed by the enzyme formylmethanofuran dehydrogenase. The prosthetic group of this enzyme contains a molybdo- or tungstopterin (1 in Fig. 10.1) belonging to the sulfite reductase family.[2] As shown in Scheme 10.2, the initiating step is the activation of $CO_2$ by coordination to the $Mo^{IV}$ centre in the reduced form of the molybdopterin, followed by the two-electron reduction of $CO_2$ via electron transfer from molybdenum. In turn, $Mo^{IV}$ is oxidized to $Mo^{VI}$. Re-reduction to $Mo^{IV}$ is achieved by NADH; for $NAD^+$/NADH, see Sidebar 9.2.

In step (b), the formyl group is transferred to tetrahydro-methanopterin and reduced, in two successive two-electron reduction steps, (c), to the methyl group. These processes are catalysed by hydrogenases based on iron-nickel, 2. Under nickel limitation, an iron-based hydrogenase is expressed by some archaea, 3 in Fig. 10.1. From methylmethanopterin thus formed, the methyl group is then transferred to mercaptomethanesulfonate, coenzyme-M HSCoM, (d). This transfer is mediated by a cobalamin {Co} (4), a factor akin to vitamin $B_{12}$. {Co} differs from the cobalamin of vitamin $B_{12}$ (see Fig. 13.4 in Section 13.1) in as far as the axial

---

[2] For molybdopterins—short for molybdo-pyranopterins—, see also Section 7.1.

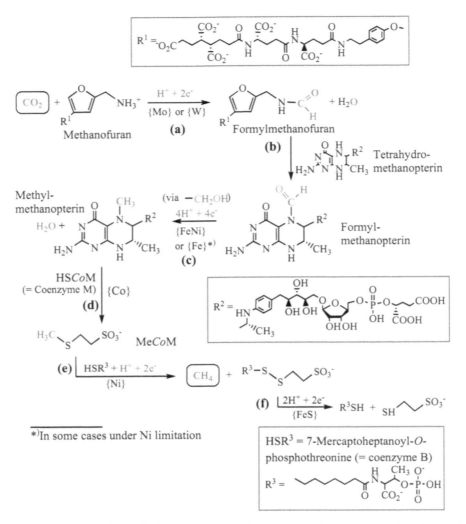

Scheme 10.1 Overview of the reductive conversion of $CO_2$ to $CH_4$ by methanogens. HSCoM and MeCoM in steps (d) and (e) are coenzyme M and methyl-coenzyme M, respectively. For the cofactors {Mo/W}, {FeNi}, {Fe}, {Co}, and {Ni}, see Figure 10.1.

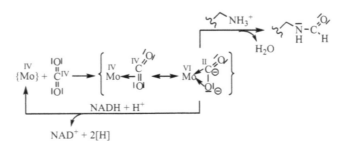

Scheme 10.2 Catalysis of the reductive transfer of $CO_2$ to methanofuran (upper right; for details of formylmethanofuran see Scheme 10.1). {Mo} represents the active centre of a molybdopterin, 1 in Fig. 10.1.

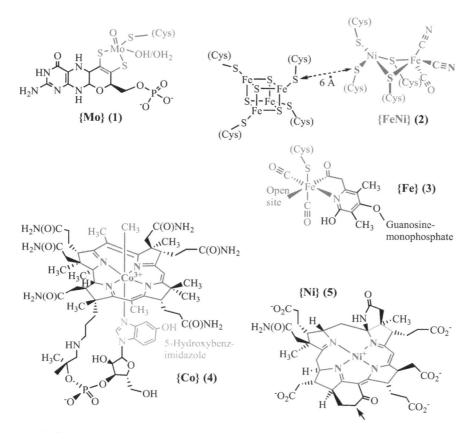

**Figure 10.1** The active centres of enzymes that are involved in methanogenesis (here: the reductive conversion of $CO_2$ into $CH_4$). The central units involved in the transfer of reduction equivalents are highlighted in blue. {Mo} (**1**) is a molybdopterin belonging to the sulfite reductase family. Mo can be replaced by W. The iron-nickel centre of the iron-nickel hydrogenase {FeNi} (**2**) directly interacting with the substrate is coupled, through electron transfer, to a [4Fe,4S] cuboidal core. In some archaea, {FeNi} can be replaced by the iron-only hydrogenase {Fe} (**3**) under constrained Ni supply. The cofactor {Co} (**4**) differs from cobalamin (vitamin $B_{12}$; Fig. 13.4 in section 13.1) in as far as the axial 5,6-dimethylbenzimidazole of cobalamin is replaced by 5-hydroxybenzimidazole in {Co}. Factor $F_{430}$ ({Ni}, **5**), with a partially saturated porphinoid skeleton, provides the final reduction equivalents ($Ni^+ \rightarrow Ni^{3+} + 2e^-$) for the formally $CH_3^+$ in MeCoM through coordination of $CH_3^-$ to $Ni^{3+}$; see also Scheme 10.3. At the position indicated by an arrow, a $CH_3S-$ substituent is bound (in the $S$ configuration) to the cofactor $F_{430}$ of the respective MeCoM reductase in *aerobic methane oxidation*; Section 10.3.

base is 5-hydroxybenzimidazole, instead of the 5,6-dimethylbenzimidazole in $B_{12}$. HSCoM[3] is thus transformed to methylcoenzyme-M, MeSCoM.

In step (**e**), MeSCoM and coenzyme-B (also a mercapto compound) react to form a hetero-disulfide, providing the final two electrons for the formation and release of methane. This

---

[3] *Co*—short for coenzyme—is denoted in italics throughout to distinguish it from the chemical symbol for cobalt.

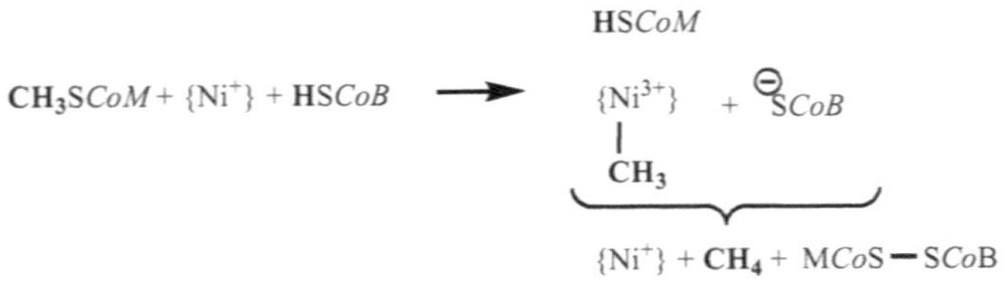

HSCoM

$$CH_3SCoM + \{Ni^+\} + HSCoB \longrightarrow \{Ni^{3+}\} + {}^{\ominus}SCoB$$

$$CH_3$$

$$\{Ni^+\} + CH_4 + MCoS - SCoB$$

Scheme 10.3  Release of methane by the interaction of methyl-coenzyme M with coenzyme B and the nickel centre of the factor $F_{430}$ of methyl-coenzyme M reductase. The methyl group is at a distance of 2.1 Å from the nickel centre, the sulfur of HSCoM at 2.4 Å [7]. HSCoB is situated ~8.7 Å away from the Ni.

step is catalysed by MeCoM-reductase, containing the nickel-based porphinoid cofactor $F_{430}$, **5**. In the reduced state of $F_{430}$, nickel is in the +I state.

The recovery of HSCoM and HSCoB by two-electron reduction, (**f**), is catalysed by heterodisulfide reductase. This reductase harbours a cuboidal $[Fe_4S_4]$ ferredoxin in its active centre. As shown in Scheme 10.3, the methyl group delivered by MeSCoM becomes intermittently coordinated as methylide $H_3C^-$ to $Ni^{+III}$, followed by the release of methane in the course of a redox interaction of the nickel centre with the two sulfidic cofactors, HSCoM and ${}^-$SCoB [7].

## 10.3  Biogenic oxidation of methane

The reverse of methanogenesis, the oxidative conversion of methane to carbon species in higher oxidation states ($CO_2$; via methanol, formaldehyde, formate or CO), is carried out by methanotrophic prokaryotic bacteria, also referred to as methanotrophs or methanophiles. Oxidation of methane requires the breaking of the thermodynamically highly stable C—H bond ($\Delta H = -435\,kJ\,mol^{-1}$), and can occur aerobically (with oxygen) with methanol as the primary oxidation product, Eq. (10.4), or non-aerobically. Suitable electron acceptors in anaerobic methane oxidation are nitrate, nitrite, sulfate, sulfite, $Mn^{IV}$, and $Fe^{III}$, Eqs. (10.5)–(10.7). In many cases, these reactions are performed by bacteria living in symbiosis with, and providing the oxidation equivalents for, methanotrophic archaea.

$$CH_4 + O_2 + 2H^+ + 2e^- \rightarrow CH_3OH + H_2O \tag{10.4}$$

$$3CH_4 + 8NO_2^- + 8H^+ \rightarrow 3CO_2 + 4N_2 + 10H_2O \tag{10.5}$$

$$CH_4 + SO_4^{2-} + H^+ \rightarrow CO_2 + HS^- + 2H_2O \tag{10.6}$$

$$CH_4 + 4MnO_2 + 8H^+ \rightarrow CO_2 + 4Mn^{2+} + 6H_2O \tag{10.7}$$

The aerobic oxidation of methane by $O_2$, Eq. (10.4), and the nitrite-driven non-aerobic oxidation of methane [8], Eq. (10.5), are examples of the $CH_4$ oxidation as mediated by non-symbiotic methanotrophic bacteria, hence in the *absence* of archaea. Equations (10.5)–(10.7) reflect the *net* reactions. Thus, the final oxidant for $CH_4$ in Eqn. (10.5) is $O_2$, generated along the sequence $NO_2^- \rightarrow NO \rightarrow O_2 + N_2$.

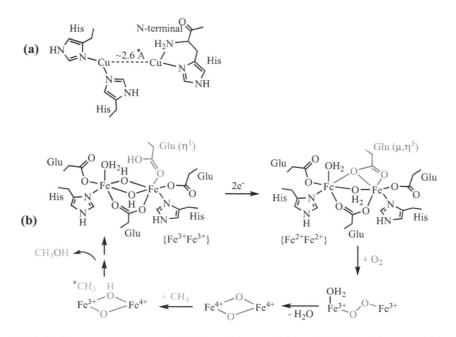

Figure 10.2 Methane monooxygenases (MMO) in methanotrophic bacteria. (a) The active centre of particulate (i.e. membrane-bound) MMO. The active intermediate, possibly a $\mu$-$O^{2-}$ bridged species, has not yet been identified. (b) Soluble MMO, including key steps of the oxidation of methane to methanol.

Interestingly, symbiotic communities of sulfate-reducing bacteria and methanotrophic archaea that oxidize $CH_4$ to $CO_2$ employ a close homolog of the nickel enzyme, namely methylcoenzyme M reductase, used by methanogenic archaea in the final step of methanogenesis [9], thus catalysing the reaction reverse to that formulated in Scheme 10.3. The factor $F_{430}$ of the MeCoM reductase in the archaea of these symbiotic methanotrophic communities contains the substituent -$SCH_3$ on the cyclohexanone ring, denoted by the arrow alongside structure **5** in Fig. 10.1.

The oxidation of methane to methanol by non-symbiotic methanotrophic bacteria, Eq. (10.4), is catalysed by methane monooxygenases (MMO) containing a di-copper [10] or a di-iron centre [11] in the active site, (a) and (b) in Fig. 10.2. The membrane-bound, Cu-based MMO is expressed under conditions of plentiful copper by all methanotrophs, while the Fe-based soluble MMO is expressed under conditions of copper limitation in a few methanotrophs only [12]. Both enzymes also catalyse the oxidation of organic substrates other than $CH_4$.

## 10.4 Nickel enzymes not involved in methane metabolism

The following nickel containing enzymes will briefly be discussed in this section: hydrogenases, CO dehydrogenase and acetyl-coenzyme-A synthetase, urease, and superoxide dismutase.

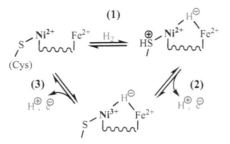

Scheme 10.4 Schematic and simplified representation of the reversible stepwise oxidation of $H_2$ (clockwise) and reduction of $H^+$ (anticlockwise) at the active site of {FeNi} hydrogenases; cf. also Eq. (10.8b); for {FeNi} see **2** in Fig. 10.1. Step (1) of the oxidation path: heterolytic addition of $H_2$; step (2): oxidation of $Ni^{2+}$ to $Ni^{3+}$ and release of $H^+ + e^-$; step (3): reduction of $Ni^{3+}$ to $Ni^{2+}$ by $H^-$ and release of another equivalent of $H^+ + e^-$.

**Hydrogenases** catalyse the reversible conversion of molecular hydrogen to protons plus electrons, Eq. (10.8a). The reaction proceeds via heterolytic cleavage of $H_2$, as depicted in Eq. (10.8b) and Scheme 10.4, as proven by the formation of HD and HDO in the presence of heavy water $D_2O$, Eq. (10.8c)

$$H_2 \rightleftharpoons 2H^+ + 2e^- \tag{10.8a}$$

$$H_2 \rightarrow H^+ + H^- \rightarrow 2H^+ + 2e^- \tag{10.8b}$$

$$H_2 + D_2O \rightleftharpoons HD + HDO \tag{10.8c}$$

Typical terminal acceptors for the electrons in Eqs. (10.8a) and (10.8b) are $O_2$, $NO_3^-$, $SO_4^{2-}$, formaldehyde (HCHO), and $CO_2$. An example of the stepwise reduction of the formyl to the methyl group has been discussed in the context of methanogenesis in Section 10.2; see step (**c**) in Fig. 10.1. The bacterial conversion of hydrogen + oxygen to water, Eq. (10.9), corresponds to the classical Knallgas (oxyhydrogen) reaction. The reaction is, for example, carried out by the hyperthermophilic bacterium *Aquifex aeolicus* [13a], one of the oldest species of bacteria with a growth optimum at 85–95 °C.

$$2H_2 + O_2 \rightarrow 2H_2O \tag{10.9}$$

Hydrogenases contain either two iron centres ({FeFe} hydrogenases; see Fig. 13.2(a) in Section 13.1), or one iron centre ({Fe}, **3** in Fig. 10.1), or an iron plus a nickel centre ({FeNi}, **2** in Fig. 10.1). The stepwise oxidation of $H_2$/reduction of $H^+$ catalysed by an {FeNi} centre in an iron–nickel hydrogenase is depicted in Scheme 10.4.

The {FeNi} centres responsible for the interconversion of $H_2/2H^+$ are coupled to one to three ferredoxins. These ferredoxins serve as mediators of the electron transport to the redox partner of the hydrogenase and, in some cases, can also exert a protective function against oxygen damage, thus providing $O_2$ tolerance for the organism [13b].

A special class among the {FeNi} hydrogenases is represented by the {FeNiSe} hydrogenases. In these enzymes, which have been isolated from sulfate reducing bacteria, one of the cysteinate residues coordinating to nickel is replaced by selenocysteinate [14], Fig.

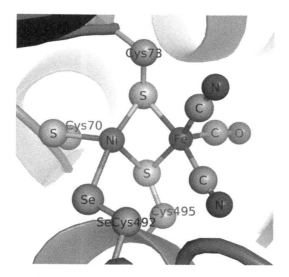

**Figure 10.3** Active site of the {FeNiSe} hydrogenase from the sulfate reducing bacterium *Desulfomicrobium baculatum*. The ribbon structures represent the polypeptide chain environment. Reproduced with kind permission from Springer Science and Business Media: Baltazar CSA, et al. *J. Biol. Inorg. Chem.* 2012; 17: 543–555. Image kindly supplied by Dr Carla Baltazar. ***Also reproduced in full colour in Plate 8.***

10.3. Again, this enables the generation of $H_2$ under oxic conditions—that is, in the presence of $O_2$.

The modes of function of the nickel enzymes **CO-dehydrogenase (CODH) and acetyl-coenzyme-A synthetase (ACS)** are closely interconnected. CODH is a CO-oxidoreductase, meaning that this enzyme catalyses the reversible redox interconversion of carbon monoxide and $CO_2$, Eq. (10.10). The term 'CO-dehydrogenase' is somewhat misleading, since water (and not CO) is dehydrogenated. Eq. (10.10), as read from left to right, is the biological equivalent of the water gas shift reaction, an industrially important reaction catalysed by, e.g. magnetite $Fe_3O_4$. The CO generated reductively from $CO_2$ and $H_2$ in the reverse reaction of Eq. (10.10) is channelled through a 70 Å long tunnel to coenzyme A, HSCoA, where it combines with a methyl group delivered by methyl-cobalamin $CH_3Cb$, to form acetyl-CoA. This net reaction is represented by Eq. (10.11).

$$CO + H_2O \rightleftharpoons CO_2 + H_2 \text{ (or 2 [H])} \tag{10.10}$$

$$CO + HSCoA + CH_3Cb(Co^{+III}) \rightarrow CH_3C(O)SCoA + H^+ + Cb(Co^{+I})^- \tag{10.11}$$

The cobalamin acts out of its 'base off' form, i.e. the otherwise axially coordinated benzimidazole (see Fig. 13.4 in Section 13.1) is detached from cobalt. Further, a nearby $[4Fe,4S]^{2+/+}$ ferredoxin is involved in electron transfer.[4] The catalytically relevant domains of the enzymes CODH and ACS are shown in Fig. 10.4.

[4] The complex $[Fe_4S_4] \sim Cb$ is commonly referred to as 'corrinoid iron sulfur protein'. Corrinoid refers to the tetra-pyrrole corrin skeleton of cobalamin.

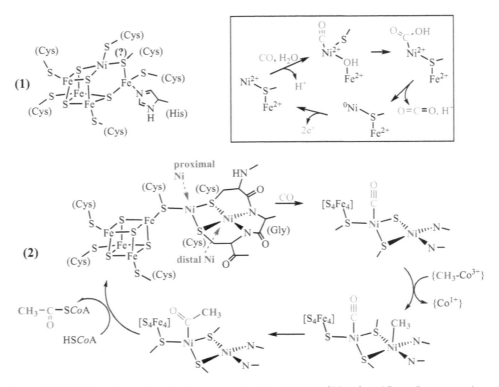

Figure 10.4 (1): The active centre of carbon monoxide dehydrogenase (CODH), and (boxed) a proposed reaction sequence for the conversion of CO and $H_2O$ into $CO_2$ and 2[H] (Eq. (10.10)). The bridging cysteinate indicated by '?' appears not to be present in all of the CODHs. (2): The core of acetylcoenzyme-A synthetase (ACS), and a plausible sequence for the formation of acetyl from methyl, generated from methyl-cobalamin {$CH_3$-Co}, and coordinatively activated CO.

The active centre of CODH contains an [$Fe_3NiS_4$] cluster in contact with an external, tetra-hedrally coordinated iron centre [15]. Two additional iron–sulfur clusters (not shown in Fig. 10.4) transfer electrons to external redox proteins. A simplified reaction sequence for the process represented by Eq. (10.10) is included in Fig. 10.4. The active centre of ACS consists of an [$Fe_4S_4$] ferredoxin linked, via a bridging cysteinate, to a dinickel centre, with the dis-tal nickel in a square-planar coordination environment of two bridging cysteinates and two nitrogen functionalities from the protein backbone.

In a first step of the probable reaction course, CO coordinates to the reduced nickel ($Ni^0$) adjacent to the iron–sulfur cluster (the 'proximal' nickel). Subsequently, methyl-cobalamine transfers its methyl group to the 'distal' nickel. The methyl then migrates onto the carbonyl ligand. In the final step, the acyl thus formed is transferred to coenzyme A.

Another redox-active enzyme, **superoxide dismutase** (SOD), catalyses the dispropor-tionation—or dismutation—of superoxide, Eq. (10.12). The radical anion superoxide $O_2^-$ is a reactive oxygen species generated during $O_2$ metabolism. It is part of the cellular defence system, but is also responsible for oxidative damage. There are several metal ion based SODs, containing Cu, Fe and Mn in their active centre. Ni-based SOD has been found in bacteria

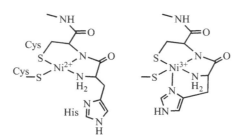

Figure 10.5 Coordination environment in the $Ni^{2+}$ and $Ni^{3+}$ states of superoxide dismutase [16]. The superoxide coordinates to $Ni^{2+/3+}$ (see Eqs. (10.13a) and (10.13b)) into the apical position, where it is additionally stabilized by hydrogen bonding interaction with the adjacent amino acids of the protein matrix.

(*Streptomyces*) and cyanobacteria, but also in a eukaryotic organism, namely *Ostreococcus*, a marine unicellular green alga.

$$2O_2^- + 2H^+ \rightarrow H_2O_2 + O_2 \tag{10.12}$$

Depending on the oxidation state ($Ni^{2+/3+}$), the Ni centre in SOD is in a planar tetra-coordinate ($Ni^{2+}$) or a tetragonal-pyramidal environment ($Ni^{3+}$), Fig. 10.5. The tetragonal plane constitutes two sulfides provided by cysteinate, and two N functions from the protein backbone. The fifth, axial ligand in the $Ni^{3+}$ state is the Nδ of a side-chain histidine. Disproportionation of $\cdot O_2^-$ involves two steps [16]: in the first step, $\cdot O_2^-$ coordinates to the $Ni^{2+}$, where it is reduced in a one-electron transfer by $Ni^{2+}$, protonated, and released as hydrogen peroxide, Eq. (10.13a). A potential proton donor is a nearby tyrosine. In the second step, a second superoxide coordinates to the $Ni^{3+}$, reducing $Ni^{3+}$ back to $Ni^{2+}$, and concomitantly becomes released as oxygen $O_2$, Eq. (10.13b).

$$Ni^{2+} + O_2^{\cdot-} \rightarrow Ni^{2+}(O_2^-), \ Ni^{2+}(O_2^-) + 2H^+ \rightarrow Ni^{3+} + H_2O_2 \tag{10.13a}$$

$$Ni^{3+} + O_2^- \rightarrow Ni^{3+}(O_2^-) \rightarrow Ni^{2+} + O_2 \tag{10.13b}$$

By contrast with the redox-active nickel enzymes introduced above, the enzyme **urease**[5] catalyses a *hydrolytic* process, namely the hydrolysis of urea to ammonia and carbamic acid/carbamate, Eq. (10.14a). In aqueous media, carbamic acid is labile and is further hydrolysed non-enzymatically to form carbonic acid and another equivalent of ammonia, Eq. (10.14b). The *non*-enzymatic degradation of urea generates ammonia and hydrocyanic acid, Eq. (10.15),[6] an extremely slow reaction with a half-life of 33 years. This compares to a half-life of $10^{-9}$ seconds for the *enzymatic* hydrolysis.

$$O=C(NH_2)_2 + H_2O \rightarrow O=C(OH)NH_2 + NH_3 \tag{10.14a}$$

---

[5] Urease was the first enzyme to be structurally characterized (in 1926 by J.B. Sumner). However, the presence of nickel was only established in 1976.

[6] The 'reverse' reaction, the formation of urea from ammonia cyanate ($NH_4OCN$) at elevated temperature, is known as the 'Wöhler urea synthesis' (1828).

$$O=C(OH)NH_2 + H_2O \rightarrow H_2CO_3 + NH_3 \qquad\qquad (10.14b)$$

$$O=C(NH_2)_2 \rightarrow NH_3 + HNCO \qquad\qquad (10.15)$$

Urea is the main terminal metabolite in the degradation of organic nitrogen compounds; its hydrolysis to ammonia and $H_2CO_3/HCO_3^-$ thus is an important input to the global nitrogen cycle (Chapter 9). Urease is present in plants, fungi, and bacteria, including the bacterium *Helicobacter pylori*. About half of the world's population is infected with *H. pylori*, a bacterium that thrives in the gastric mucosa and can cause inflammatory processes (gastritis), ulcers, and carcinomas in the stomach and the duodenum, necessitating treatment with antibiotics. The degradation of urea to ammonia with the help of the *H. pylori* urease not only screens the bacterium—by formation of a protective ammonia layer—against the harsh acidic conditions in the stomach (pH around 2) via neutralization of HCl through $NH_3$, but can also help to shield stomach tissues against acid damage. For most infected people, medication against *H. pylori* is therefore dispensable.

The urease enzyme [17] is a tetramer of trimers, $\{(\alpha\beta)_3\}_4$, with a molecular mass of 1.1 MDa. Each subunit of the active centre contains two $Ni^{2+}$, bridged by $OH^-$ plus a carbamate ligand with lysine as the amine constituent (Fig. 10.6). One of the nickel ions is five-coordinated (two His, one $H_2O$, the bridging $OH^-$ and $LysNHCO_2^-$), while the second $Ni^{2+}$ contains an additional aspartate, coordinated via the carboxylate group in a monodentate fashion. The $\mu$-$OH^-$, the two aqua ligands, plus an external water/$OH^-$ are tetrahedrally arranged.

In a first step of the catalytic turnover (**a** in Fig. 10.6), urea bridges the two $Ni^{2+}$ centres, coordinating through the carbonyl oxygen and one of the amino groups. The second $NH_2$ is hydrogen bonded to backbone carbonyls and a histidine in the protein pocket. Subsequently

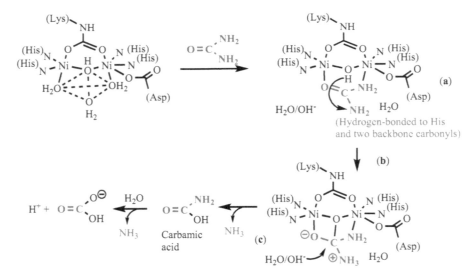

Figure 10.6 The active centre of urease from the soil bacterium *Bacillus pasteurianum*, and the likely course of the hydrolytic degradation of urea as catalysed by the two-nickel unit. See text for the successive steps (**a**) to (**c**).

(step **b**), a proton is transferred from the $\mu$-OH$^-$ to this 'outside' NH$_2$, and the carbon of urea binds to the $\mu$-O$^{2-}$. The final step, step **c**, is the attack of an external OH$^-$ onto the carbon of urea, and the concomitant release of ammonia and carbamic acid.

## Summary

Methane formation and annihilation can occur with the help of microorganisms (methanogens and methanotrophs, respectively) or abiotically. An example of abiotic methane formation is serpentinization, the reductive conversion of $CO_2$ to $CH_4$ supported by ferrous minerals. An increasingly important source for the greenhouse gas methane is methane hydrates (clathrates) from permafrost and deep-sea areas, as well as the exploitation of fossil fuels. The largest sink for atmospheric $CH_4$ is its elimination by reaction with OH radicals.

Methanogenic bacteria reductively convert substrates such as $CO_2$, formate and acetate to methane. The conversion of $CO_2$ to $CH_4$ and $H_2O$ affords eight reduction equivalents. The first step is the reductive transfer of a formyl fragment to methanofuran, mediated by a molybdopterin cofactor, and followed by propagation of formyl to methanopterin. The formyl group is then further reduced, in two successive two-electron steps, to the methyl group, a process catalysed by an iron–nickel hydrogenase. Mediated by cobalamine, the methyl group is further transmitted to mercaptoethanesulfonate (HSCoM) to form $CH_3$SCoM. In the final step, methylide $CH_3^-$ becomes coordinated to the Ni$^{3+}$ centre of factor $F_{430}$, the porphinoid cofactor of methyl-CoM reductase, and is finally released as $CH_4$.

The reverse of methanogenesis, the biogenic oxidation of $CH_4$, is carried out by methanotrophs either aerobically, i.e. by direct oxidation of $CH_4$ to $CO_2$ (via methanol) with $O_2$, or anaerobically. In the latter case, nitrite, sulfate, or high-valent transition metal ions are the electron acceptors. In several cases, the archaea involved in these processes form symbiotic consortia with chemolithotrophic bacteria providing the oxidation equivalents. Communities of sulfate-reducing bacteria/methanotrophic archaea rely on a nickel enzyme closely related to the factor $F_{430}$ of methanogens.

Nature provides three types of hydrogenases for the reversible conversion $H_2 \leftrightarrows H^+ + H^- \leftrightarrows 2H^+ + 2e^-$: iron-only hydrogenases with one or two Fe centres, and iron–nickel hydrogenases {FeNi}. In the latter, iron carries two CN$^-$ and one CO ligand, and is bridged to Ni via cysteinates, Cys$^-$. The coordination sphere of Ni is complemented by two additional Cys$^-$ or, in {FeNiSe} hydrogenase, by Cys$^-$ and selenocysteinate.

Nickel-based cofactors are also present in a couple of other redox enzymes, in a synthase and a hydrolase. Prominent examples of redox enzymes are CO-dehydrogenase (CODH) and superoxide dismutase (SOD). CODH catalyses the interconversion $CO + H_2O \leftrightarrows CO_2 + H_2$. Its function is closely connected to acetylcoenzyme-A synthase (ACS). ACS mediates the synthesis of the acetyl group from CO (provided by CODH) and a methyl delivered by methylcobalamine, and the linkage of acetyl to the sulfhydryl of coenzyme-A. The active centre of CODH constitutes an iron-nickel-sulfur cluster; the active centre of ACS two nickel centres linked to a [Fe$_4$S$_4$] ferredoxin through a cysteinate bridge. SOD directs the disproportionation of superoxide to $H_2O_2$ and $O_2$; its nickel centre switches between the planar S$_2$N$_2$ and the tetragonal-pyramidal S$_2$N$_3$ coordination.

The nickel-dependent hydrolase is a urease, which efficiently catalyses the hydrolytic break down of urea to carbamic acid and ammonia. Urease contains two Ni$^{2+}$ centres, bridged by OH$^-$ and a carbamate with lysine as the amide constituent. Activation of urea is initiated by its $\mu_2$-coordination to the two nickel ions through the carbonyl-oxygen and one of the NH$_2$ groups, and intermittent linkage of the carbon to the bridging hydroxyl.

Suggested reading

**Ferry JC. CO in methanogenesis.** *Ann. Microbiol.* **2010; 60: 1–12.**
The pathways of CO metabolism are reviewed.

**Thauer RK. Functionalization of methane in anaerobic microorganisms.** *Angew. Chem. Int. Ed.* **2010; 49: 6712–6713.**
A brief review of the mechanisms of the anaerobic oxidation of methane by methanotrophic bacteria/archaea.

**Himes RA, Barnese K, Karlin KD. One is lonely and three is a crowd: two coppers are for methane oxidation.** *Angew. Chem. Int. Ed.* **2010; 49: 6714–6716.**
Mechanistic pathways for bacterial soluble and particulate methane monooxygenases are elucidated, in particular with respect of the role of the active copper and iron centres.

**Ragsdale SW. Nickel-based enzyme systems.** *J. Biol. Chem.* **2009; 284: 18571–18575.**
A review that focuses on the catalytic mechanism of eight nickel-based enzyme systems, and structural features of the metal sites.

## References

1. (a) Kamat SS, Williams HJ, Raushel FM. Intermediates in the transformation of phosphonates to phosphate by bacteria. *Nature* 2011; 480: 570–573;. (b) Kamat SS, Williams HJ, Dangott LJ, et al. The catalytic mechanism for aerobic formation of methane by bacteria. *Nature* 2013; 497: 132–136.

2. (a) Lenhart K, Bunge M, Ratering S, et al. Evidence for methane production by saprophytic fungi. *Nat. Commun.* 2012; 3: 1046–1054;. (b) Metcalf WW, Griffin BM, Cicchillo RM, et al. Synthesis of methylphosphonic acid by marine microbes: A source for methane in the aerobic ocean. *Science* 2012; 337: 1104–1107.

3. Lane N and Martin WF. The origin of membrane bioenergetics. *Cell* 2012; 151: 1406–1416.

4. Thauer RK, Kaster A-K, Goenrich M, et al. Hydrogenases from methanogenic archaea, nickel, a novel cofactor, and $H_2$ storage. *Annu. Rev. Biochem.* 2010; 79: 507–536.

5. Kim YJ, Lee HS, Kim ES, et al. Formate-driven growth coupled with $H_2$ production. *Nature* 2010; 467: 352–356.

6. Ferry JG. Methanogenesis biochemistry. In: *Encyclopedia of life science*. Macmillan Publ., 2002, pp.1–9.

7. Cedervall PE, Dey M, Li X, et al. Structural analysis of a Ni-methyl species in methyl-coenzyme M reductase from *Methanothermobacter marburgensis. J. Am. Chem. Soc.* 2011; 133: 5626–5628.

8. Ettwig KF, Butler MK, Le Paslier D, et al. Nitrite-driven anaerobic methane oxidation by oxygenic bacteria. *Nature* 2010; 464: 543–548.

9. (a) Shima S, Krueger M, Weinert T, et al. Structure of a methyl-coenzyme M reductase from Black Sea mats that oxidize methane anaerobically. *Nature* 2012; 481: 98–101;. (b). Thauer RK. Anaerobic oxidation of methane with sulfate: on the reversibility of the reactions that are catalyzed by enzymes also involved in methanogenesis from $CO_2$. *Curr. Opin. Microbiol.* 2011; 14: 292–299.

10. Balasubramanian R, Smith SM, Rawat S, et al. Oxidation of methane by a biological dicopper centre. *Nature* 2010; 465: 115–119.

11. (a) Kovaleva EG, Neibergall MB, Chakrabarty S, et al. Finding intermediates in the $O_2$ activation pathways of non-heme iron oxygenases. *Acc. Chem. Res.* 2007; 40: 475–483; (b) Lee SJ, McCormick MS, Lippard SJ, et al. Control of substrate access to the active site in methane monooxygenase. *Nature* 2013; 494: 380–384.

12. Culpepper MA and Rosenzweig AC. Architecture and active site of particulate methane monooxygenase. *Crit. Rev. Biochem. Mol. Biol.* 2012; 47: 483–492.

13. (a) Pandelia M-E, Nitschke W, Infossi P, et al. Characterization of a unique [FeS] cluster in the electron transfer chain of the oxygen tolerant [NiFe] hydrogenase from *Aquifex aeolicus. Proc. Natl. Acad. Sci. USA* 2011; 108:

6097–6102;. (b) Goris T, Wait AF, Saggu M, et al. A unique iron-sulfur cluster is crucial for oxygen tolerance of a [NiFe]-hydrogenase. *Nat. Chem. Biol.* 2011; 7: 310–318.

14. Marques MC, Coelho R, De Lacey AL, et al. The three-dimensional structure of [NiFeSe] hydrogenase from *Desulfivibrio vulgaris* Hildenborough: a hydrogenase without a bridging ligand in the active site in its oxidised, "as-isolated" state. *J. Mol. Biol.* 2010; 396: 893–907.

15. Gong W, Hao B, Wei Z, et al. Structure of the $\alpha_2\varepsilon_2$ Ni-dependent CO dehydrogenase of the *Methanosarcina barkeri* acetyl-CoA decarbonylase/synthase complex. *Proc. Natl. Acad. Sci. USA* 2008; 105: 9558–9563.

16. Barondeau DP, Kassmann CJ, Bruns CM, et al. Nickel superoxide dismutase structure and mechanism. *Biochemistry* 2004; 43: 8038–8047.

17. Zambelli B, Musiani F, Benini S, et al. Chemistry of $Ni^{2+}$ in urease: sensing, trafficking, and catalysis. *Acc. Chem Res.* 2011; 44: 520–530.

# 11 Photosynthesis

Earth's primordial atmosphere contained but traces of oxygen, a situation which lasted for about two billion years. Then, photosynthetically active cyanobacteria began to evolve, increasingly enriching the atmosphere and the oceans with oxygen and thus forcing organisms to adapt to this novel challenge, or to retreat into anoxic niches. These cyanobacteria contained organelles—thylakoids—capable of converting $CO_2$ and water into biomass and $O_2$, using light as an energy source. The incorporation of these organelles into more developed organisms (algae and plants) by a process referred to as endosymbiosis, and their development into what is known as chloroplasts, initiated a further surge of oxygen up to today's level.

Photosynthesis depends on metals, in particular magnesium (in the light-harvesting chlorophylls), manganese (in the oxygen evolving complex (OEC), where $H_2O$ is oxidized to $O_2$), and iron and copper, both in the electron transfer chain where the electrons freed in the course of water oxidation are ultimately shuttled to the substrate $CO_2$, which is finally reductively converted to organic compounds.

In this chapter, the overall involvement of metal centres in photosynthesis is evaluated, specifically discussing the central role of the chlorophylls on the one hand, and of the oxygen evolving centre on the other hand. The exploitation of sunlight as an energy source in the conversion of $CO_2$ plus water into organics plus $O_2$ by cyanobacteria, algae and plants has also initiated research directed towards mimics of this natural process, i.e. towards the establishment of an 'artificial photosynthesis'. This aspect will also be addressed here.

## 11.1 Overview

The atmosphere of the primordial Earth was a reducing one (Chapter 2), containing only traces of free oxygen as a result of the splitting of molecules such as water and carbon dioxide by electric discharge and high-energy radiation (UV, cosmic rays). This situation changed with the 'Great Oxygenation Event' about 2.4 Ga ago, when photosynthetically active cyanobacteria, also called blue-green algae, started to produce—and to release into the atmosphere—more $O_2$ than could be eliminated by oxidation processes, such as oxidative decay of organics, or the conversion of ferrous to ferric iron, sulfide to sulfate, ammonia to nitrate, and so forth. This change forced organisms that failed to adapt to the novel situation into oxygen-free niches.

A second surge of photosynthetic oxygen supply, 0.8–0.55 Ga ago, finally provided today's atmospheric oxygen contents of 20.95% (by volume, dry air), an indispensable source for energy production and thus sustenance of life for aerobic organisms. The average concentration of $O_2$ in sea water at 1 bar and 15 °C is 6 mL $O_2$ per 1 L of water, which corresponds to a concentration of 0.25 mM. Today, the global solar energy conversion by photosynthesis amounts to 125 TW per year; by contrast, the present global energy consumption by industry and households has been approximately 20 TW in 2013; 1 TWyr = 31.54 EJ.

In the stratosphere, $O_2$ is partially converted to ozone, $O_3$, by UV-C radiation (<242 nm). Ozone in turn decays to $O_2$ and O by absorption of UV-B (around 310 nm). These conversions effectively filter out UV-C and most of UV-B, thus protecting life on our planet's surface—a process which has also helped to drive evolution.

The oxidation of water to oxygen, Eq. (11.1), requires a redox potential of +0.815 V (at pH 7); the overall process is complex and involves a couple of successive electron transfer steps. The common reaction path by which oxygen is produced is through photosynthesis, represented by Eq. (11.2). Here, the oxidation of water to oxygen is coupled to the reduction of $CO_2$ to carbohydrates ('carbon fixation') such as glucose $C_6H_{12}O_6$ or starch, symbolized in Eq. (11.2) by $\{CH_2O\}$.[1] The asterisk in Eq. (11.2) indicates that the oxygen produced in the course of photosynthesis derives from water.

The energy source for photosynthesis, also referred to as autotrophic carbon fixation/ assimilation, is light, $h\nu$, which is absorbed by 'antenna' molecules (such as chlorophylls and carotenoids), with absorption maxima in the blue-violet, red and yellow-orange. Green light is not absorbed—the reason why most plant leaves are green to our eyes. The ability to imitate photosynthesis, i.e. exploiting sunlight as an energy source, increasingly motivates research groups to elucidate the many intertwined paths underlying this naturally occurring process. 'Understanding the fundamental factors that control this complex four-electron, four-proton reaction is [...] essential for the development of efficient artificial photosynthetic machines to convert sunlight into stored chemical energy' [1].

$$2H_2O \rightarrow O_2 + 4H^+ + 4e^- \tag{11.1}$$

$$CO_2 + 2H_2O^* + h\nu \rightarrow \{CH_2O\} + O_2^* + H_2O \tag{11.2}$$

Global carbon fixation by land plants amounts to ca. $120 \times 10^{12}$ kg C per year. A 100 year old beech tree produces about $10^3$ L of $O_2$ and 12 kg of carbohydrates every day, corresponding to 100 mL of $O_2$ and 1.2 g of glucose per 1 m$^2$ of foliage. The reaction complementary to photosynthesis is respiration (aerobic dissimilation), Eq. (11.3). This is an exergonic process by which organics such as carbohydrates are degraded (catabolized) to hydrogencarbonate and protons which, with an additional molecule of $O_2$ and NADH, end up as water. Alternatives for $CO_2$, used by a variety of archaea and eubacteria, are hydrocarbons (such as $CH_4$ and $C_4H_{10}$), formate (Eq. (11.4)), and $CS_2$. In the latter case, Eq. (11.5), $CO_2$ is generated by *hydrolysis* of $CS_2$, catalysed by a $CS_2$ hydrolase (section 12.2.2).

$$O_2 + \{CH_2O\} \rightarrow HCO_3^- + H^+ + energy \tag{11.3}$$

$$HCO_2^- + H_2O \rightarrow HCO_3^- + H_2 \tag{11.4}$$

$$CS_2 + 2H_2O \rightarrow CO_2 + 2H_2S \tag{11.5}$$

[1]  The German botanist Julius Sachs showed in 1862 that green plants produce starch via photosynthesis.

Photosynthesis depends on the presence of chlorophylls as part of the light-harvesting machinery. In cyanobacteria, algae, and higher plants, chlorophyll is confined to cell organelles known as chloroplasts. For most algae and all land plants, these organelles have descended from cyanobacterial ancestors, and hence are the product of endosymbiosis. Algal chloroplasts can also be incorporated symbiotically into other organisms, where they are termed zoochlorellae. An example is *Paramecium bursaria*, a ciliate protozoa. But even vertebrates can undergo a symbiotic relationship with algae, as demonstrated by the embryos of the salamander *Ambystoma maculatum* [2].

In photosynthetic bacteria, which resort to hydrogen sulfide instead of water as the electron source, Eq. (11.6), chlorophyll (here referred to as bacteriochlorophyll) is an integral part of the chlorosomes in the bacterial cell membrane. Bacterial photosynthesis is also referred to as anoxygenic photosynthesis.

$$CO_2 + 2H_2S + h\nu \rightarrow \{CH_2O\} + 1/nS_{2n} + H_2O \ (n = 3,4) \tag{11.6}$$

Instead of resorting to light as an energy source, 'chemical energy' (i.e. energy liberated in the course of a chemical reaction) can be employed in carbon fixation in the absence of chloroplasts. Furthermore, sources other than $CO_2$, e.g. CO or acetate, can be utilized. We can distinguish between the following categories, depending on the energy and carbon source (cf. also Scheme 2.2 in Chapter 2): phototrophic (light energy) vs. chemotrophic (chemical energy), and autotrophic ($CO_2$ as carbon source) vs. heterotrophic (other carbon sources).

The overall reaction as summarized by Eq. (11.1) proceeds via successive light-dependent and light-independent (or dark) reactions. In the light reactions, two photosystems are at work: photosystem II (PSII) and photosystem I (PSI), both embedded in the thylakoid membrane of the chloroplasts. The photosystems are surrounded by antenna complexes for the light to be captured, referred to as 'light-harvesting' complexes (LHCs). These antennae, comprising up to 200 pigments, are mainly chlorophyll-a (yellow-green) and chlorophyll-b (blue-green), but other pigments, such as anthocyanins (red, purple, or blue, depending on the pH), xantophylls (yellow), and carotinoids (orange), are also present. *Bacterial* photosynthesis relies on a single photosystem only, with bacteriochlorophylls as antenna molecules.

From a bioinorganic point of view, the chlorophylls are the more distinguished pigments: they represent coordination compounds with $Mg^{2+}$ in the coordination centre, plus a porphyrinogenic ligand system; thus representing a prominent role for a non-transition, non-redox-active metal ion in a biological context. For a more general view of the role of magnesium in biological systems see Section 3.4.

## 11.2 The reaction pathway

More than one hundred proteins are involved in the global regulation of the photosynthetic machinery, steering more than 50 distinct chemical transformations. Scheme 11.1 illustrates the primary steps in the light-induced generation of an electron in PSII, and Fig. 11.1 provides an overview of selected key reactions in the shuttle of electrons from PSII to PSI, and further to the oxidized form of *nicotine-adenine dinucleotide phosphate*, NADP⁺. The overall reactions are summarized, in a somewhat simplified form, in Eq. (11.7) for the light reaction of

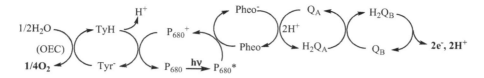

**Scheme 11.1** The initial steps of the light-driven electron transfer from a tetramer of chlorophyll-a ($P_{680}$, Fig. 11.2) in PSII, coupled to the oxidation of water to $O_2$ by the oxygen-evolving complex OEC (Fig. 11.4). The quinone/hydroquinone pool employs two electrons (plus two protons). For further details of the electron shuttle, see Fig. 11.1. Tyr=tyrosine, Pheo=pheophytin (a chlorophyll where $Mg^{2+}$ is replaced by two $H^+$), Q=quinone, $H_2Q$=hydroquinone.

PSII, in Eq. (11.8) for the light reaction of PSI, and in Eq. (11.9) for the dark reaction (Calvin cycle). The electrons shuttled between PSII and PSI, Eqs. (11.7a) and (11.8b), are highlighted in bold. The electron transfer from PSII to PSI (Fig. 11.1) also provides protons, and thus the $H^+$ gradient afforded for the synthesis of the energy storage system adenosine triphosphate (ATP) from adenosine diphosphate (ADP) and inorganic phosphate.

The primary reaction centres $P_{680}$ and $P_{700}$ in the two photosystems are tetramers of molecules of chlorophyll-a (Fig. 11.2) in slightly different environments, with absorption maxima at 680 and 700 nm, respectively. OEC (Eq. (11.7b)) is short for *oxygen evolving complex* (vide infra); for NADPH (the reduced form of $NADP^+$) see Sidebar 9.2.

$$\text{PSII}: P_{680} + h\nu \rightarrow \{P_{680}\}^* \rightarrow [P_{680}]^+ + \mathbf{e^-} \tag{11.7a}$$

$$[P_{680}]^+ + \tfrac{1}{2}H_2O \rightarrow P_{680} + \tfrac{1}{4}O_2 + H^+ \text{(catalysed by the OEC)} \tag{11.7b}$$

$$\text{PSI}: P_{700} + h\nu \rightarrow [P_{700}]^+ + e^- \tag{11.8a}$$

$$[P_{700}]^+ + \mathbf{e^-} \rightarrow P_{700} \tag{11.8b}$$

$$NADP^+ + 2e^- + H^+ \rightarrow NADPH \tag{11.8c}$$

$$\text{Dark reaction}: 2(NADPH + H^+) + CO_2 \rightarrow \{CH_2O\} + 2NADP^+ + H_2O \tag{11.9}$$

Accordingly, chlorophyll-a ($P_{680}$) is excited, by absorbing light, to $P_{680}^*$, which in turn transfers an electron to pheophytin, a chlorophyll-a molecule devoid of the central $Mg^{2+}$. The electron is further transported to plastoquinone Q. Plastoquinones are 5,6-dimethylquinones with an additional substituent in position 2, comprising 6–10 isoprene residues, Fig. 11.1. Transfer of two electrons with concomitant transfer of two protons generates the hydro-plastoquinone $H_2Q$. The electron transfer between pheophytin and quinone is catalysed by an enzyme containing an active iron centre of composition $\{Fe(His)_4Glu\}$ (not directly participating in the $e^-$ transfer). Two $Q/H_2Q$ pairs are involved. The $e^-$ originally stemming from $P_{680}$ (Chl-a) are finally shuttled – with the help of a Rieske centre, cytochromes and plastocyanin – to oxidized $P_{700}$ (Chl-a'). The electrons generated photochemically from Chl-a' (Eq. (11.8a)) go into the dark reaction. The oxidized form of chlorophyll-a, $P_{680}^+$, is finally reduced back to $P_{680}$ by tyrosine. The oxidation of tyrosine to a tyrosine radical ($TyrH \rightarrow Tyr^\bullet + H^+ + e^-$) goes along with the transfer of the proton to a nearby histidine. The tyrosine radical thus formed takes up an electron delivered by water oxidation in the OEC.

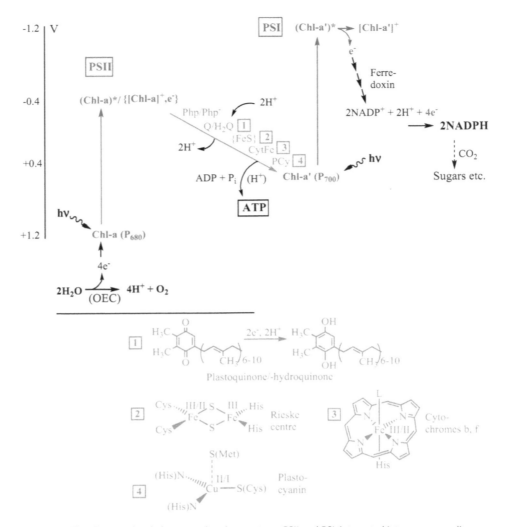

Figure 11.1 The electron shuttle between the photosystems PSII and PSI, integrated into an energy diagram (the so-called Z-scheme; blue, light blue and grey). Large positive potentials correspond to systems at low energy and vice versa. For the first steps of the photochemical release of electrons from $P_{680}$, see also Scheme 11.1. Functional groups of enzymes that take part in the electron shuttle are drawn in light blue. $2 \times 4$ quanta of light are afforded for the transport ($H_2O \rightarrow\rightarrow NADP^+$) of $4e^-$. The ATP synthesis is driven by the $H^+$ motive force; the protons are delivered by the OEC and from the reaction $H_2Q \rightarrow Q + 2 H^+ + 2 e^-$. For details concerning the Rieske centre, cytochromes, and iron–sulfur proteins/ferredoxin, see Sidebars 5.1 and 5.2.

The active centres of several key enzymes participating in the electron shuttle from PSII to PSI are shown in Fig. 11.1. In the final step of this shuttle, $Cu^{2+}$ is reduced to $Cu^+$. In the active centre of the respective enzyme, plastocyanin (a type I, or 'blue' [in its oxidized form], copper protein; cf. sidebar 6.2), copper is in a trigonal pyramidal environment of two histidines and a cysteinate, as illustrated in Fig. 11.3. The apical position is occupied, at a rather long distance of 2.90 Å, by the thioether sulfur function of methionine.

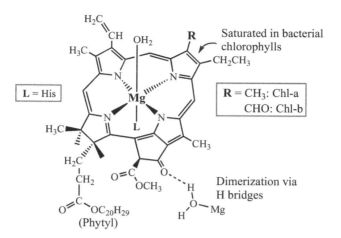

Figure 11.2 The structure of chlorophylls.

In bacterial photosynthesis, plastocyanin is replaced by azurin. In azurin, copper is in a strongly distorted trigonal-bipyramidal environment. Here, the second axial ligand is a protein side-chain carbonyl at a distance of 3.12 Å.

As noted previously, the oxidation of water to oxygen—Eq. (11.10) and the starting situation in Scheme 11.1—is catalysed by PSII, containing a complex cluster, the oxygen evolving complex (OEC), as its active site for water-oxidation catalysis. This complex is based on four manganese ions and one calcium ion. In the reduced form, the metal centres are connected by five oxido groups and six carboxylato groups of amino acid (aspartate and glutamate) side-chains (Fig. 11.4, left), five of which are coordinated in the bidentate mode [4]. Accordingly, the reduction equivalents released from the oxidation of water are taken up by a tyrosyl radical to form tyrosine, which then delivers the electron to $P_{680}^{+}$. This redox-active tyrosine (Tyr 161) is in the proximity of, but not directly coordinated to, the $CaMn_4O_5$ cluster. The cluster constitutes a somewhat distorted $CaMn_3O_4$ cube, linked, via μ-O, to an 'external' Mn. Two water molecules are coordinated to these external manganese and to the calcium ions.

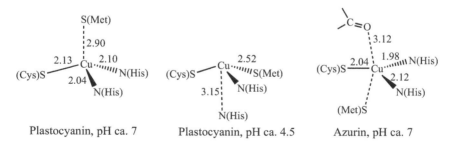

Figure 11.3 Coordination environments of copper in the electron transport centres plastocyanin (plants, algae, and cyanobacteria) and azurin (bacteria). The coordination environment of Cu in the cuprous form of plastocyanin is almost trigonal planar. The coordination environment is pH dependent [3]. Bond lengths in Å.

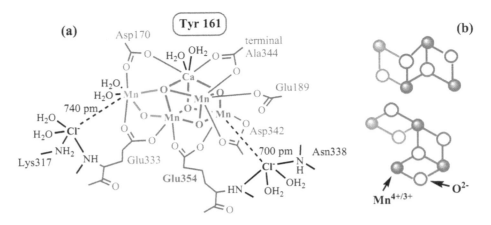

Figure 11.4 (a) The oxygen evolving complex (OEC) within photosystem II [4a]. The $CaMn_4O_5$ cluster is high-lighted in blue (with the central $CaMn_3O_4$ cluster in dark blue), the chloride environments in black, and amino acid residues in grey. The tyrosine (Tyr 161, $Y_Z$) is involved in the redox changes. Glu354 is provided by the chlorophyll binding protein. (b) $\{Mn_4(\mu\text{-}O)_{4/5}\}$ clusters extracted as constituents from the lattice of the $MnO_2$-based tunnel-type mineral hollandite (see end of this section).

In addition, there are two chloride binding sites at non-bonding distances of approximately 7 Å. Each of these chlorides carries two water molecules, and they are additionally linked to backbone nitrogens of amino acids (Glu333 and Glu354) directly coordinated, via the carboxylate side-chain, to $Mn^{n+}$. This structure suggests that they enhance the stability of the cluster during redox turnover and/or act as shuttles for water molecules to the active centre.

$$\{OEC\}$$
$$4\,TyrO^\bullet + 2H_2O \rightarrow 4\,TyrOH + O_2 \qquad (11.10)$$

In the course of the water splitting process, the OEC runs through five successive states, $S_0$ to $S_4$, as illustrated in Fig. 11.5. Four of these changeovers ($S_0 \rightarrow S_1 \rightarrow S_2 \rightarrow S_3 \rightarrow S_4$) require light, while the fifth, $S_4 \rightarrow S_0$, goes along with the oxidation of $O^{2-}$ in water and the release of oxygen. During turnover, the manganese centres change between different oxidation states, most likely involving the oxidation states +III, +IV (and, perhaps, +V) [5]. The coordination of water to a $Mn^{n+}$ centre considerably facilitates its deprotonation through the labilization of the O—H bonds (stabilization of the $\{Mn\text{-}OH\}$ and $\{Mn=O\}$ moieties), an effect which becomes increasingly prominent as the oxidation state of manganese increases. The crucial step, allowing for the formation of an O—O bond, likely is the generation of a $Mn^{IV}$-oxyl radical, $Mn^{IV}O^\bullet$, in the S3 state. Model studies (Section 11.3) suggest that the presence of $Ca^{2+}$ in the cluster further facilitates, via its Lewis acidic character, the activation of the substrate $H_2O$.

Each of the steps in the sequence $S_0 \rightarrow S_1 \rightarrow S_2 \rightarrow S_3 \rightarrow S_4$ requires one light quantum h$\nu$, provides one reduction equivalent (e$^-$), and releases, in most cases [4d], one H$^+$. The proposed oxidation states of the manganese centres are based on electron paramagnetic resonance, EPR; cf. Sidebar 11.1. Also of interest in this context is the ability of simple $MnO_2$ to catalyse the disproportionation of $H_2O_2$ into $O_2$ and $H_2O$, Eq. (11.11), a process otherwise performed

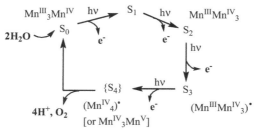

Figure 11.5  S state transitions (Kok cycle; modified from [6]) of the OEC in the course of the oxidation of water to $O_2$. $O_2$ is released as the metastable $\{S_4\}$ state is reached. The subscripts of S indicate the number of accumulated oxidation equivalents, with $S_0$ ($Mn^{III}_3Mn^{IV}$) as the starting situation. The release of four protons is associated with the oxidation steps $S_0 \rightarrow S_1 \rightarrow S_2 \rightarrow S_3$ and $S_3 \rightarrow \{S_4\} \rightarrow S_0$. ($Mn^{III}Mn^{IV}_3$)• in the $S_3$ state represents a reactive manganese-oxyl ($Mn^{IV}$-O•), i.e. a species with radical character on the oxyl oxygen.

by catalases (Sidebar 6.1). Note, however, that there are distinct differences between the catalase activity of $MnO_2$ on the one hand, and the OEC on the other hand: Catalases enable the $2e^-$ transfers in already existing $\{O-O\}$ systems, while the OEC effects a $4e^-$ transfer in conjunction with the *formation* of an O—O bond.

$$H_2O_2 \rightarrow H_2O + \tfrac{1}{2}O_2 \tag{11.11}$$

How did a rather complex system such as the OEC evolve? An evolutionary origin from manganese superoxide dismutase (containing one Mn centre in a histidine dominated environment) or manganese catalase (with two manganese centres, bridged by $O^{2-}/OH^-$ and glutamate) has been proposed [6a]. A possible inorganic progenitor for the OEC is the mineral hollandite [7], found around submarine hot springs. Hollandite is a tunnel-type mineral, the rough stoichiometry of which corresponds to $MnO_2$. Due to inherent disorder in the lattice, several $Mn^{IV}$ sites are occupied by $Mn^{III}$, charge-balanced by $Mn^{2+}$ and $Ca^{2+}$ in the tunnels. The structure contains clusters, with Mn···Mn distances of 2.8 and 3.4 Å (Fig. 11.4 right), that conform with the $\{Mn_4(\mu\text{-}O)_{4/5}\}$ core of the OEC.

This similarity has provoked speculation that oxidic manganese minerals were progenitors in the evolution of the water oxidase. However, the direct incorporation of mineralized metal oxide cluster fragments into a protein structure of a living organism is not very likely to occur. Rather, the spontaneous formation of the respective metal oxide fragments, here mixed manganese–calcium oxides, formed in an aqueous medium with the 'correct' conditions and mediated by cellular activity could provide the scenario for the generation of an inorganic-organic hybrid structure such as present in PSII. We will explore these scenarios, and related ones, in Section 11.3.

---

### Sidebar 11.1    EPR spectroscopy

Electron paramagnetic resonance (EPR) spectroscopy detects paramagnetic species such as organic radicals, and metal ions with one or more unpaired valence shell electrons. The tyrosyl radical Tyr• in PSII is an example for the former, $Mn^{n+}$ in the OEC of PSII are examples for the latter.

Organic radicals usually contain one unpaired electron which, in a magnetic field $B$, can align with the field, symbolized by $\uparrow$ (spin quantum number $M_s = -1/2$; ground state), or be arranged in a direction opposite to the field (spin quantum number $M_s = +1/2$, symbolized by $\downarrow$; excited state). The energy difference between these two states is $\Delta E$. Electromagnetic radiation $h\nu$ can promote the electron from its ground state to the excited state, if the resonance condition $\Delta E = h\nu = g\beta B$ is fulfilled (where $\beta$ is the Bohr magneton and $g$ the so-called $g$-factor).

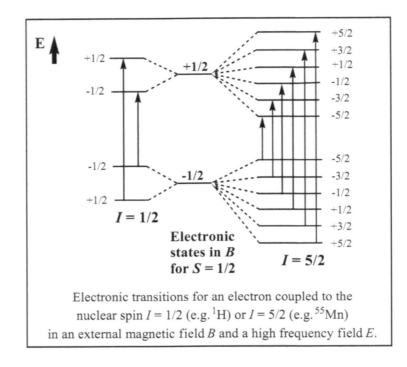

Electronic transitions for an electron coupled to the nuclear spin $I = 1/2$ (e.g. $^1$H) or $I = 5/2$ (e.g. $^{55}$Mn) in an external magnetic field $B$ and a high frequency field $E$.

In a typical EPR experiment, the magnetic field strength is $B \approx 0.5\,\text{T}$, the radiation frequency $\approx$ 10 GHz. The $g$ value for $S = \frac{1}{2}$ states of organic radicals is always close to the $g$ value for an undisturbed electron ($g = -2.0023$). In metal complexes, where the chemical environment is largely influenced by the extent of covalency in the metal-to-ligand bonds, $g$ values and the spectral pattern are modulated by the nature (and oxidation state) of the metal centre, particularly so in the case of quadrupolar metal nuclei, i.e. nuclei with a nuclear spin $> \frac{1}{2}$, such as $^{55}$Mn (spin $= 5/2$) and $^{59}$Co (spin $= 7/2$).

If the spin of a single electron (spin state $S = 1/2$) is in contact with the nuclear spin $I$, for example of the proton ($I = 1/2$), coupling between the two spins occurs. In a magnetic field $B$, splitting of the magnetic ($M$) states thus created occurs, e.g. in the case of $I = 1/2$: $M_s = -1/2 \rightarrow M_I = +1/2$ and $-1/2$; $M_s = +1/2 \rightarrow M_I = -1/2$ and $+1/2$, as shown in the left part of the above scheme. Taking into account the selection rules $\Delta M_s = 1$ and $\Delta M_I = 0$, two transitions are allowed. The result is a doublet structure of the EPR signal, where the extent of coupling—the spacing between the two components—is quantified by the hyperfine coupling constant $A = B_2 - B_1$, commonly expressed in units of $10^{-4}\,\text{cm}^{-1}$.

With a nuclear spin $> 1/2$ (a quadrupolar nucleus), multiplets are obtained. In the case of $^{55}$Mn ($I = 5/2$), six transitions are allowed; right part in the scheme. The situation becomes even more complicated if there is more than one unpaired electron. Examples are high-spin $Mn^{3+}$ (four unpaired electrons, $S=2$) and $Mn^{4+}$ (three unpaired electrons, $S=3/2$). Here, along with the multiline system at $g \approx 2$, EPR signals at $g \gg 2$ appear (vide infra). For the simplest case, where two of the three electrons of $Mn^{4+}$ are paired, and hence a resulting electronic state $S=1/2$, a six-line pattern will arise, as depicted in the right part of the scheme—provided that the local environment of the manganese nucleus is isotropic, as in a complex of $O_h$ symmetry.

In the case of deviations from the ideal symmetry, e.g. in a complex of composition $[MnL_2L'_4]$ (L and L' are two different ligands) with $C_{4v}$ symmetry, an axial spectrum is obtained, with two sets of six (partially overlapping) lines and hence two different coupling constants, $A_z$ and $A_{x,y}$ (if the $z$ axis is the fourfold axis of the molecule). At even lower symmetry, a rhombic spectrum arises, with three sets of lines, hence $A_z$, $A_x$, and $A_y$. The size of EPR spectroscopic coupling constants varies with the nature (donor/acceptor strength) of the ligands in a complex and, in the case of cyclic ligands such as the imidazole in the side chain of histidine, with the geometrical orientation of the ligand. (For spin states, see also Sidebar 4.3; for symmetry, Sidebar 4.4 in Chapter 4.)

Coupling patterns become more complex, and $g$ values $\gg 2$ arise, for spin states $> \frac{1}{2}$, and in cases where coupling in multinuclear centres occurs, as in the OEC. Thus, the $S_1$ state of the OEC (cf. Fig. 11.5), with an integer spin value, exhibits a multiline signal centred at $g=12$ [8]. For the $S_2$ state ($Mn^{III}Mn^{IV}_3$), with three of the Mn centres (the in-cube Mn) coupled ferromagnetically and two of the Mn centres (the outside Mn and one of the in-cube Mn) coupled antiferromagnetically, the overall spin is $S=5/2$, and a broad EPR signal at $g \approx 4.1$ arises along with the resolved multiline system at $g \approx 2$ [9].

## 11.3 Modelling photosynthesis

The resemblance of the oxidomanganese cluster of the OEC to structural units in minerals such as hollandite addressed in the preceding chapter has been a basis for the development of biomimetic water-oxidation catalysts for 'artificial photosynthesis'. Similarly, the development of coordination compounds, based on both manganese and other transition metals, continues to be a major subject area in bio-inspired coordination chemistry directed towards practical applications—here the conversion of sunlight into other forms of energy, or into primary 'fuels' (such as hydrogen) that can store energy.

Minerals belonging to the perovskite family, if specifically designed, have been shown to effectively catalyse the water-splitting reaction, although mechanistically distinct from the OEC, viz. likely via the oxidation of hydroxide $OH^-$ to the hydroxyl radical $OH\cdot$. Genuine perovskite is calcium titanate $CaTiO_3$. Both $Ca^{2+}$ and $Ti^{4+}$ can be exchanged for other alkaline earth and transition metal ions, respectively. An example is $Ba_{0.5}Sr_{0.5}Co_{0.8}Fe_{0.2}O_{3-\delta}$, where the index $\delta$ denotes the sub-stoichiometric (with respect to $CaTiO_3$) oxygen. In this specific perovskite, the $\sigma$-bonding $e_g$ orbital[2] of the transition metal exposed to the surface has near unity occupancy, i.e. it is occupied by 1 electron. This specific occupancy provides greater covalency in the metal-to-oxygen bond between the metal ion and the oxygen in $H_2O$ or $OH^-$, which promotes the charge transfer between $M^{4+}$ (M=Co, Fe) and $O^{2-}$ (from water or $OH^-$), and thus the oxidation of $O^{2-}$ [10].

While these 'photo-active' minerals are of considerable interest for future developments towards artificial materials as catalysts in solar energy production, they do not provide a basis

---

[2] An $e_g$ orbital is a doubly degenerate ($e$), even ($g$, for gerade) orbital, commonly the set $d(z^2)+d(x^2-y^2)$.

for understanding the evolutionary process that led to the employment of such a complex system as the OEC in nature. This is different for pre-mineral systems produced *in situ* from appropriate precursor compounds. Thus, it has been shown that, from alkaline aqueous solutions containing $Mn^{2+}$, permanganate $Mn^{VII}O_4^-$ and $Ca^{2+}$, precipitates of calcium–manganese oxides form with the approximate composition $CaMn^{III}_{0.4}Mn^{IV}_{1.6}O_{4.5}(OH)_{0.6}(\cdot xH_2O)$ [11]. This specific composition, based on structure evaluation by X-ray absorption spectroscopy (see Sidebar 11.2), is reminiscent of the naturally occurring mineral family of birnessites (Na,Ca, K)$_{0.6}$($Mn^{III}$,$Mn^{IV}$)$_2O_4\cdot1.5$ $H_2O$.

These mixed-valence oxides with partially unsaturated Mn centres for the intermittent coordination of water molecules are active in the catalysis of water oxidation, and so represent functional models for the {$CaMn_4O_5$} cluster of the OEC. Even more so, the amorphous precipitates contain motifs reminiscent of the OEC, viz. cubane structures {$CaMn_3O_4$} where one $Ca^{2+}$ and three Mn ions are bridged by oxido anions; **1** in Fig. 11.6. The oxidant in natural water oxidation, catalysed by the Ca-Mn-oxido cluster, is the tyrosyl radical. Its equivalents in model systems for the oxidation of water are one-electron oxidants such as $Ce^{4+}$, Eq. (11.12), or [Ru(bpy)$_3$]$^{3+}$.

$$4\,Ce^{4+} + 2\,H_2O \rightarrow 4\,Ce^{3+} + 4\,H^+ + O_2, \quad cat. \approx CaMn^{IV/III}_2O_5 \tag{11.12}$$

Isolated cubane clusters containing the [$Mn^{IV}_3CaO_4$]$^{6-}$ core can also be stabilized with appropriate multidentate ligands [12]. An example is the complex [$Mn_3CaO_4(THF)(Ac)_3L$], **2** in Fig. 11.6. THF is tetrahydrofuran (coordinating to $Ca^{2+}$), Ac is acetate bridging $Ca^{2+}$ to the three $Mn^{4+}$ centres, and L is a trianionic hexadentate ligand connecting the three $Mn^{4+}$ via three {N-O$^-$} functions, where N is a pyridine nitrogen, and O$^-$ an alkoxide oxygen. One—but just only one—of the $Mn^{4+}$ can reversibly be reduced to $Mn^{3+}$. Hence, both of the manganese oxidation states present in the OEC are mimicked. The presence of $Ca^{2+}$ in the cluster facilitates redox interconversion between $Mn^{IV}$ and $Mn^{III}$. Symproportionation[3] of $Mn^{2+}$ and $MnO_4^-$

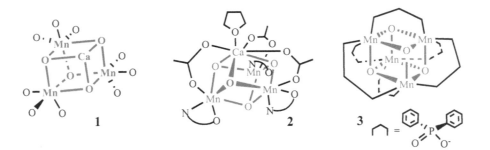

Figure 11.6 Models for the cubane cluster in the oxygen evolving complex of photosystem II that are active in water oxidation catalysis. **1** [11] is a cut-out of an artificial, amorphous material resembling, in composition, the mineral birnessite. **2** [12] and **3** [13] are synthetic cluster compounds. The central cubanes are high-lighted in blue. For details, see text.

[3] When two educts containing the same element in different oxidation states react to form a product with the element in an oxidation state intermediate to those of the educts, this is referred to as *symproportionation* (or *comproportionation*). The reverse situation is known is *disproportionation*.

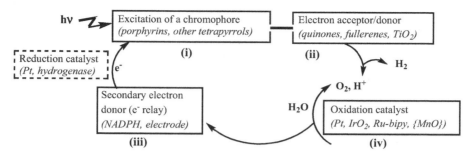

**Figure 11.7** Key steps in artificial photosynthesis. Selected examples for the realization by chemical species are shown in italics. Ru-bipy is a bipyridyl–ruthenium complex such as the dimeric cationic [{Ru(bipy)$_2$H$_2$O}$_2$μ-O]$^{4+}$. Hydrogen can be replaced by other 'fuels' such as methanol and ammonia, however, the electron source in all such schemes is always water.

coupled to self-assembly in the presence of diphenylphosphinate Ph$_2$PO$_2^-$ yields cubane clusters Mn$^{III}_2$Mn$^{IV}_2$(O$_2$PPh$_2$)$_6$, **3** in Fig. 11.6 [13a]. Impregnating **3** into a Nafion[4] matrix causes a reorganization of **3** into birnessite-like nanoparticles dispersed in this matrix. This disordered Mn$^{III/IV}$-oxide phase is catalytically active in water oxidation [13b].

In the light of efforts to replace energy production based on fossil fuels, the perusal of artificial photosynthesis is an attractive goal, but also a major challenge. The core reaction in photosynthesis is the generation and delivery of reduction equivalents (electrons), stemming from water, for the production of reduced species like H$_2$, CH$_3$OH, etc, which might be used as alternative fuels. So far, a robust and efficient system mimicking photosynthesis is not yet available, although a couple of promising approaches towards this goal are being attended to in various laboratories. Following the biological blueprint, artificial photosynthesis requires (i) excitation of a chromophore, also acting as an antenna for harvesting sunlight, (ii) an electron acceptor which is linked to and reduced by the chromophore, and transfers electrons to the substrate to be reduced (e.g. H$^+$), (iii) a secondary electron donor which restores the chromophore to its ground state, and (iv) a catalyst promoting the water oxidation. Figure 11.7 provides a schematic view of the component parts involved in such a process, including selected chemical species which can be employed to model the successive functions in natural photosynthesis [14].

---

**Sidebar 11.2**   X-ray absorption spectroscopy (XAS, XANES, EXAFS)

X-ray absorption spectroscopy (XAS) provides structure information on non-crystalline (amorphous) samples. In particular, bond distances are available with an accuracy approaching that of X-ray diffraction (XRD) spectroscopy of single crystals and microcrystalline powders, while bond angles and coordination numbers, and thus coordination geometries, are less precisely determined. XAS is an element-specific method, i.e. it requires the choice of a specific energy window of a synchrotron radiation source for the excitation of a specific element in a complex compound. Depending on the region under investigation, one distinguishes between X-ray absorption near edge structure (XANES) and extended X-ray absorption fine structure (EXAFS), where 'edge' refers to the energy necessary to

---

[4] Nafion is a sulfonated tetrafluoroethene polymer.

excite an electron, commonly from the K shell (i.e. a 1s electron), to the outermost valence state and beyond, i.e. the ionization limit. An advantage of XAS with respect to XRD is its less invasive—and hence potentially less destructive—nature, sometimes a prerequisite for the structural characterization of metalloproteins.

For a transition metal such as Mn, the inflection point of the K-edge (the binding energy threshold of the 1 s electrons) corresponds to the dipole allowed $1s \rightarrow 4p$ electronic transition. In ideally octahedral complexes, this is the only edge feature observed. Any deviations from $O_h$ symmetry provides a more or less intense pre-edge feature arising from the $1s \rightarrow 3d$ transition. The energetic position of the K-edge of a metal ion in a coordination compound is a measure for the oxidation state of the metal ion, but also intimately correlates with the sum of the electronegativities of the ligand functions attached to the metal. Less energetic electronic transitions can be excited from 2p (into 3d) levels, and are referred to as L-edge spectra.

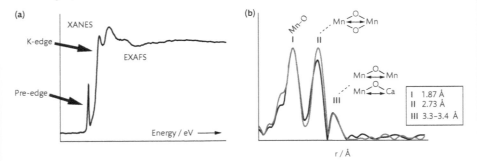

(a) The XAS spectrum of a vanadate-dependent bromoperoxidase (Section 7.2). (b) Fourier-transformed EXAFS range of the XAS spectrum of the $Mn_4CaO_5$ cluster ($S_1$ state in black, $S_2$ state in grey) of the oxygen evolving complex in photosystem II (adapted from: Yano J, et al. *J. Biol. Chem.* 2011; 286: 9257). For details, see below and Fig. 11.4a.

As the K-electron originating from the K-shell of a central transition metal is excited beyond the ionization limit, it interacts with the valence electron shells of the ligands bonded to the central atom. This is referred to as backscattering or, with the wave nature of the electron in mind, interference between the electron wave corresponding to the electron released by ionization of the metal 1s state and the electron waves in the metal-to-ligand bonding sphere. This kind of interaction provides information on the distance between the metal M and ligands L and thus on the nature of the ligand atom (e.g. O, N, or S) directly bound to M, as well as inter-metal distances and distances between M and ligand atoms in the second (and third) coordination sphere, up to about 5 Å. In addition, the (approximate) numbers of atoms in the 1st and 2nd coordination sphere are obtained. This part of the spectrum to high energy of the K-edge, the EXAFS region, is commonly represented as Fourier transformation.

The spectrum for the OEC (b) in the above figure) exemplifies the situation: Peak I at 1.87 Å reflects all of the Mn–O bonds, assuming hexa-coordination for Mn. The Mn···Mn distances within the central $Mn_3CaO_4$ cube (dark blue in Fig. 11.4a) are reflected by peak II, while peak III corresponds to the Mn··· Mn distance to the extant Mn (light blue in Fig. 11.4a) plus the Mn···Ca distances.

## ⊕ Summary

Apart from a few anaerobic niches, our planet has become enriched with oxygen during the past 2.4 billion years, culminating in today's atmospheric content of ca. 21% $O_2$ by volume. This oxygen is provided by photosynthesis—the oxidation of water by plants, algae, cyanobacteria, and several genera of eubacteria, coupled to the reductive fixation of $CO_2$. The herbal photosynthetic apparatus is located in the plants' chloroplasts, which contain two photosystems, PSII and PSI. Light is captured by antenna molecules, mainly chlorophyll-a and -b, which contain $Mg^{2+}$ coordinated to a porphyrinogenic (tetrapyrrole) system. The respective photo-active chlorophyll-a centres in PSII and PSI are $P_{680}$ and $P_{700}$.

Absorption of light by $P_{680}$ causes its activation to $P_{680}^*$ followed by oxidation to $P_{680}^+$. The electron is picked up by pheophytin (a chlorophyll-a without $Mg^{2+}$), which in turn is oxidized by plastoquinone Q. The hydroquinone $H_2Q$ formed in this redox process delivers the electron to an electron shuttle chain connecting PSII and PSI. $P_{680}^+$ becomes re-reduced to $P_{680}$ by tyrosinate $Tyr^-$. The $Tyr\cdot$ radical thus formed is reduced back to $Tyr^-$ by an electron delivered by water oxidation, a process which is catalysed by the oxygen evolving complex, OEC, the manganese containing active site for water-oxidation within PSII.

Redox enzymes involved in the electron shuttle between PSII and PSI comprise a Rieske centre, cytochromes-b and -c, and plastocyanin. Plastocyanin is a 'blue' copper protein with $Cu^{+/2+}$ in a trigonal environment of Cys and two His, and is also weakly coordinated to an axial methionine. $P_{700}^+$ in PSI picks up the electron from the cuprous form of plastocyanin. The electron released by light from $P_{700}$ is taken by $NADP^+$ (to generate NADPH), and further channelled into the dark reaction for the reductive fixation of $CO_2$.

The OEC is a cluster of composition $\{Mn_4Ca(\mu\text{-}O)_5(H_2O/OH)_4\}$ (with some uncertainty still with respect to the number of $H_2O/OH$), containing the central cuboidal $Mn_3CaO_4$ core. The metal ions are further coordinated by the side-chain carboxylates of 3 Glu and 2 Asp, and a terminal Ala. During turnover, four successive oxidation states are passed through, from $S_0$ ($Mn^{III}_3Mn^{IV}$) to $S_4$ ($Mn^{IV}_4$)·, each one involving the release of an electron and a proton. The final step, $S_4 \rightarrow S_0$, is coupled to the release of $O_2$. Structural units resembling the OEC are also present in manganese oxide minerals like hollandite and birnessite.

Photosynthesis has inspired efforts directed towards the conversion of sunlight into energy or primary fuels. Artificial catalysts include synthetic 'minerals' belonging to the perovskite family, and *in situ* generated birnessite, for example of composition $CaMn^{III}_{0.4}Mn^{IV}_{1.6}O_{4.5}(OH)_{0.6}$, as well as molecular, manganese based (cubane) clusters stabilized by multidentate ligands. *Mechanistically*, these mineral-based catalysts are, however, not comparable to the OEC.

## Suggested reading

**Spiegel FW. Contemplating the first plantae.** *Science* 2012; 335: 809–810.

Introduces the way by which chloroplasts spread through the eukarya after having been colonized by a photosynthetic cell (a cyanobacterium).

**Berg IA, Kockelkorn D, Ramos-Vera WH, et al. Autotrophic carbon fixation in archaea.** *Nat. Rev. Microbiol.* 2010; 8: 447–460.

A general overview of $CO_2$ fixation pathways, in consideration of the specific archaean autotrophic carbon fixation strategies that fundamentally differ from the Calvin cycle in eukarya.

**Barber J. Photosynthetic generation of oxygen.** *Phil. Trans. R. Soc. B* 2008; 363: 2665–2674.

In this review, the molecular architecture of photosystem II and its oxygen evolving centre as based on a recent high resolution X-ray study is featured.

**Gust D, Moore TA, Moore AL. Solar fuels via artificial photosynthesis.** *Acc. Chem. Res.* 2009; 42: 1890–1898.

The account illustrates that solar fuel production based on artificial reaction centres that mimic the photosynthetic reaction centre is feasible—though (not yet) sufficiently efficient for practical use.

Wiechen M, Berends H-M, Kurz P. Water oxidation by manganese compounds: from complexes to 'biomimetic rocks'. *Dalton Trans.* 2012; 41: 21–31.
This perspective presents recent progress for promising strategies that have been followed to model the oxygen evolving centre both with respect to structural and functional analogy.

 ## References

1. Eisenberg R and Gray HB. Preface on making oxygen. *Inorg. Chem.* 2008; 47: 1697–1699.

2. Kerney R, Kim E, Hangarter RP, et al. Intracellular invasion of green algae in a salamander host. *Proc. Natl. Acad. Sci. USA* 2011; 108: 6497–6502.

3. Sas KN, Haldrup A, Hemmingsen L, et al. pH-dependent structural change of reduced spinach plastocyanin studied by perturbed angular correlation of γ-rays and dynamic light scattering. *J. Biol. Inorg. Chem.* 2006; 11: 409–418.

4. (a) Umena Y, Kawakami K, Shen J-R, et al. Crystal structure of oxygen-evolving photosystem II at a resolution of 1.9 Å. *Nature* 2011; 473: 55–61; (b) Barber J. Photosynthetic generation of oxygen. *Phil. Trans. R. Soc. B* 2008; 363: 2665–2674; (c) Grundmeier A and Dau H. Structural models of the manganese complex in photosystem II and mechanistic implications. *Biochim. Biophys. Acta* 2012; 1817: 88–105. Part of a special issue, vol. 1817 (1), on 'photosystem II', containing 21 subject-related articles; (d) Dau H, Limberg C, Reier T, et al. The mechanism of water oxidation: from electrolysis via homogeneous to biological catalysis. *ChemCatChem* 2010; 2: 724–761.

5. Armstrong FA. Why did nature choose manganese to make oxygen? *Phil. Trans. R. Soc B* 2008; 363: 1263–1270.

6. (a) Najafpour MM. A possible evolutionary origin for the $Mn_4$ cluster in photosystem II: from manganese superoxide dismutase to oxygen evolving complex. *Orig. Life Evol. Biosph.* 2009; 39: 151–163; (b) Zaharieva I, Wichmann JM, Dau H. Thermodynamic limitations of photosynthetic water oxidation at high proton concentrations. *J. Biol. Chem.* 2011; 286: 18222–18228.

7. Sauer K and Yachandra VK. A possible evolutionary origin for the $Mn_4$ cluster of the photosynthetic water oxidation complex from natural $MnO_2$ precipitates in the early

oceans. *Proc. Natl. Acad. Sci. USA* 2002; 99: 8631–8636.

8. (a) Haddy A. EPR spectroscopy of the manganese cluster of photosystem II. *Photosynth. Res.* 2007; 92: 357–368; (b) Boussac A, Sugiura M, Rutherford AW, et al. Complete EPR spectrum of the $S_3$-state of the oxygen-evolving photosystem II. *J. Am. Chem. Soc.* 2009; 131: 5050–5051.

9. Pantazis DA, Ames W, Cox N, et al. Two interconvertible structures that explain the spectroscopic properties of the oxygen-evolving complex of photosystem II in the $S_2$ state. *Angew. Chem. Int. Ed.* 2012; 51: 9935–9940.

10. Suntivich J, May KJ, Gasteiger HA, et al. A perovskite oxide optimized for oxygen evolution catalysis from molecular orbital principles. *Science* 2011; 334: 1383–1385.

11. Zaharieva I, Najafpour MM, Wiechen M, et al. Synthetic manganese–calcium oxides mimic the water-oxidizing complex of photosynthesis functionally and structurally. *Energy Environ. Sci.* 2011; 4: 2400–2408.

12. Kanady JS, Tsui EY, Day MW, et al. A synthetic model of the $Mn_3Ca$ subsite of the oxygen-evolving complex in photosystem II. *Science* 2011; 233: 733–736.

13. (a) Dismukes GH, Brimblecombe R, Felton GAN, et al. Development of bioinspired $Mn_4O_4$-cubane water oxidation catalysts: lessons from photosynthesis. *Acc. Chem. Res.* 2009; 42: 1935–1943. (b) Hocking RK, Brimblecombe R, Chang L-Y, Water-oxidation catalysis by manganese in a geochemical-like cycle. *Nat. Chem.* 2011; 3: 461–466.

14. Hammerström L and Hammes-Schiffer S. (Eds.) Artificial photosynthesis and solar fuels. *Acc. Chem. Res.* 2009; 42: 1859–2029. Contains 17 subject-related articles by research groups working in related fields (including ref. [13a]).

# 12     The biochemistry of zinc

Next to iron, zinc is the most abundant transition metal in all living organisms. A few thousand proteins relying on zinc for their function and/or structure have been characterized. In contrast to iron, zinc is redox-inactive under physiological conditions, a fact that restricts zinc's role to structural functions, and those catalytic functions where changes of the metal oxidation state are not needed.

Of the numerous zinc proteins, we will select a few examples that highlight the diverse modes of operation that are pertinent to zinc ions in coordination environments dominated by amino acid residues. We will provide examples for the following five main categories of proteins in which zinc attains a structural function and/or mediates catalytic processes: (i) Enzymatic activity in hydrolytic processes; (ii) substrate activation for oxidative detoxification; (iii) interconversion between carbon dioxide and hydrogencarbonate; (iv) transcription of the genetic information contained in deoxyribonucleic acid (DNA) for protein synthesis; and (v) demethylation—and thus repair—of DNA damaged by methylation.

The following questions will thus be addressed: (i) how do zinc ions mediate the breakdown of proteins in our food, making available amino acids for resorption and thus usage in the synthesis of our body's own proteins; (ii) how is ethanol converted to acetaldehyde (co-responsible for the hangover after drinking alcoholic beverages); (iii) in what way is metabolically released carbon dioxide processed for the transport into the lungs; (iv) what is the primary step of the transcription of information, encoded in DNA, into an amino acid; and (v) how do organisms cope with DNA damage stemming from unwanted methylation ('hypermethylation') caused by (for example) methylating agents present in nutrients?

Finally, we will address storage, transport, and eventually release of zinc ions by small proteins termed thioneins, and thus the question of how thioneins contribute to controlling zinc homeostasis in the body. Thioneins contain a particularly high percentage of cysteine residues; another issue in the context of thioneins is therefore their ability to scavenge toxic ions with a high affinity to sulfide, such as $Cd^{2+}$ and $Hg^{2+}$.

## 12.1   An overview of zinc

The essential biological role of zinc was first noted by Jules Leonard Raulin in 1869. Raulin worked with cultures of the mould *Aspergillus niger* and discovered that the multiplication of the fungus was inhibited whenever zinc was absent in the culture medium. The essential nature of zinc is now well established. Zinc is present in all body tissues, with a majority (around 85%) in muscle tissue and bones. The highest local concentrations are found in the prostate and in eye tissues (retina and choroids). An average human individual with a body mass of 70 kg contains ca. 2 g of zinc. Zinc is thus the second-most abundant transition metal in human beings, outclassed only by iron, of which we contain 4–5 g.

After resorption, and prior to distribution in the various tissues, $Zn^{2+}$ is primarily bound to and transported by serum albumin; the mean serum zinc concentration is 0.6 mM. There are two binding sites for $Zn^{2+}$ in serum albumin: Site 1 provides three nitrogen donors stemming from two histidines and one asparagine, plus a carboxylate-O from aspartate; site 2 provides one histidine and three oxygen-functional groups. The strength of $Zn^{2+}$ coordination is modulated by concomitant binding of fatty acids to albumin [1].

The daily requirement for zinc amounts to 3–25 mg, 10–15 mg of which is provided by food. Hence, normal nutritional behaviour ensures a balanced supply of zinc, the absence of dysfunctions of zinc resorption and body distribution provided. In developing countries, however, zinc deficiency is fairly prevalent as a consequence of malnutrition or unbalanced diet [2a], resulting in growth retardation. Further, as we age, the biorecovery of $Zn^{2+}$ decreases, making the elderly susceptible to zinc deficiency symptoms (see below). In either case, the zinc imbalance can be counteracted by zinc supplementation. Impaired zinc utilization has also been associated with cardiovascular diseases and diabetes mellitus [2b].

Zinc deficiency is responsible for a plethora of somatic malfunctions, including cell-mediated immunity and apoptosis.[1] Normal immune response can be restored, in many cases, by nutritional additives such as zinc acetate or gluconate. $Zn^{2+}$ also indirectly counteracts reactive oxygen species (ROS) by inhibiting NADPH oxidase[2], by decreasing oxidative stress through copper-zinc superoxide dismutase (Section 6.2), and by the induction of thioneine synthesis. Metallothioneines are cysteine-rich low molecular mass proteins that can remove ROS. This is illustrated in Eq. (12.1) for the annihilation of the superoxide radical anion. Ointments based on zinc oxide are used in skin protection and wound healing, and have enjoyed great popularity in this respect throughout the decades.

$$3RSH + O_2^{\cdot-} + H^+ \rightarrow 1\tfrac{1}{2}RS-SR + 2H_2O \tag{12.1}$$

Zinc homeostasis is maintained by *zinc importers* (abbreviated Zip) and *zinc transporters* (ZnT).[3] *Importers* channel $Zn^{2+}$ across the cell wall *into* the cytosol, and also mediate its efflux out of intracellular organelles. The zinc binding domains of the respective membrane-bound proteins are rich in amino and carboxy termini, such as histidine (His) and glutamate (Glu). *Transporters*, or zinc-chaperones, mediate $Zn^{2+}$ *efflux* across cell membranes, as well as influx into cellular organelles, including zinc sequestering vesicles and synaptic vesicles of neurons. The metal binding domain of these chaperones is characterized by a His-rich loop. Another

---

[1] Apoptosis is the *programmed*, and hence biologically regulated, cell death of, for example, damaged or worn out cells. Apoptosis is thus distinct from necrotic cell death, which results from injury or parasite infestation.

[2] NADPH is a membrane-bound enzyme complex that generates the reactive oxygen species superoxide $O_2^{\cdot-}$ by electron transfer to $O_2$. For the molecular structure of NADPH, see Sidebar 9.2.

[3] The terms 'metal ion transporter' and 'metal ion importer' are not employed consistently in the literature. Importers are also referred to as transporters (of ions across cell membranes into the cytosol); for a transporter of ions within cells to, e.g. cellular compartments, the term *metallochaperone* has become customary.

group of zinc binding proteins involved in $Zn^{2+}$ homeostasis are the above mentioned thioneins. Thioneins bind $Zn^{2+}$ via cysteine (Cys) residues.

The first zinc-dependent enzyme—carboanhydrase—became characterized as late as 1940 [3]. Carboanhydrase is present in red blood cells and catalyses the hydration of $CO_2$ and the dehydration of $H_2CO_3$; for details see Section 12.2.1. Some 14 years passed until the second zinc enzyme, carboxipeptidase, was discovered. Given that zinc is the second-most abundant transition metal in our body, and more than three thousand zinc-containing proteins are known today, the late and (initially) slow process of their detection and characterization appears surprising at first sight.

The reason for the initially sluggish progress in the discovery of zinc-based proteins stems from the intrinsic chemical and physical properties of zinc ions: $Zn^{2+}$ is redox-inactive in physiological environments; and the closed $d^{10}$ shell generally excludes detection by electron absorption spectroscopy (UV-vis) and electron paramagnetic resonance (EPR). In addition, the nuclear properties of the only magnetically active zinc isotope, $^{67}Zn$ (natural abundance 4.1%, nuclear spin 5/2, nuclear quadrupole moment 15 fm$^2$), are far from ideal for the detection of distinct zinc signals by nuclear magnetic resonance, NMR (Sidebar 14.2). Zinc enzymes have therefore often been characterized after exchange of $Zn^{2+}$ for $Cd^{2+}$ or $Co^{2+}$. The nuclei $^{111}Cd$ and $^{113}Cd$ are versatile NMR probes, and $Co^{2+}$, a paramagnetic $d^7$ system, is susceptible to EPR (Sidebar 11.1) and UV-vis spectroscopy. In addition, structural information for $Co^{2+}$ substituted zinc proteins can be yielded from paramagnetic $^1H$ NMR of organic moieties in the coordination periphery of the metal.

Zinc ions adopt a variety of different physiological roles, including catalytic, structural, and regulatory functions. With respect to function, the following categories can be distinguished; we discuss details in the context of selected zinc proteins in later sections.

(i)  Catalytic function: carboanhydrase; hydrolases and lipases, synthases, isomerases, ligases; oxidases/reductases.

Hydrolases (esterases, peptidases, phosphatases) and lipases catalyse the breaking of bonds by water. Ligases catalyse bond formations using a biochemical energy source (such as ATP), while synthases catalyse syntheses without the use of an energy source.

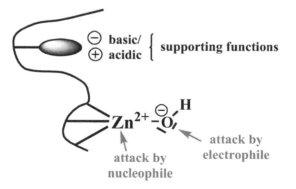

Figure 12.1 Schematic illustration of a zinc centre in a hydrolase. The looped line represents the protein backbone. $Zn^{2+}$ is preferentially linked to Glu/Asp, Cys and His; see also Fig. 12.3. The p$K_a$ for the equilibrium $\{Zn^{2+}\text{-}OH_2\} \leftrightarrows \{Zn^{2+}\text{-}OH^-\} + H^+$ is $\approx 7$.

Redox-active zinc enzymes such as the *catalytic* centre in alcohol dehydrogenase require a redox-active cofactor, here NADH/NAD⁺. A catalytic $Zn^{2+}$ centre in a hydrolase is typically tetrahedrally coordinated to three amino acid residues (glutamate or aspartate, cysteinate, histidine) and $H_2O/OH^-$. Its enzymatic action is based on the fact that the $\{Zn^{2+}-OH^-\}$ moiety can activate nucleophiles as well as electrophiles (see Fig. 12.1). An example of a *mixed metal* hydrolase is purple acid phosphatase, containing a dinuclear $\{Fe^{2+}(\mu\text{-Asp},\mu\text{-OH})Zn^{2+}\}$ active centre.

(ii) Structural function

(a) Structural function on the molecular level: stabilization of the tertiary structure of oligomers of proteins and of domains.

The coordination environment of structural $Zn^{2+}$ is dominated by cysteinate. Examples for structural $Zn^{2+}$ centres in enzymes are Cu,Zn superoxide dismutase, cytochrome-c oxidase,[4] and the *structural* centre in alcohol dehydrogenase. In zinc finger transcription factors, involved in the process of information transfer from DNA, $Zn^{2+}$ stabilizes a protein loop and thus enables recognition of a base triplet of DNA to be transcribed into a messenger-RNA. Here, $Zn^{2+}$ is commonly coordinated to two cysteinyl and two histidine residues of the protein (class I) or, less commonly, to four cysteines (class II). The coordination number six—three histidines and three water molecules in a facial coordination arrangement—is achieved in the hexameric storage form of insulin, where three insulin dimers are linked by two or four $Zn^{2+}$.

(b) Structural function on the level of 'materials': the foam nest for fertilized eggs of the Malaysian frog *Polypedates leucomystax* contains a blue dimeric 26 kDa protein, ranasmurfin, with the two identical subunits linked by cystine bridges and $Zn^{2+}$ in a penta-coordinate environment (Fig. 12.2a) [4].

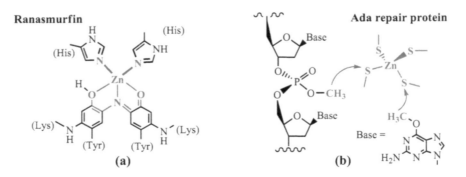

Figure 12.2 (**a**) The $Zn^{2+}$ centre of ranasmurfin, a protein from the foam nest of a tropical tree frog, links the two subunits (represented here by the His residues) of the protein dimer. (**b**) The zinc centre of the Ada DNA repair protein demethylates the phosphoester linkage between nucleobases, in particular guanine (shown) by transfer of the methyl group to a coordinated cysteinate sulfur.

[4] For Cu,Zn-superoxide dismutase and cytochrome-c oxidase, see Sections 6.2 and 5.3, respectively.

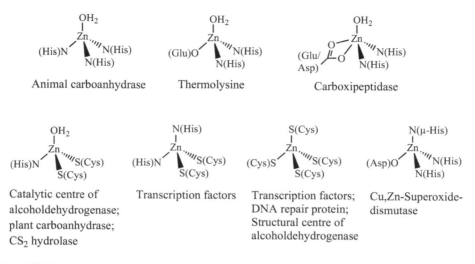

Figure 12.3 Typical coordination environments of tetrahedral zinc centres in zinc proteins.

(iii) DNA repair: DNA damage by alkylation of the nucleobases (guanine, thymine) and/or the phosphate linkers is counteracted by the enzyme 'Ada DNA repair protein'. 'Ada' stands for 'adaptive', referring to the induction of the enzyme when cells are treated with alkylating agents. $Zn^{2+}$ is coordinated to four cysteinyl residues. Repair comes about by transfer of the alkyl group—in most cases a methyl group—from the base and/or phosphate to cysteinate [5], as shown in Fig. 12.2 b).

(iv) Modulation of (intra-)cellular signal response and transduction, including synaptic transmission: Zinc ions modulate a broad variety of physiological responses [6]. Examples are the regulation of cell proliferation and the modulation of the synaptic activity of brain cells by free and labile (weakly coordinated) zinc ions.

(v) Zinc storage: cysteine-rich, small proteins (molecular mass ca. $6000\,g\,mol^{-1}$) termed thioneins can coordinate up to seven thiophilic cations such as $Zn^{2+}$, $Cu^+$ and $Cd^{2+}$. They probably serve as zinc and/or cysteine storage proteins, and can be active in the redox annihilation of hyperoxide and peroxide.

The common coordination behaviour of $Zn^{2+}$ in biological systems is characterized by the coordination number 4 in a tetrahedral environment, as shown in Fig. 12.3. This arrangement enables minimization of strains and thus represents an energy-minimized geometry in a protein pocket [7].

The coordination environment is multifaceted: hard (O-functional), medium (N-functional), and soft (S-functional) ligands are employed, where the N- and S-functional ligands predominantly are histidine and cysteinate. The general coordination chemistry of zinc is more flexible, i.e. the coordination numbers 3, 5, and 6 are also seen, as is square planar coordination for the coordination number 4. Fig. 12.4 displays respective examples.

In the next two sections, zinc's role in biocatalysis on the one hand, and as a structural motif on the other, will be outlined in more detail.

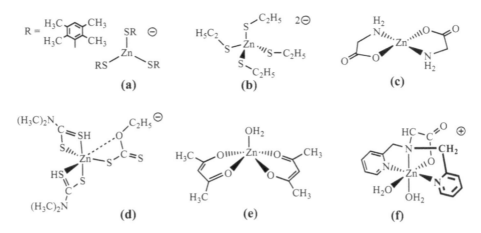

Figure 12.4 Examples for zinc complexes with coordination numbers 3 (**a**, trigonal planar), 4 (**b**, tetrahedral; **c**, planar), 5 (**d**, distorted trigonal-bipyramidal; **e**, tetragonal-pyramidal), and 6 (**f**, octahedral). The ligands are tetramethylphenylthiolate (**a**), ethylthiolate (**b**), glycinate (**c**), dithiocarbamate and dithiocarbonate (**d**), acetylacetonate (**e**), and dipicolylglycinate (**f**). The dashed line in (**d**) indicates a weak interaction between $Zn^{2+}$ and the oxygen of the ethyl-dithiocarbonato ligand.

## 12.2  Zinc enzymes

### 12.2.1  Carboanhydrase

As noted at the start of the preceding section, carboanhydrase (also referred to as carbonic anhydrase) was the first zinc protein to be structurally characterized. The human body contains 1–2 g of this 29.3 kDa protein. Its task is the rapid setting of the equilibrium between hydrogencarbonate and carbon dioxide+hydroxide, Eq. (12.2), a reaction which, in the absence of a catalyst, proceeds with rate constants $k_{\rightarrow}=8.5\times10^3$ $M^{-1}$ $s^{-1}$ and $k_{\leftarrow}=2\times10^{-4}$ $s^{-1}$—that is, it goes on very slowly. The process is accelerated by a factor of $10^7$ in the presence of the enzyme.

$$CO_2 + OH^- \rightleftharpoons HCO_3^- \tag{12.2}$$

Carboanhydrase thus rapidly converts $CO_2$ *in statu nascendi* (i.e. *instantaneously* in metabolizing tissue where oxidative degradation of organics occurs) to $HCO_3^-$. Hydrogencarbonate is then transported in the blood stream to the pulmonary alveoli. Here, protonation of $HCO_3^-$ by haemoglobin (Hb) takes place, with concomitant formation of oxyhaemoglobin, Eq. (12.3). The carbonic acid thus formed is broken down to water and $CO_2$, which is exhaled.

$$HCO_3^- + Hb\cdot H^+ + O_2 \rightarrow Hb\cdot O_2 + H_2CO_3(\rightarrow H_2O + CO_2) \tag{12.3}$$

The zinc centre in carboanhydrase is coordinated to the Nε of three histidine residues, and water/$OH^-$. As shown in Scheme 12.1, $CO_2$ is activated by an electrophilic attack of the carbon of $CO_2$ on the oxygen of the hydroxido ligand, followed by the formation of a hydrogencarbonato complex via redistribution of the proton, a process which is mediated by a

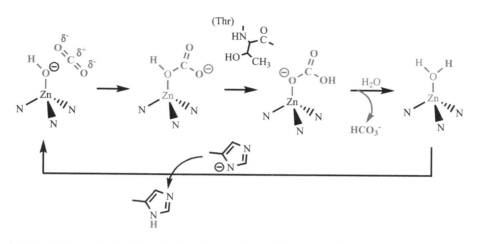

Scheme 12.1 The mode of action of carboanhydrase. N stands for the Nε of histidine residues of the protein, and Thr is a threonine residue at the active site protein pocket, mediating the proton shift.

nearby threonine, Thr. In the last step, hydrogen carbonate is detached and replaced by water. The active, hydroxidic form of the enzyme is restored by transfer of a proton to a base, most likely deprotonated histidine. Tripodal tridentate nitrogen donors such as *tris*(pyrazolyl) borate form complexes with $Zn^{2+}$ that mimic carboanhydrase [8], Scheme 12.2.

In higher plants, where carboanhydrase is involved in the acquisition of inorganic carbon (conversion of $H_2CO_3 / HCO_3^-$ to $CO_2$) for photosynthesis, two of the histidines are replaced by cysteines, forming a {Zn(His)(Cys)$_2$OH} motif, as in the catalytic centre of alcohol dehydrogenase, Fig. 12.3. Such a coordination environment is also present in a carboanhydrase containing *cadmium* instead of zinc, isolated from the marine diatom *Thalassiosira weissflogii* [9]. Cadmium is commonly classified as a highly toxic metal, because $Cd^{2+}$ (like $Hg^{2+}$) can replace $Zn^{2+}$ in zinc proteins and thus defunctionalize zinc-dependent enzymes. In marine environments, zinc concentrations are very low, and algae such as diatoms apparently have adapted to this situation by resorting to cadmium.

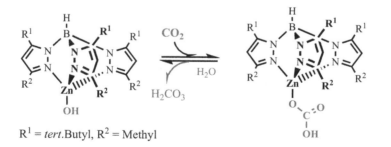

$R^1 = tert.$Butyl, $R^2 =$ Methyl

Scheme 12.2 The *tris*(pyrazolyl)boratozinc complex is a structural and functional mimic of carboanhydrase.

## 12.2.2 Hydrolases

Hydrolases are enzymes that fragment chemical bonds with the help of water by transfer of $OH^-$ and $H^+$ to the originally connected molecular fragments. This is represented for a molecule AB, consisting of the fragments A and B, in Eq. (12.4). A large number of zinc-dependent hydrolases are known, four of which, including peptidases and phosphatases, will be described in more detail in the following.

$$A-B + H_2O \rightarrow A-H + HO-B \tag{12.4}$$

The protein-based components of our food are broken down in the gastrointestinal tract by peptidases to smaller fragments (peptides[5]) and finally amino acids, which then become resorbed and are employed as building blocks for our body's own peptides and proteins. *Endo*peptidases catalyse the hydrolysis of the peptide bond *within* a polypeptide strand, while *exo*peptidases catalyse the hydrolytic detachment of an amino acid at the N- or the C-terminus of the protein. Hydrolytic bond cleavage is illustrated in Eq. (12.5).

$$\text{—N—C—C—NH—C—C} + H_2O \longrightarrow \text{—N—C—C} + H_2N\text{—C—C} \tag{12.5}$$

*Amino-* and *carboxi*peptidases—both *exo*peptidases—are distinguished according to whether they start from the N- or the C-terminal end of the peptide. A particularly well investigated exopeptidase is carboxipeptidase A from bovine pancreas, with a molecular mass of 34.6 kDa. The $Zn^{2+}$ centre (see also Fig. 12.3) is coordinated to the Nε of two histidines, a carboxylate (in the $\eta^2$ mode) of glutamate, and a water molecule. A second glutamate and a tyrosine, both in the proximity to the active centre, participate in the catalytic process. An example of an *endo*peptidase is thermolysin, isolated from the thermophilic *Bacillus thermoproteolyticus*, a bacterium that thrives at temperatures around 65 °C. In thermolysin, the active centre glutamate is coordinated in the $\eta^1$ mode. The tertiary structure of thermolysin is stabilized by four $Ca^{2+}$ against denaturation at high temperatures; see Section 3.5.

General features of the mechanistic cycle of the hydrolysis of a peptide bond by a peptidase are illustrated in Fig. 12.5: In step (1), the protein is activated by replacement of the water attached to zinc through the peptide (coordinating via the carbonyl-O), while the water molecule becomes hydrogen-bonded to the external glutamate. *Activation* here refers to the effective increase of the polarity within the carbonyl group. In step (2), hydroxide is transferred from the water molecule (hydrogen-bonded to and thus activated by glutamate) to the carbonyl carbon. Concomitantly, glutamate and the peptide amide function become protonated. The 'extra' proton is delivered by the nearby tyrosine of the enzyme. The intermediate thus formed rearranges, releasing—step (3)—the new C- and N-terminal peptide fragments of the original protein.

Another hydrolase, isolated from an extremophile, is carbon disulfide hydrolase [10]. The 24 kDa subunit of the octamer contains two cysteines and one histidine coordinated to $Zn^{2+}$. $CS_2$ hydrolase is used by the acido- and hyperthermophilic archaea *Acidianus* to generate

---

[5] There is no sharp limit for the use of the terms 'protein' and 'peptide'. A peptide is commonly an assembly of a few amino acids (oligopeptides) up to several dozen amino acids (polypeptides). Still larger assemblies are referred to as proteins. In any case, the linkage formed by condensation of the carboxylate function of an amino acid (the C-terminus) and the amine function (the N-terminus) of an adjacent amino acid is termed '*peptide* bond', irrespective of the number of amino acids involved in the build-up of a peptide or protein.

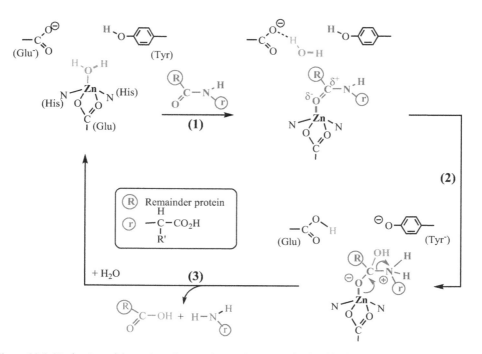

**Figure 12.5** Mechanism of the action of a typical peptidase with a {Zn(Glu)(His)$_2$H$_2$O} active centre. Glu and Tyr are two of the nearby amino acid residues participating in the hydrolytic cleavage of the peptide bond. For a description of paths (1), (2), and (3), see the text.

energy from the hydrolytic conversion of CS$_2$ to CO$_2$ and H$_2$S by way of COS, Eq. (12.6). H$_2$S is further (non-enzymatically) oxidized to sulfuric acid, providing the highly acidic medium of the volcanic solfataras, the habitat of these archaea.

$$CS_2 + H_2O \rightarrow COS + H_2S; \quad COS + H_2O \rightarrow CO_2 + H_2S \tag{12.6}$$

The immediate coordination environment of zinc is identical to that in carboanhydrase, Section 12.2.1, suggesting that the activation of (hydrophobic) CS$_2$ by carbon disulfide hydrolase is identical to the first step of the activation of CO$_2$ by carboanhydrase, Fig. 12.1.

Enzymes that catalyse the hydrolysis of the phosphoester bond, Eq. (12.7), are termed phosphatases. Phosphoesters play a pivotal role in a plethora of physiological processes, including energy storage, the activation of substrates for catabolic processes, and the linkage of nucleotides in DNA and RNA. Sidebar 12.1 provides an overview of functional phosphoesters and -amides in biological systems, along with some basic information on the structurally similar phosphate 'competitors' vanadate and arsenate. Alkaline and acid phosphatases cleave ester bonds under (slightly) alkaline and acidic conditions, respectively. An example of a zinc-dependent alkaline phosphatase is illustrated in Fig. 12.6a. The active centre contains two cooperative Zn$^{2+}$ and a structural Mg$^{2+}$ (not shown in Fig. 12.6).

$$H_2PO_3(OCH_3) + H_2O \rightarrow H_2PO_4^- + CH_3OH + H^+ \tag{12.7}$$

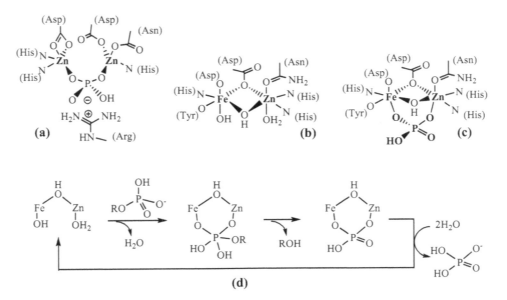

Figure 12.6 (a) The active site of the dizinc centre in a zinc-based alkaline phosphatase from the human placenta. (b) The active centre of purple acid phosphatase in kidney beans, its phosphate adduct (c), and the catalytic cycle of phosphoester hydrolysis (d). The pentavalent transition state of phosphorus is shown in the centre of the sequence (d).

Purple acid phosphatase is representative of a *hetero*-metal enzyme, where both metal centres, here $Fe^{3+}$ and $Zn^{2+}$, again cooperate.[6] The enzyme catalyses the hydrolytic conversion of phosphoesters in the pH range 4–7. Purple acid phosphatase from kidney beans is a homodimer with a molecular mass of 55 kDa per monomeric subunit [11]. Both metal centres, $Fe^{3+}$ in its high-spin state and $Zn^{2+}$, act as Lewis acids. Coordinating the phosphoester to the two metal ions in the bidentate bridging mode induces hydrolysis of the ester bond via an intermediate pentavalent state. The coordination environment of the active centre and the operating sequence of the hydrolytic process are illustrated in Fig. 12.6b–d.

---

**Sidebar 12.1    Phosphate and the phospho-ester/amide bond in biological processes**

Phosphates and phosphate derivatives, esters of orthophosphoric acid ($H_3PO_4$) in particular, have numerous functions in biological systems. The average concentration of phosphate in body fluids is ca. 1 mM; at pH 7, inorganic phosphate is present as $H_2PO_4^-$ and $HPO_4^{2-}$ in about equal amounts (the $pK_a$ for $H_2PO_4^- \rightleftharpoons HPO_4^{2-} + H^+$ is 7.21). The main body pool for inorganic phosphate is hydroxyapatite $Ca_5(PO_4)_3(OH)$. Hydroxyapatite adopts a supporting function in bones; Section 3.5.

Major functions of *organic* phosphate derivatives include:

---

[6] Phosphatases containing Fe(III)–Mn(II) and Fe(III)–Fe(II) centres are also known.

- Linkers between ribose/deoxyribose moieties in RNA/DNA (**a**).

- Energy storage: Examples of energy storage molecules are phosphates of nucleosides such as adenosine triphosphate (ATP, Fig. 3.7 in Section 3.4) and guanosine triphosphate (GTP). Other molecules for energy storage and transfer/release include creatine phosphate (Scheme 3.3 in Section 3.4), acylphosphates (**b**), phosphoarginine (**c**), and phosphoguanidine (**d**). Energy is released upon hydrolytic cleavage of the phosphoester bond; examples ($\Delta G$=free reaction enthalpy, or Gibbs energy):

$$ATP^{3-} + H_2O \longrightarrow ADP^{2-} + H_2PO_4^- \qquad \Delta G = -30.5 \text{ kJ mol}^{-1}$$

$$H_2C=C\begin{smallmatrix}CO_2^-\\OPO_3^{2-}\end{smallmatrix} + H_2O \longrightarrow H_3C-C\begin{smallmatrix}CO_2^-\\O\end{smallmatrix} + HPO_4^{2-} \qquad \Delta G = -61.9 \text{ kJ mol}^{-1}$$

Phosphoenolpyruvate              Pyruvate

- Integral part of coenzymes such as nicotine-adenine dinucleotide phosphate (NADPH/NADP$^+$, Sidebar 9.2 in Chapter 9).

- Phospholipids in membrane structures (**e**).

- Activation of substrates in metabolic processes, e.g. glucose-6-phosphate (**f**).

- Signalling and messenger molecules such as cyclic adenosyl monophosphate cAMP (**g**) and inositol phosphates (**h**).

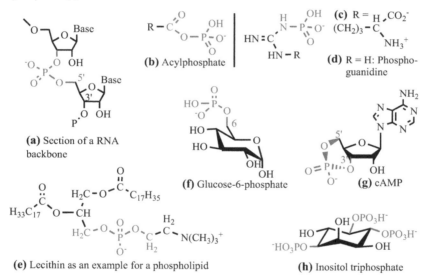

(**a**) Section of a RNA backbone

(**b**) Acylphosphate

(**c**) R = $\begin{smallmatrix}H\\(CH_2)_3\end{smallmatrix}-C\begin{smallmatrix}CO_2^-\\NH_3^+\end{smallmatrix}$

(**d**) R = H: Phospho-guanidine

(**e**) Lecithin as an example for a phospholipid

(**f**) Glucose-6-phosphate

(**g**) cAMP

(**h**) Inositol triphosphate

The phosphate–vanadate antagonism: Monovanadate $H_2VO_4^-$, the predominant form of vanadate at physiological conditions, is a structural analogue of phosphate $H_nPO_4^{(3-n)-}$ ($n$=1 and 2); see also Scheme 14.4 in Section 14.3.4. Vanadate can compete with, and/or regulate, phosphate in its various physiological functions. The ionic radii of V$^{5+}$ (50 pm, for tetrahedral coordination) and P$^{5+}$ (52 pm) are almost identical; vanadate can thus substitute for phosphate in processes that depend on phosphate. In contrast with phosphorus, the transition metal vanadium can easily attain the stable *penta*-coordination (which is just a transition state in phosphate-dependent metabolic processes). The usually *inhibitory* effect of vanadate towards phosphatases and kinases stems from these specific characteristics. Another striking difference between vanadate and phosphate is the accessibility of vanadate, but not phosphate, to reduction (to vanadium(IV)) under physiological conditions.

Arsenate vs. phosphate: Arsenate and phosphate esters are structurally and thermodynamically strikingly similar. This fact has initiated a lively discussion on the potential ability of bacteria to build arsenate rather than phosphate into ester linkages, when these bacteria thrive in arsenate-rich habitats, such as the Mono Lake in California with $c(As) \approx 200\ \mu M$. However, *kinetically* arsenate esters are extremely short-lived: half-lives for diesters (as in DNA linkages) are $6 \times 10^{-2}$ seconds, as compared to $3 \times 10^{7}$ years for the respective phosphoesters. Another difference between arsenate and phosphate is the more pronounced susceptibility of the former to reduction.

How do organisms discriminate between arsenate and phosphate, and thus protect themselves against arsenate poisoning? The periplasm of the bacterial cell contains phosphate-binding proteins (PBPs) that deliver phosphate to a transporter for transport across the membrane into the cell. These PBPs can discriminate phosphate $HPO_4^{2-}$ over arsenate $HAsO_4^{2-}$ with a selectivity of about three orders of magnitude. The differentiation has been attributed to symmetrical ($HPO_4^{2-}$) vs. asymmetrical ($HAsO_4^{2-}$) P/As-O-H$\cdots$O(Asp) interaction [12].

Phosphate-dependent enzymes, in particular phosphatases and kinases, can be inhibited by the phosphate analogue vanadate (Sidebar 12.1), which is of interest in the context of the potential medicinal use of vanadate and vanadium compounds (Section 14.3.4). 'Physiological' vanadate concentrations might contribute to the regulation of phosphate-dependent enzymes and processes. Sidebar 12.1 also contains a brief account of the phosphate-arsenate antagonism.

### 12.2.3 Alcohol dehydrogenase

Alcohol dehydrogenase (ADH) is an 80 kDa homodimer that catalyses the dehydrogenation of alcohols, ethanol in particular, to aldehydes. Each of the two subunits of ADH contains a structural zinc centre with $Zn^{2+}$ tetrahedrally coordinated to four Cys, plus a catalytic zinc centre where $Zn^{2+}$ is coordinated to two Cys, one His, and an aqua/hydroxido ligand. The task of the catalytic centre is to activate the substrate alcohol, thus enabling hydride transfer from the alcohol to the nearby cofactor $NAD^+$. Fig. 12.7 reflects the arrangement of the catalytic site, and Eq. (12.8) the net reaction.

Further reduction—and thus detoxification—of the aldehyde is achieved by the NAD cofactor of aldehyde dehydrogenases, Eq. (12.9). Aldehyde dehydrogenases do not contain a metal centre for substrate activation, but can come with $Mg^{2+}$ for structure stabilization.

$$RCH_2OH + NAD^+ \rightarrow RCHO + NADH + H^+ \tag{12.8}$$

$$RCHO + NAD^+ + H_2O \rightarrow RCO_2H + NADH + H^+ \tag{12.9}$$

Figure 12.8 illustrates the dehydrogenation of alcohols at the catalytic centre of ADH: In a first step, (**a**), the alcohol attacks the hydroxido-zinc centre nucleophilically, followed by deprotonation of the alcohol to alcoholate, (**b**), and removal of water. Hydride $H^-$ is then transferred from the alkoxido ligand to the nicotinamide ring of the adjacent coenzyme $NAD^+$, (**c**). The intermittently generated carbocation is stabilized by rearrangement to the aldehyde; the product aldehyde and NADH finally dissociate from the active site, (**d**), which then picks up a water molecule, (**e**). The cycle closes by deprotonation of the aqua ligand and reentry of an $NAD^+$, (**f**). A structural model for the active site zinc with ethanol coordinated to the forth position is also provided in Fig. 12.8.

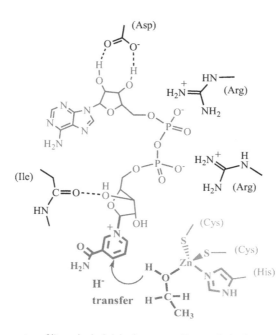

**Figure 12.7** The active centre of liver alcohol dehydrogenase. The catalytic zinc centre is shown in light blue, the substrate alcohol in grey, and the cofactor NAD$^+$ in blue. For NAD$^+$/NADH, see also Sidebar 9.2. Along with the three amino acid residues coordinated to Zn$^{2+}$, several amino acids of the protein matrix, stabilizing the active centre, are also shown. Modified from Lippard SJ and Berg JM. *Principles of bioinorganic chemistry.* Mill Valley, CA: University Science Books, 1994.

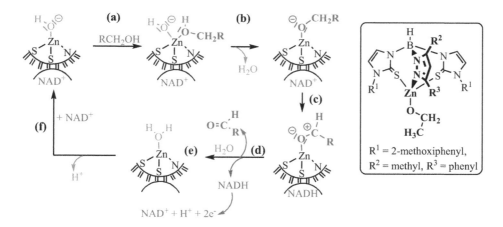

**Figure 12.8** The catalytic cycle of the oxidation (dehydrogenation) of alcohols by alcohol dehydrogenase. See text for details. Boxed: A structural model of the active site zinc. The tridentate ligand is pyrazolyl*bis*(thioimidazolyl)borate(1−) [13]. For steps **(a)** to **(f)**, see text.

## 12.3  The role of zinc in the transcription of genes

Whenever a peptide or protein is to be assembled, amino acids have to be provided. The information to direct the supply of the correct amino acids during the synthesis of a peptide is encoded in DNA in the form of hereditary units, or genes. In the initial step of information transfer, the information about the amino acid is *transcribed* into a messenger RNA, mRNA. The information captured in the mRNA is then 'decoded' by the ribosomes, the intracellular factories for protein synthesis, which *translate* the information, and tie the correct amino acid to the end of a growing peptide/protein. The *transport* of the amino acid to the ribosome is carried out by transfer RNA, tRNA. The overall procedural sequence is displayed in Scheme 12.3. The synthesis of the mRNA from a DNA template is catalysed by the enzyme RNA polymerase. This enzyme is directed towards the 'correct' part of the DNA by a *transcription factor*.

A prominent class of these transcription factors is represented by the zinc finger proteins [14]. Zinc finger proteins contain finger-like loops, stabilized by coordination of $Zn^{2+}$ at the base of the loop. The specific amino acid sequence in the loops (Fig. 12.9a) allows for docking of the zinc finger—via hydrogen bonds—to the DNA site to be transcribed (Fig. 12.10), i.e. to the site containing the segment of information within the DNA that is to be transcribed into complementary mRNA. Zinc fingers thus virtually 'pilot' the RNA polymerase to the correct location. Each loop in a zinc finger protein recognizes three bases of DNA.

For specific applications, artificial zinc finger proteins with nuclease activity have been developed. These zinc finger nucleases (ZFN) contain, along with the DNA binding domain, a catalytic nuclease domain that targets DNA, and breaks the DNA double strand at the target site, thus creating a gene knock-out. ZFNs have been engineered to manipulate the genomes of various plants and animals, and are also being developed towards clinical trials for the knockout of gene sequences responsible for, e.g. lung cancer [15] and HIV/AIDS.

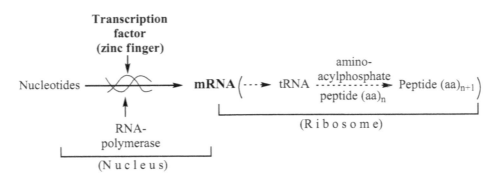

Scheme 12.3 Simplified schematic representation of the transcription of information (DNA → mRNA) and translation (peptide/protein synthesis via tRNA). The DNA is symbolized by the double helix. The extension of the peptide/protein (aa)$_n$ in the ribosome by an amino acid leads to the new peptide/protein (aa)$_{n+1}$. In this transfer reaction, both the amino acid, connected to acetylphosphate ((b) in Sidebar 12.1), and the peptide are activated by intermittent linkage to the 3′ position of the ribose moiety of the tRNA.

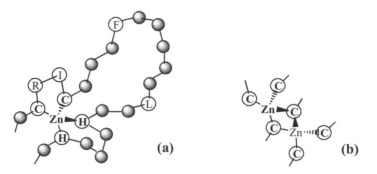

**Figure 12.9** (**a**) Cartoon of a zinc finger protein. Typically, $Zn^{2+}$ is in a tetrahedral environment of two cysteines (C) and two histidines (H). Amino acids are symbolized by spheres. The 12-residue pattern between Cys and His is a motif frequently found in zinc fingers. Arginine (R), isoleucine (I), phenylalanine (F), and leucine (L) are constituents of the *zinc finger Zif-268*. Zif-268 regulates neurological genes and genes responsible for cell proliferation. (**b**) The dinuclear $Zn_2(\mu\text{-}Cys)_2Cys_4$ motif is present in zinc finger proteins of yeasts such as *Saccharomyces cerevisiae*.

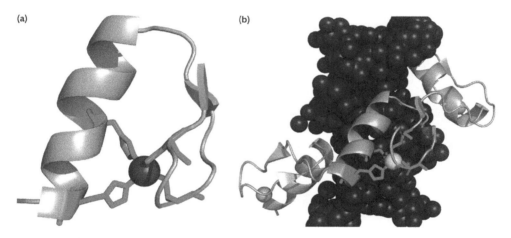

**Figure 12.10** (**a**) Section of a zinc finger sub-domain, with $Zn^{2+}$ (grey sphere) coordinating to a helical subunit via two His, and to the loop via two Cys. (**b**) Docking of a zinc finger into the major groove of DNA (blackish spheres; $Zn^{2+}$: light grey sphere). The $(His)_2(Cys)_2$ motif is shown in blue. Generated from PDB 1AAY by J. Crowe.

## 12.4 Thioneins

Metallothioneins form a superfamily of ubiquitous, low molecular mass proteins (62–68 amino acids in human thioneins), often devoid of aromatic amino acids and particularly rich in cysteine residues (up to one third) that preferentially bind $Zn^{2+}$ and $Cu^+$, and thus play an intriguing role in the homeostasis of these two metal ions. They also function as xenobiotics by efficiently complexing $Cd^{2+}$, $Hg^{2+}$, and $Ag^+$, and eventually protect against oxidative stress through free radical scavenging, Eq. (12.1) in Section 12.1. Further, they play an as yet to be

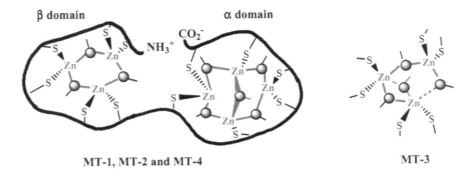

**β domain**          **α domain**

**MT-1, MT-2 and MT-4**          **MT-3**

Figure 12.11 Human zinc-loaded thioneins. The bridging cysteine residues of the $Zn_3$ and $Zn_4$ clusters are represented by grey spheres. The $Zn_3$ cluster of the MT-3 thioneins is shown to the right. The dashed line represents a weak interaction (2.90 Å).

clarified role in neurodegenerative diseases. The binding sites for the metal ions in vertebrate metallothioneins are provided by cysteines, but histidine can contribute in bacterial and plant thioneins.

Four human metallothioneins, MT-1 to MT-4, have been characterized [16]. They are all very similar with respect to the amino acid sequence, 20 of which are cysteines. And they all contain two $\{Zn^{2+}-S(Cys)\}$ clusters—namely a $Zn_3(Cys)_9$ and a $Zn_4(Cys)_{11}$ system in the β and α domain, respectively, of the protein. The clusters are displayed in Fig. 12.11: The central bicyclic $Zn_4(\mu-S)_5S_6$ system of the tetranuclear zinc cluster is arranged in a manner reminiscent of adamantane. In the cyclic $Zn_3(\mu-S)_3S_6$ system of the trinuclear cluster of MT-1, MT-2 and MT-4, the $Zn_3(\mu-S)_3$ ring adapts the chair conformation as in cyclohexane, while in the MT-3 subgroup, two of the $Zn^{2+}$ ions are doubly bridged and one of the $Zn-S(\mu-Cys)$ interactions is particularly weak.

## ➕ Summary

In proteins, zinc can adopt structural and catalytic functions. A *structural* function implies structural support of a protein domain—in particular stabilization of a nearby catalytically active site, commonly a redox active centre. In these structural zinc centres, $Zn^{2+}$ is predominantly coordinated to 4 cysteinates. An example of an enzyme with a structural zinc centre $\{Zn(Cys)_4\}$ *and* a catalytic zinc domain $\{Zn(OH)(Cys)_2His\}$ is alcohol dehydrogenase ADH. ADH catalyses the oxidative conversion, and thus detoxification, of an alcohol (in particular ethanol) to an aldehyde by activating the transfer of $\{H^- + H^+\}$ to the enzyme's cofactor $NAD^+$.

The catalytic function of zinc centres reflects the bipolar character of the $\{Zn^{2+}-OH^-\}$ moiety present in many of these enzymes. $Zn^{2+}$ can be attacked by a Lewis base (a nucleophile), while the hydroxido ligand is targeted by a Lewis acid (electrophile). The dominating amino acid residues in the coordination environment of $Zn^{2+}$ in catalytically active zinc enzymes are histidine, cysteinate and aspartate/glutamate.

Examples of catalytic functions of zinc-dependent proteins are carboanhydrase and $CS_2$ hydrolase, thermolysin and carboxipeptidases, as well as alkaline and acid phosphatases. Carboanhydrase enables the interconversion of water + carbon dioxide into hydrogencarbonate/carbonic acid and vice versa, and thus plays a central role (i) in the off-transport of $CO_2$ formed by oxidative degradation of organic matter in tissues, and (ii) in the acquisition of $CO_2$ for photosynthesis by converting carbonic acid into $CO_2$. $CS_2$ hydrolase is used by the extremophilic archaea *Acidianus* to generate energy from the hydrolytic conversion of $CS_2$ to $CO_2$ and $H_2S$. Thermolysin and carboxipeptidases are enzymes that catalyse the hydrolytic rupture of intra-molecular peptide bonds (an example is the *endo*peptidase thermolysin), or the peptide bond at the C-terminus of a protein (carboxipeptidase, an *exo*peptidase). Zinc-dependent phosphatases with two cooperative metal centres can be homonuclear (2 $Zn^{2+}$ in acid phosphatase) and heteronuclear ($Zn^{2+}+Fe^{2+}$ in alkaline purple phosphatase). The enzymes enable the hydrolysis of the phosphoester bond in various substrates where phosphate acts as an energizer (in, for example, glucose-6-phosphate and ATP), or as a linker and thus with a structure function (as in DNA).

Other fields where zinc proteins feature include the repair of methylated DNA, gene transcription, and zinc homeostasis. In all of the respective functional proteins, $Zn^{2+}$ is in a tetrahedral environment of four cysteinyl residues. The Ada repair protein demethylates methylated DNA bases and phosphate linkers in DNA by transfer of the methyl group to a coordinated cysteinate of the repair protein. Transcription factors pilot RNA polymerase to those sections of the DNA where the information for a specific amino acid (to be transported by transfer-RNA into the ribosomes) is transcribed into a messenger-RNA. Zinc homeostasis is controlled by thioneins. These are small, cysteine-rich proteins that can incorporate up to 7 $Zn^{2+}$, partitioned in two domains harbouring 3 and 4 $Zn^{2+}$, respectively. Thioneins also strongly coordinate $Cu^+$, as well as toxic metal ions, in particular $Cd^{2+}$, $Hg^{2+}$, and $Ag^+$. Further, the Cys residues function as anti-oxidants.

## Suggested reading

**Chasapis CT, Loutsidou AC, Spiliopoulou CA, et al. Zinc and human health: an update. *Arch. Toxicol.* 2012; 86: 521–534.**
Provides a recent and comprehensive overview of health impairment caused by zinc deficiency, and health problems that can be counteracted or ameliorated by zinc supplementation.

**Parkin G. Synthetic analogues relevant to the structure and function of zinc enzymes. *Chem. Rev.* 2004; 104: 699–767.**
An overview of the structure and function of zinc enzymes, and model complexes mimicking the active centres in these enzymes structurally and functionally.

**Metallothioneins: chemical and biological challenges. *J. Biol. Inorg. Chem.* 2011; 16: no. 7.**
This is a special issue of the journal, providing insight into the various aspects of thioneins.

## References

1. Lu J, Stewart AJ, Sleep D, et al. A molecular mechanism for modulating plasma Zn speciation by fatty acids. *J. Am. Chem. Soc.* 2012; 134: 1454–1457.

2. (a) Prasad AS. Zinc: role in immunity, oxidative stress and chronic inflammation. *Curr. Opin. Clinic. Nutr. Metab. Care* 2009; 12: 646–652; (b) Foster M and Samman S. Zinc and redox signaling: perturbation associated with cardiovascular disease and diabetes mellitus. *Antiox. Redox Signal.* 2010; 13: 1549–1573.

3. Keilin D and Mann T. Carbonic anhydrase. Purification and nature of the enzyme. *Biochem. J.* 1940: 34: 1163–1176.

4. Oke M, Ching RTT, Carter LG, et al. Unusual chromophores and cross-links in ranasmurfin: A blue protein from the foam nests of a tropical frog. *Angew. Chem. Int. Ed.* 2008; 47: 7853–7856.

5. Myers LC, Terranova MP, Ferentz AE, et al. Repair of DNA methylphosphotriesters through a metalloactivated cysteine nucleophile. *Science* 1993; 261: 1164–1167.

6. Fukada T, Yamasaki S, Nishida K, et al. Zinc homeostasis and signaling in health and disease. *J. Biol. Inorg. Chem.* 2011; 16: 1123-1134.

7. Dudev T and Lim C. Tetrahedral vs. octahedral zinc complexes with ligands of biological interest: a DFT/CDM study. *J. Am. Chem. Soc.* 2000; 122: 11146-11153.

8. Looney A, Parkin G, Alsfasser R, et al. Zinc pyrazolylborate relevant to the biological function of carbonic anhydrase. *Angew. Chem. Int. Ed.* 1992; 31: 92-93.

9. (a) Lane TW, Saito MA, George GN, et al. A cadmium enzyme from a marine diatom. *Nature* 2005; 435: 42; (b) Xu Y, Feng L, Jeffrey PD, et al. Structure and metal exchange in the cadmium carbonic anhydrase of marine diatoms. *Nature* 2008; 452: 56-61.

10. Smeulders MJ, Barends TRM, Pol A, et al. Evolution of a new enzyme for carbon disulphide conversion by an acidothermophilic achaeon. *Nature* 2011; 478: 412-416.

11. Schenk G, Gahan LR, Guddat LW. Crystal structure of a purple acid phosphatase, representing different steps of this enzyme's catalytic cycle. *BMC Struct. Biol.* 2008; 8: 6.

12. Elias M, Wellner A, Goldin-Azulay K, et al. The molecular basis of phosphate discrimination in arsenate-rich environments. *Nature* 2012; 491: 134-137.

13. Seebacher J, Shu M, Vahrenkamp H. The best structural model of ADH so far: a pyrazolylbis(thioimidazolyl)borate zinc ethanol complex. *Chem. Commun.* 2001; 1026-1027.

14. (a) Pabo CO, Peisach E, Grant RA. Design and selection of novel $Cys_2His_2$ zinc finger proteins. *Ann. Rev. Biochem.* 2001; 70: 313-340; (b) Razin SV, Borunova VV, Maksimenko OG, et al. $Cys_2His_2$ zinc finger protein family: classification, function and major members. *Biochemistry (Moscow)* 2012; 77: 217-226.

15. Sigma-Aldrich. ZFN Technology – have your genomic work cut out for you. *Biowire* 2010; 10: 7-11.

16. Vašák M and Meloni G. Chemistry and biology of mammalian metallothioneins. *J. Biol. Inorg. Biochem.* 2011; 16: 1067-1078.

# 13 Metal– and metalloid–carbon bonds

As shown in the preceding chapters, there are many biologically active metal complexes in which the central metal ion is coordinated to a more or less complex organic ligand system, in most cases via oxygen, nitrogen, and sulfur functions of the ligand. This specific type of coordination compound is commonly referred to as a metal–organic compound. In contrast, if a ligand coordinates to a metal centre directly via a carbon functionality, the compound is referred to as an organometallic compound. Several non-metals, e.g. iodine and arsenic, and elements with a status in between metal and non-metal, such as selenium, also form element-to-carbon bonds in biologically active molecules; these elements are referred to as 'metalloids'. For metals such as mercury, and metalloids such as arsenic, the making and breaking of the M–C bond plays a pivotal role in determining the toxicity and detoxification.

In this chapter we consider the role of the transition metal–carbon bond in the activation of substrates such as CO, $CO_2$, $N_2$, $CH_4$, alkenes, and alkynes, focusing in part on selected examples of metalloenzymes which have already been dealt with in previous chapters. In this context, the specific bond characteristics of the metal–carbon bond will also briefly be commented on. We will then explore the special role of adenosyl- and methyl-cobalamin (vitamin $B_{12}$), the latter in the frame of the broad range of physiologically important methyl transfer reactions, and briefly set out the physiological implications of the selenium–carbon bond. Finally, the biogeochemical making and breaking of the metal- and metalloid–carbon bond in poisoning by and detoxification of mercury, lead, and arsenic will be addressed.

## 13.1 Organometallic compounds of transition metals

**The iron–carbon bond**: the central trigonal-prismatic $Fe_6$ unit of the $M$ cluster $\{Fe_7MoS_9\}$ in molybdenum nitrogenase has incorporated within it a light atom which likely is a carbon [1]; Fig. 13.1a. The mean Fe–C distance is 2.04 Å, a common value for an $Fe^{2+}$–carbide ($C^{4-}$) σ bond. Two of the Fe–C distances extend to ca. 2.6 Å in the course of $N_2$ binding and $N_2$ reduction. For details, cf. Scheme 9.2 in Section 9.2. The net reaction catalysed by molybdenum nitrogenase is depicted in Eq. (13.1).

$$N_2 + 10H^+ + 8e^- \rightarrow 2NH_4^+ + H_2 \tag{13.1}$$

Side reactions of molybdenum- and vanadium-dependent nitrogenases (Section 9.1) include the reduction of CO to methane and other hydrocarbons, Eq. (13.2), the reductive

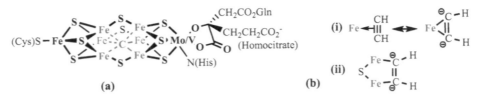

Figure 13.1 (a) The M cluster of the enzyme nitrogenase, with a light atom, most likely carbon (blue), in the centre of the Fe$_6$ cluster (grey). (b) Presumed modes of the coordination of ethyne prior to its protonation to ethene: (i) side-on, (ii): bridging.

protonation of isonitriles to primary amines, Eq. (13.3), and the reductive protonation of ethyne to ethene, Eq. (13.4). All of these reactions presuppose activation of the substrate by coordination to—presumably—the iron centres. The formation of Z-dideuteroethene in the case of the reductive protonation of ethyne presupposes that ethyne coordinates in the side-on $\eta^2$ mode to one iron centre ((i) in Fig. 13.1b), or bridges two iron centres ((ii) in Fig. 13.1b). We consider the different bonding modes ($\sigma$ and $\pi$, end-on and side-on) in Sidebar 13.1.

$$CO + 3H_2 \rightarrow CH_4 + H_2O \tag{13.2}$$

$$RNC + 6H^+ + 6e^- \rightarrow RNH_2 + CH_4 \tag{13.3}$$

$$C_2H_2 + 2D^+ + 2e^- \rightarrow Z - C_2H_2D_2 \tag{13.4}$$

Examples of iron centres coordinated to cyano and carbonyl ligands are iron-only hydrogenase [2] (a in Fig. 13.2), and iron–nickel hydrogenase [3] (b in Fig. 13.2).[1] Biosynthesis of CO and CN$^-$ is commonly carried out by cleavage of tyrosine, Eq. (13.5) [4], which generates p-cresol together with CO and CN$^-$.

$$\tag{13.5}$$

Both CO and CN$^-$ are strong ligands and thus secure the low-spin state of iron essential for efficient binding of H$_2$. The CO ligands are strong $\pi$ acceptors and thus increase the Lewis acidity of the iron centre, favouring binding and hence activation of molecular hydrogen. The increased Lewis acidity also supports heterolytic cleavage of H$_2$, as illustrated by the hydrogenase reaction in Eq. (13.6).

$$H_2 \leftrightarrows H^+ + H^-, \text{ and} : H^- \rightarrow H^+ + 2e^- \tag{13.6}$$

Carbon monoxide (CO) and cyanide (CN$^-$) are otherwise highly toxic when entering the organism from external sources. This toxicity can be ascribed to the effective coordination of these ligands to the iron centre of the haem group in haemoglobin (CO) and cytochrome-c oxidase (CN$^-$); such binding causes the blockage of O$_2$ transport in the bloodstream and disruption of O$_2$ reduction, respectively (Chapter 5). Other molecules delivering organic

---

[1] For iron-only and iron–nickel hydrogenases, see Sections 10.2 (Fig. 10.1) and 10.4.

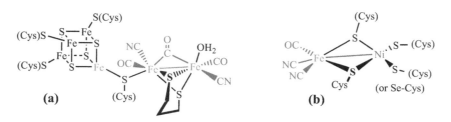

**Figure 13.2** Iron-containing enzymes with iron–carbon bonds in (**a**) iron-only hydrogenase and (**b**) iron–nickel hydrogenase. Se-Cys is selenocysteinate. There is evidence, from recent model studies [2b], that the bridge-head group in the dithiolene is NH rather than $CH_2$.

fragments can also interfere with the metabolism of $O_2$. An example is phenylhydrazine, the toxicity of which goes back to the transfer of a phenyl group ($C_6H_5$) to the iron centre of haemoglobin, Eq. (13.7), thus inhibiting its function as an $O_2$ transporter.

$$Hb(Fe^{II}) + C_6H_5NH-NH_2 + H^+ + e^- \rightarrow Hb(Fe^{III}-C_6H_5) + H_2N-NH_2 \qquad (13.7)$$

**Sidebar 13.1**   Ligand-to-metal bonding in organometallic compounds

Bonding modes for five types of organic ligands are pictured. Left: Valence-bond (VB) formulations; right: (simplified) orbital representations. Positive orbital lobes are shaded, negative lobes are open. Arrows: $\leftarrow \sigma$ donor bond; $\leftrightarrow$ mesomeric arrow.

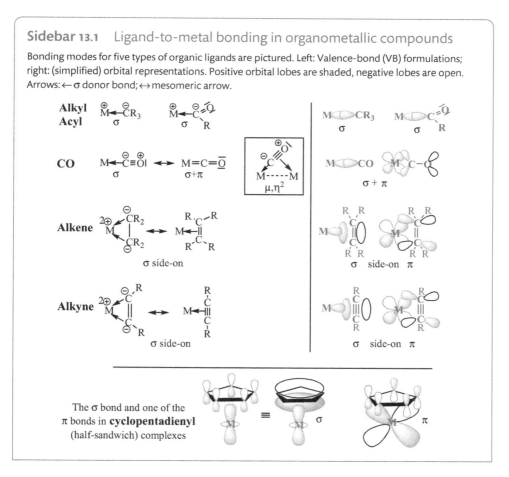

The $\sigma$ bond and one of the $\pi$ bonds in **cyclopentadienyl** (half-sandwich) complexes

Alkyl- and acyl–metal compounds: the anionic alkyl $R_3C^-$ and acyl $O=C(R)^-$ (where R=H or any alkyl or aryl residue) binds to the cationic metal centre via a $\sigma$-donor bond. The donating orbital at the carbon typically is an $sp^3$ (alkyl) or $sp^2$ hybrid (acyl), the accepting orbital on the metal a metal hybrid such as $sp^3$ (tetrahedral complexes), $dsp^2$ (trigonal-bipyramidal) or $d^2sp^3$ (octahedral).

Carbonyl–metal compounds: CO coordinates to the (commonly low-valent) metal by a combination of a $\sigma$ donor and a $\pi$ acceptor bond, thus providing double bond character to the M–C bond. In the case of $\sigma$ bonds, the electron density has a maximum along the M–C bond axis; in the case of $\pi$ bonds–with maxima above and below the bond axis–the electron density along this axis is zero. Boxed: Carbonyl groups can also connect two metal centres ('hapto-2', $\eta^2$) in the bridging ($\mu$) mode. Commonly, there is (weak) bonding interaction between the two metals.

Alkenes and alkynes coordinate in the side-on mode. In the VB description, this binding mode can be described as a hybrid of a triangular structure with two formal charges on the metal (positive) and the ligand (negative) on the one hand, and a donor-acceptor interaction between ligand and metal on the other hand, where the donor is the ligand $\pi$ system and the acceptor a $\sigma$ type hybrid orbital at the metal. The bond thus formed is a $\sigma$ bond. In addition, $\pi$ bonding interaction between metal and alkene/alkyne takes place, as shown in the orbital overlap representation for a metal-d and a ligand $\pi$ orbital.

The cyclopentadienide ligand ($\eta^5 - C_5H_5^- \equiv Cp$) and other aromatic molecules commonly coordinate in the side-on (half-sandwich) mode. The bonding interaction between the metal and Cp is best described by a triple bond ($\sigma + \pi + \pi$), where the bonding orbitals at the metal are $d_{z^2}$ (and $p_z$) for the $\sigma$ bond, and $d_{xz}$ and $d_{yz}$ for the two $\pi$ bonds. The bonding orbitals at Cp are formed by appropriate linear combinations of the $p_z$ orbitals of the five carbons.

**The nickel–carbon bond**: Ni–C bonds are formed in reactions catalysed by nickel–iron carbon monoxide dehydrogenase (CODH) and acetylcoenzyme-A synthase (ACS) [5]; see also Section 10.4. In CODH, a functional nickel centre is linked, through a cysteinate, to a [3Fe-4S] cluster (Fig. 13.3a). The overall (reversible) reaction, reminiscent of the water-gas shift reaction, is represented by Eq. (13.8).

$$CO + H_2O \rightleftharpoons CO_2 + 2H^+ + 2e^- \tag{13.8}$$

Equation (13.9a) illustrates the course of the reaction catalysed by ACS synthase; the net reaction of the process is represented by Eq. (13.9b). In a first step, the CO delivered by catalytically conducted reduction of $CO_2$ according to the reverse reaction in Eq. (13.8) is transferred to the nickel centre {Ni} of ACS synthase. Next, a methyl group is linked to the nickel centre ((b) in Eq. (13.9a) and in Fig. 13.3). The transfer of this methyl group is carried out by methyl-cobalamin $CH_3Cb$, which we discuss later in this section. Then, the methyl fragment is shifted to CO to form an acetyl group $\sigma$-bonded to Ni. In a final step, the acetyl group is linked to coenzyme-A.

$$CO + HSCoA + Cb \cdot CH_3 \rightleftharpoons CH_3C(O)SCoA + H^+ + Cb^- \tag{13.9b}$$

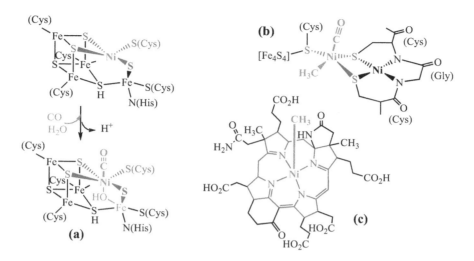

Figure 13.3 (a) The active centre of the nickel–iron CO dehydrogenase, CODH. The lower part shows the interim active site structure in the process of the conversion of CO into $CO_2$. (b) The active centre of nickel–iron acetylcoenzyme-A synthase (ACS), reflecting the intermediate state (b) in Eq. (13.9a). (c) Factor $F_{430}$ of methyl-coenzyme M reductase, with the methyl group coordinated to $Ni^{III}$. In (a), (b), and (c), the active site moieties are highlighted in blue.

Another example for the Ni—C bond formation is provided by methyl coordination to the nickel centre of the porphinogenic factor $F_{430}$ of methyl-coenzyme M reductase (Fig. 13.3c). Methyl-coenzyme M is $CH_3S—CH_2—CH_2SO_3^-$; the methyl group is transferred to the nickel centre of factor $F_{430}$, situated 2.1 Å proximal of $Ni^{3+}$ [6a]. Subsequently, reductive elimination of methane takes place, the final step in methanogenesis. (We discuss methanogenesis in Section 10.2). Eq. (13.10) illustrates the net reaction which, in principle, is reversible [6b].[2]

$$\{Ni^{III}-CH_3\}+H^++2e^- \rightleftharpoons \{Ni^{I}\}+CH_4 \tag{13.10}$$

**The molybdenum–carbon bond**: the first step in methanogenesis is the activation of carbon dioxide by side-on coordination of $CO_2$ to molybdenum, followed by the shift of $O^{2-}$ from $CO_2$ to Mo and concomitant protonation and formation of a $\sigma$-formyl complex, as depicted in Scheme 13.1. This reaction sequence is accompanied by an intramolecular oxidation of $Mo^{IV}$ to $Mo^{VI}$. In a next step, the formyl group is transferred to methanofuran to form formylmethanofuran. The overall process is catalysed by methylmethanofuran dehydrogenase, the cofactor of which, {Mo}, belongs to the sulfiteoxidase family of molybdopterin cofactors (Section 7.1). Reduction of $Mo^{VI}$ back to $Mo^{IV}$ is carried out by $FADH_2$.[3]

**The cobalt–carbon bond**: the 'antipernicious anaemia factor', or vitamin $B_{12}$ (cyanocobalamin), discovered in 1926 in raw liver, is a cobalt coordination compound in which $Co^{3+}$

---

[2] Oxidant in the back reaction ('reverse methanogenesis') is sulfate, oxidation product $CO_2$. This reaction is carried out by methanotrophic bacteria.

[3] $FADH_2$ is the reduced form of flavin-adenine-dinucleotide; see Sidebar 9.2.

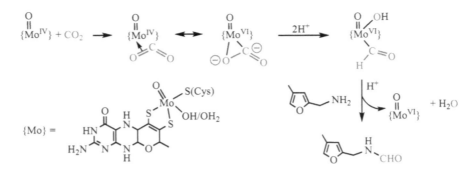

Scheme 13.1 Activation of $CO_2$ and transfer of the formyl group by the molybdenum cofactor {Mo} of methylmethanofuran dehydrogenase.

is coordinated equatorially into the plane formed by the four nitrogen donors of a corrinoid ligand (a ligand system related to haems), and additionally to an axial nitrogen of 5,6-dimethylbenzimidazole and to an axial cyanide. For $CH_3Cb$ and AdCb see Fig. 13.4. In the physiologically relevant forms of cobalamin (Cb), the cyanide ligand is replaced by $H_2O/OH^-$, the methyl group ($CH_3Cb$), or 5′-deoxiadenosyl (adenosyl-cobalamin AdCb, coenzyme $B_{12}$). In many bacteria and archaea, 5,6-dimethylbenzimidazole is exchanged for other benzimidazoles or by adenosine. These variants are referred to as cobamides.

Cobalamins and cobamides are important cofactors in various enzymatic processes. Examples are isomerases and methyl transferases (for details see below). In these processes, an intermediate state forms in which $Co^{III}$ is in the +II or +I states, referred to as cob(II)alamin

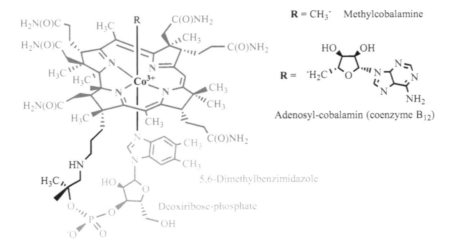

Figure 13.4 The structure of cobalamins, the cofactor in methyl transferases ($R=CH_3^-$) and isomerases ($R=5'$-deoxiadenosin(1–)). The equatorial, porphyrinogenic ligand system (high-lighted in dark blue) is termed the corrinoid system or corrin ligand.

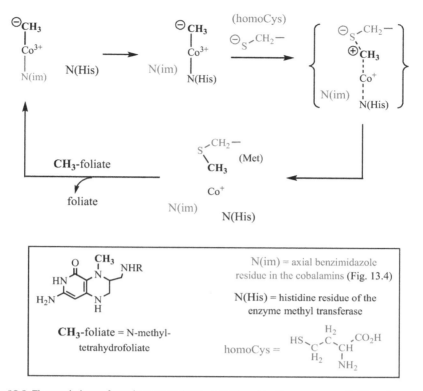

Scheme 13.2 The methyl transfer to homocysteinate as catalysed by the enzyme methyl transferase, employing CH$_3$Cb as cofactor for the delivery of the methyl group, and methyl-tetrahydrofoliate to restore CH$_3$Cb. The transition state (in curly brackets) has been proposed on the basis of density functional calculations with CH$_3$S$^-$ as a model for homoCys [7b].

and cob(I)alamin. In any of the oxidation states, +III, +II, and +I, the cobalt ion is in its low-spin configuration.[4]

Let us now explore in some detail two reactions that exemplify the role of the Co—C bond in enzymatic processes: (i) the methylation of homocysteine to methionine by methionine synthase, and (ii) the transformation of glutamate into 3-methylaspartate by glutamate mutase. Example (i) involves methyl transfer with the participation of CH$_3$Cb as a cofactor, while example (ii) addresses the role of AdCb in the rearrangement of a carbon chain.

(i) The cobalamin-dependent methionine synthase reaction: the overall reaction scheme is depicted in Scheme 13.2 [7], and the net reaction (the methylation of homocysteine to the essential amino acid methionine) is shown in Eq. (13.11), where Cb and fol are short for cobalamin and foliate, respectively, and the methyl group shifted from CH$_3$Cb to homocysteine to form methionine is highlighted in italics.

$$CH_3Cb + HS(CH_2)_2CH_2(NH_2)CO_2H + CH_3 - fol \rightarrow$$
$$CH_3Cb + CH_3S(CH_2)_2CH_2(NH_2)CO_2H + Hfol \qquad (13.11)$$

---

[4] See Sidebar 4.3 for high- and low-spin configurations.

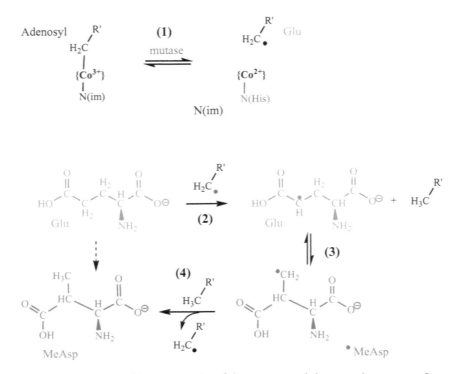

Scheme 13.3 Reaction paths of the isomerization of glutamate to methylaspartate by coenzyme $B_{12}$ dependent glutamate mutase. Black: coenzyme $B_{12}$ (adenosyl-cobalamin) and the adenosyl radical; grey: glutamate mutase, symbolized by N(His); light blue: glutamate; dark blue: methylaspartate. N(im) is benzimidazole. The broken arrow (pointing from Glu to MeAsp) indicates the net reaction.

(ii) An example for an enzyme employing the cofactor 5'-deoxiadenosyl cobalamin (coenzyme $B_{12}$) is glutamate mutase, which converts glutamate (Glu) to methylaspartate (MeAsp). The net (reversible) reaction is indicated by the broken arrow in the lower left of Scheme 13.3. Anaerobic bacteria that are present in the gastrointestinal tract, such as those belonging to the genus *Clostridium*, exploit this isomerization as the initiating step in the fermentation of MeAsp. MeAsp is then further degraded to butyrate, $CO_2$, $NH_4^+$, and $H_2$ [8a].

In a first step of the overall reaction cycle, the mutase binds cofactor $B_{12}$, {$Co^{3+}$} in Scheme 13.3, by replacing benzimidazole N(im) for a histidine residue of the enzyme. Concomitantly, the cobalt-carbon bond is split homolytically, leaving Co in the oxidation state +II and furnishing an adenosyl radical [8b]. This is depicted in step (1) in Scheme 13.3. In step (2), the radical character is interchanged between adenosyl and glutamate, i.e. a hydrogen atom on $C_\gamma$ of glutamate is transferred to the adenosyl radical. Step (3) reflects the isomerization of the glutamate radical to a methylaspartate radical. In step (4), a hydrogen atom is exchanged between the MeAsp radical and adenosine to generate methylaspartate and restore the adenosyl radical.

## 13.2 Carbon bonds to main group metals and metalloids

In this section, we consider the involvement of the carbon-to-metal and carbon-to-metalloid bond in the environmental and physiological speciation of Hg, Pb, As, and Se, with special emphasis upon the conversion of these elements into toxic species and their detoxification.

### 13.2.1 Mercury

The present day level of global mercury emission is $2 \times 10^6$ kg per year [9a]. Nonetheless, from a global perspective, there is no dramatic contamination of the environment by mercury: natural and anthropogenic inputs of mercury into the environment essentially balance each other, as does re-deposition of mercury. Locally, however, environmental mercury contamination can become a severe problem. As an example, industrial wastes dumped into the Minamata Bay in Japan resulted in the bioaccumulation of highly toxic methylmercury in shellfish and fish, causing—beginning in the mid-1950s—massive poisoning of the local populace, including a death toll of ca. 3000 and prenatal damages. Natural mercury sources include native mercury droplets in rock, and cinnabar (HgS).

Volcanic emissions, and decontamination of $Hg^{2+}$ compounds by certain bacterial strains (see below) and transgenic plants, contribute to the natural release of elemental mercury into the atmosphere. Important anthropogenic sources include the chlorine–alkali electrolysis by the amalgamation process, the extraction of gold via amalgamation, metallurgical and smelting processes, and emissions from waste incinerating plants (mercury batteries, electric light bulbs) and crematories (dental amalgam fillings).

Once in the atmosphere or in the hydrosphere, elemental mercury and mercury compounds are subjected to speciation. An overview is provided in Fig. 13.5. A prominent mercury species in the atmosphere is gaseous elemental mercury, with an average concentration of 1.6 ng m$^{-3}$ in the Northern hemisphere, and a residence time in the atmosphere of about one year. Starting from this species, the following speciation steps can be noted:

1. Atmospheric depletion of elemental mercury, Hg, is achieved by oxidative addition of methyl iodide (released from the oceans) to form methylmercury iodide $CH_3HgI$. Additionally, Hg can be oxidatively converted into mercurous/mercuric bromide in the presence of bromine (again stemming from sea water) plus ozone. The reactive species formed from bromine and ozone is likely to be BrO. Methylmercury can also form from $HgBr_2$ (and other mercuric species) plus acetate.

2. After deposition of $CH_3HgI$ into the ocean water, further methylation—formation of $Hg(CH_3)_2$—occurs. Exchange of iodide by chloride, the dominating anion in sea water, generates methylmercury chloride ($CH_3HgCl$).

3. Light-induced homolytic splitting of the Hg–CH$_3$ bond, with concomitant formation of ethane, returns Hg to the atmosphere. Alternatively, chlorine and OH radicals convert methyl mercury to inorganic mercuric compounds.

4. In an anoxic environment (for example, in sediments rich in decaying organic material) $H_2S$ converts $CH_3HgCl$ into $CH_3HgSH$. $CH_3HgSH$ can either dismutate to generate sparingly soluble HgS + volatile $Hg(CH_3)_2$, or $(CH_3Hg)_2S$.

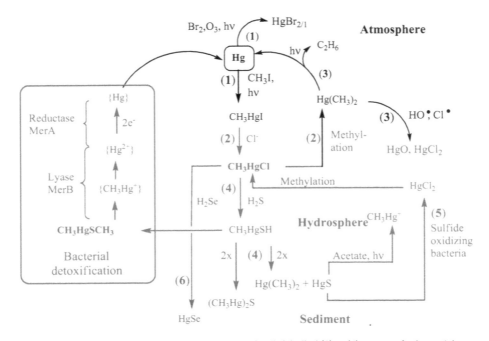

Figure 13.5 The biogeochemical cycle of mercury. For details labelled (1) to (6) see text; for bacterial detoxification (blue box) see also Scheme 13.4. Speciation is shown that predominantly takes place in the atmosphere (black), hydrosphere (blue), and in sediments (grey).

5. Sulfide oxidizing bacteria can remobilize mercury by converting insoluble HgS to water soluble mercuric chloride, which then re-enters the methylation cycle. Remobilization is also achieved non-biotically in the presence of acetate by light.

6. Irreversible removal of mercury and its final deposition into the sediment is accomplished by conversion into mercuric selenide HgSe in the presence of hydrogen selenide $H_2Se$.

Mercury-resistant bacteria have specialized in the detoxification of methylmercury. They do so by heterolytic rupture of the mercuric-to-methyl bond, with subsequent reduction of $Hg^{2+}$ to elemental Hg, which is not directly toxic. The net reaction is represented by Eq. (13.12); details of this bacterial detox are highlighted on the left-hand side of Fig. 13.5. For mechanistic aspects of the conversion of organic into inorganic mercury, see below.

The two enzymes in the two successive processes, MerB (an organomercurial lyase for proteolysis of the $CH_3-Hg^+$ bond, forming $CH_4$ and $Hg^{2+}$) and MerA (a mercurial reductase for the reduction of $Hg^{2+}$ to $Hg^0$), are encoded in the *merB* and *merA* operons[5] of the bacterial DNA. The high efficacy of the *mer* system has generated interest with respect to its potential use in biotechnologically modified plants, including *Arabidopsis*, tobacco and deep rooting poplar trees, for the remediation of mercury-contaminated soils in industrially polluted areas [9b,c].

$$CH_3Hg^+ + H^+ \rightarrow CH_4 + Hg^{2+} \tag{13.12}$$

[5] An operon is a functional unit of DNA, encoding several proteins with related tasks—in this case, the detoxification of mercury (*mer*).

The toxicity of mercury is due to the high affinity of $Hg^{2+}$ and $CH_3Hg^+$ for the sulfhydryl group of cysteine residues in proteins, rendering enzymes and other proteins inactive.[6] Inorganic mercuric ($Hg^{2+}$) and mercurous ($Hg_2^{2+}$) compounds, once having entered an organism, are biotically converted to $CH_3Hg^+$ through methyl transfer, mediated by methyl-cobalamin (see Section 14.1). The lipophilic $CH_3Hg^+$ is a contaminant of particular concern for animals and humans because of its ability to cross the blood-brain barrier.

Details of the mechanism of the proteolysis of the organo–mercury bond catalysed by the organomercurial lyase MerB in mercury-resistant bacteria are sketched in Scheme 13.4.

The lipophilic active centre of MerB exploits the high affinity of $CH_3Hg^+$ for thiol residues. Two cysteine residues (CysH and Cys'H in Scheme 13.4) and an active site aspartate ($Asp^-$) are involved in the Hg–C cleavage. The substrate for MerB is linear $RS-Hg-CH_3$, where R is methyl or (less common) another alkyl. The following steps, supported by the crystal structure of MerB [10a] and DFT calculations [10b] are noted:

(i)   $Asp^-$ picks up $H^+$ from HCys.

(ii)  $Cys^-$ thus formed coordinates to mercury, weakening the RS–Hg bond in $RSHgCH_3$.

(iii) Protonation of mercury-bound $RS^-$ by HCys' initiates release of the thiol RSH from Hg with concomitant coordination of $Cys'^-$ to the remaining $CH_3Hg^+$ and thus formation of trigonal planar $CH_3HgCys(Cys')$.

(iv)  HAsp transfers a proton to mercury-bound $CH_3^-$; $CH_4$ is released with simultaneous coordination of water.

The $Hg^{2+}$ integrated into the active site of MerB is finally shuttled with periplasmatic MerB to the cytosolic reductase MerA, an oxidoreductase which, by means of its cofactor $FADH_2$ (cf. Sidebar 9.2), reduces $Hg^{2+}$ to $Hg^0$, Eq. (13.13). $Hg^0$ leaves the cell via diffusion across the cell membrane.

$$Hg^{2+} + FADH_2 \rightarrow Hg + FAD + 2H^+ \tag{13.13}$$

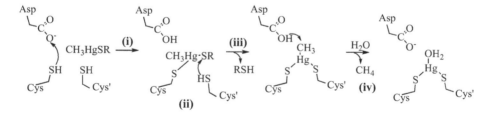

Scheme 13.4 Mechanism of the cleavage of the mercury–carbon bond as catalysed by the lyase MerB, based on [10a]. See Fig. 13.5 for the context in overall biogeochemical mercury speciation, and the text for details of steps (i)–(iv). Mercury is in the +II state throughout this sequence. Subsequent reduction to $Hg^0$ is carried out by the reductase MerA.

[6] The high affinity of $Hg^{2+}$ for thiolates is also exploited in the chelation therapy of mercury poisoning. Compounds such as dimercaptopropanesulfonic acid (Dimaval®) form 2:2 complexes with $Hg^{2+}$ which are efficiently carried out of the body.

### 13.2.2 Lead

Most lead found in the environment results from human activities. Lead is used in water pipelines, batteries, computer and TV screens, stained glass windows, and, in the form of tetraethyllead $Pb(C_2H_5)_4$, still as an additive to gasoline. In water pipelines, a coating of sparingly soluble $Pb(OH)_2\cdot2PbCO_3$ forms, from which $Pb^{2+}$ can be mobilized by slightly acidic water. The volatile anti-knocking agent in gasoline, $Pb(C_2H_5)_4$, either escapes directly into the atmosphere, where it decomposes to form tetramethyllead + ethane and/or elemental lead, or is broken down to elemental lead in the combustion engine and oxidized after emission to lead(II), as depicted in Scheme 13.5.

The lead oxide, hydroxide, and halides thus formed undergo further speciation when transported into the hydrosphere [11]: sulfide converts $Pb^{2+}$ into insoluble lead sulfide PbS, while methyl iodide oxidizes lead(II) compounds to trimethyllead(IV) $Pb(CH_3)_3^+$ which dismutates to generate tetramethyllead and PbS in the presence of $H_2S$.

The main concern with lead compounds, and organolead compounds such as $Pb(C_2H_5)_4$ in particular, is their high toxicity. Exposure to organolead compounds occurs via inhalation, ingestion and skin contact. The toxicity of lead is due to its strong affinity to the sulfhydryl groups of cysteine residues of enzymes and proteins (in this respect, the toxicity of lead resembles that of mercury), and further to the ability of lead ions to cause oxidative stress. OH radicals and peroxides thus generated induce damage of DNA and neurons. An example for the pathological effects of lead, resulting from the deactivation of enzymes, is the disruption of the haem biosynthesis, and thus anaemia.

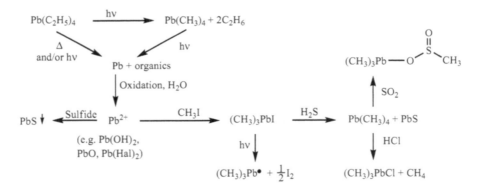

Scheme 13.5 Examples for the environmental speciation of tetraethyllead. Hal stands for Cl, Br, or I.

### 13.2.3 Selenium

Contrasting the other elements dealt with in this section (Hg, Pb, As), selenium—although also potentially toxic—is also an essential micronutrient for humans. Selenium is present in the amino acids selenocysteine SeCys and selenomethionine, and in the form of selenouridine in selenium-containing RNA. Due to its essential nature, SeCys is referred to as the 21st proteinogenic amino acid. SeCys is related to cysteine in such a way that the sulfur in Cys is replaced by Se. Elevated selenium levels are, however, toxic; there is in fact a comparatively narrow range between toxic levels (daily intake > 400 μg) and dietary deficiency (< 40 μg per day) [12]. Organic selenium compounds, in particular $Se(CH_3)_2$, are less toxic than the inorganic compounds selenite $HSeO_3^-$ and selenate $SeO_4^{2-}$; biotransformation of inorganic into organic selenium compounds is thus considered a detoxification process.

SeCys is present in several enzymes. Examples are glutathione peroxidase and formate dehydrogenase. Glutathione peroxidase catalyses the oxidation of glutathione GSH to its disulfide by peroxide, and thus counteracts oxidative damage via removal of peroxide. The respective reaction is shown in Eq. (13.14). Formate dehydrogenase catalyses the oxidation of formate to $CO_2$ by means of the oxidized form of nicotine adenine dinucleotide (NAD⁺), Eq. (13.15). Formate dehydrogenase contains the molybdopterin cofactor (Section 7.1), with selenocysteinate coordinated to molybdenum.

$$2\,GSH + H_2O_2 \rightarrow GS\text{-}SG + 2H_2O \tag{13.14}$$

$$HCO_2^- + NAD^+ \rightarrow CO_2 + NADH \tag{13.15}$$

As depicted in Scheme 13.6, line (1), the biosynthesis of SeCys proceeds via selenation of a serinyl residue linked to transfer-RNA. The selenium for this conversion of serine to SeCys is delivered by selenophosphate, line (2) in Scheme 13.6, generated from adenosine triphosphate ATP and selenide $Se^{2-}$, and catalysed by selenophosphate synthase [13].

Several aspects of the environmental speciation of selenium are outlined in Scheme 13.7. The main sources of environmental selenium are volcanic emissions ($H_2Se$), coal combustion ($SeO_2$), and biota ($Se(CH_3)_2$ and $Se_2(CH_3)_2$). Once released into the atmosphere, the oceans or the soil, low-valent selenium compounds are oxidized to selenite $HSeO_3^-$ and selenate $SeO_3^{2-}$. Inorganic selenium compounds arriving in the biosphere can reductively be reconverted to alkyl- (essentially methyl-) selenium species with $Se^{-I}$ and $Se^{-II}$, with the intermittent formation of methylselenates and –selenites [14]. The main biologically available methylation agent for selenium compounds is S-adenosylmethionine (boxed in Scheme 13.7), but eventually methyl groups can also be transferred by methylcobalamin.

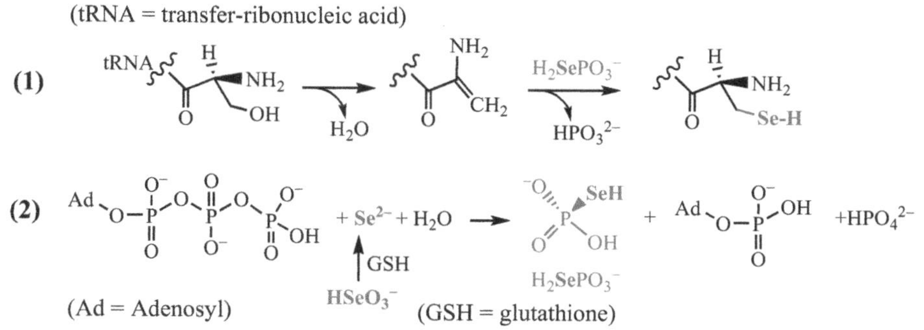

(tRNA = transfer-ribonucleic acid)

Scheme 13.6 Conversion of serine (as a constituent of serinyl-tRNA) to selenocysteine by selenophosphate (1), and generation of selenophosphate from selenide and adenosyltriphosphate (2). Reaction (1) is catalysed by selenocysteine synthase, reaction (2) by selenophosphate synthase. Actually (and not shown here), the amino group of the serinyl moiety is functionalized by pyridoxalphosphate.

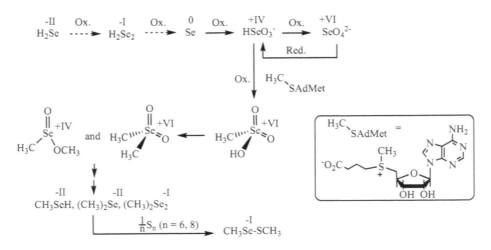

Scheme 13.7 Selected biogeochemical transformations of selenium. Roman numerals indicate selenium oxidation numbers. Ox.=oxidation (e.g. by $O_2$ and $O_3$), Red.=reduction.

### 13.2.4 Arsenic

Arsenic is a highly toxic element for most organisms. Its toxicity notwithstanding, some microorganisms metabolize both inorganic and organic forms of arsenic and/or tolerate high arsenic concentrations in aquatic environments. An example is arsenate(V) $HAsO_4^{2-}$ ($AsO_3OH^-$), which is used by specialized bacteria as the terminal electron acceptor in respiration, exploiting the reduction of arsenate to arsenite(III) $H_2AsO_3^-$ ($AsO(OH)_2^-$). Arsenic, mainly in the form of arsenite, is a main concern in drinking water in several regions of our planet: arsenic leaches into the ground water from mineral deposits such as arsenopyrite FeAsS, and from arsenical pesticides.

Once inorganic compounds of arsenic become introduced into the organism, they are metabolized to methylarsenic species, formally derived from arsenate and arsenite, by replacement of up to three OH/O$^-$ functions for methyl groups. The product primarily excreted by humans after arsenic poisoning is dimethylarsenate $(CH_3)_2As^VO_2^-$. Scheme 13.8 provides an overview of metabolic pathways exchanging between arsenate(V) and arsenite(III) on the one hand, and inorganic and organic (methylarsenous) compounds on the other hand.

Under environmental conditions, arsenate and arsenite are interconverted by common redox chemistry. Arsenate is a phosphate analogue and crosses cell membranes via phosphate transport systems. Once in the (bacterial) cell,[7] reduction to arsenite and re-oxidation to arsenate by various redox-active species in the cellular systems takes place [15]. Reduction of arsenate to arsenite is catalysed by an arsenate reductase, and the process involves interim coordination of arsenate to a cysteinyl residue and formation of a disulfide bond at the active site of the enzyme [16]. This is highlighted in dark blue in box **1** of Scheme 13.8. Box **2** in Scheme 13.8, accentuated in grey, provides details of the conversion of arsenite and methylarsenites into trimethylarsine As(CH$_3$)$_3$ ('Gosio poison'[8]).

---

[7] How bacteria can protect themselves against uptake of arsenate is briefly addressed in Sidebar 12.1.

[8] Genuine Gosio poison is a mix of methylarsines which form in damp and mouldy wallpapers containing the pigment Scheele's Green (copper arsenite, $Cu_3As_2O_6$) by bio-reduction and –methylation.

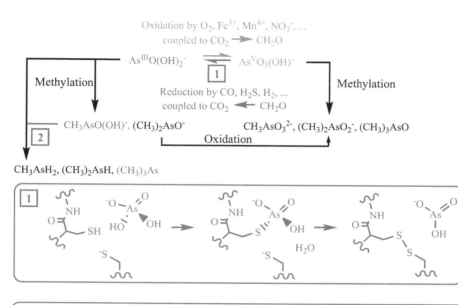

$$CH_3AsH_2, (CH_3)_2AsH, (CH_3)_3As$$

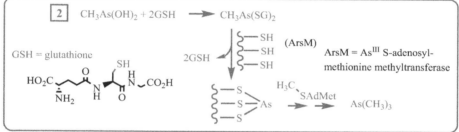

Scheme 13.8 Selected biotransformations of arsenic species. The upper part provides an overview. Box **1** exemplifies the biotransformation of arsenate into arsenite (i.e. the reduction; dark blue), box **2** the transformation of arsenite and methylarsenites into trimethylarsine $As(CH_3)_3$. For $CH_3$-SAdMet (S-adenosylmethionine), cf. the box in Scheme 13.7. Further details are explained in the accompanying text.

The methylation of arsenic likely starts by an exchange of the oxide/hydroxide groups of arsenite for the sulfhydryl of glutathione GSH, i.e. with the formation of $(CH_3)_nAs(SG)_{3-n}$, where $n=0$, 1, or 2; for $n=1$ see box **2** of Scheme 13.8. The glutathione-bound $As^{3+}$ is then transferred to three sulfhydryl groups of the enzyme $As^{III}$ S-adenosyl-methionine methyl-transferase, ArsM [17]. Methylation is finally accomplished by S-adenosylmethionine, and hence by the cofactor which has also been identified as the methylating agent in the generation of methyl–selenium compounds.

Arsenite(III) can bind to the bis(thiolate) function of lipoic acid, the cofactor of pyruvate dehydrogenase, which catalyses the final step in glycolysis, i.e. the decarboxylation of pyruvate [18]. This is demonstrated in Fig. 13.6a for the $As^{III}$ compound lewisite ('dew of death'), which had been used as a biological warfare agent.

But arsenic compounds have also been employed in chemotherapy. A famous example is salvarsan ('healing arsenic'), developed in Paul Ehrlich's institute in 1909, and employed in the

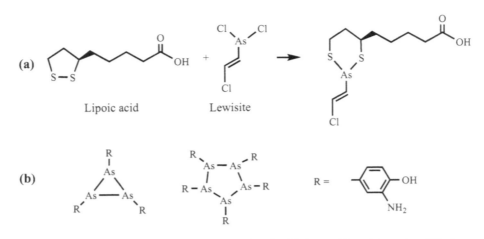

Figure 13.6 (a) Inhibition of lipoic acid by the war gas lewisite. (b) Composition of the anti-syphilis drug salvarsan.

treatment of syphilis and several other microbial diseases. Salvarsan (Fig. 13.6b) is a mixture of tri- and pentacyclic organoarsenic compound [19] with arsenic in the oxidation state +I.

## Summary

This chapter addresses the implications of organometal and -metalloid compounds for living organisms, and the processing of these compounds in biogeochemical cycles.

The trigonal-prismatic central $Fe_6$ unit of the M cluster of nitrogenases incorporates, in bonding contact with Fe, a carbon that mediates the activation of $N_2$ by modulation of Fe–C distances. Furthermore, formation of Fe–C bonds—and thus substrate activation—takes place in the course of the reductive protonation of substrates such as CO, isonitriles, alkynes, and alkenes by nitrogenase through coordination and thus activation of these substrates to iron ions of the $Fe_6$ cluster.

In iron- and nickel-based hydrogenases, the activation of $H_2$ is supported by CO and $CN^-$ ligands in the coordination sphere of Fe. In these hydrogenases, CO and $CN^-$ do not directly take part in substrate alteration. An example for direct activation of CO is the enzyme carbon monoxide dehydrogenase, based on nickel and iron. Here, methyl cobalamin transfers $CH_3^-$ to CO coordinated to nickel, forming an acetyl–nickel complex.

In methanogenesis (the biologically-conducted reduction of $CO_2$ to methane), the initial step is the activation of $CO_2$ by its coordination to the molybdenum centre of the molybdopterin cofactor of methylmethanofuran dehydrogenase, followed by reduction of $CO_2$ to Mo-centred formyl. The final step in the overall reaction cascade of methanogenesis is the transfer of $CH_3^+$ to the nickel centre in the porphinoid cofactor $F_{430}$, and the reductive protonation of the nickel-bound methyl to methane.

Many methyl ($CH_3^-$) transfer reactions are catalysed by enzymes employing methyl cobalamin as cofactor. Cobalamins contain $Co^{n+}$ ($n = 1, 2, 3$) coordinated to a corrin ligand (a porphinogenic system) and an axial benzimidazole. Adenosyl cobalamin (coenzyme $B_{12}$) is a cofactor in isomerases. An example for these specific enzymes is glutamate mutase, which reversibly isomerizes glutamate to methylaspartate via a radical mechanism, involving the interim formation of an adenosyl radical.

The metalloids selenium and arsenic undergo complex biogeochemical speciation. Selenium, a constituent of the essential amino acid selenocysteine, is released into the environment mainly in the form of $H_2Se$ and selenite, arsenic in the form of arsenite and arsenate. Methylation by S-methyladenosine converts inorganic Se and As compounds into various organic species, including methylselenates/arsenates, methylselenites/arsenites, and methylselanes/arsanes such as $(CH_3)_2Se$ and $(CH_3)_3As$. In the case of arsenic, these latter transformations are preceded by the intermittent formation of $As(SG)_3$, where HSG is glutathione.

## Suggested reading

**Gordon JC and Kubas GJ. Perspectives on how nature employs the principle of organometallic chemistry in dihydrogen activation in hydrogenases.** *Organometallics* **2010; 29: 4682–4701.**
A review which provides insight into how nature generates carbon monoxide and cyanide, and why nature employs these 'inorganics' as ligands for the activation of iron in the reactive centres of hydrogenases.

**Kräutler B and Jaun B. Metalloporphyrins, metalloporphinoids, and model systems. In: Kraatz H-B Metzler-Nolte N (eds)** *Concepts and models in bioinorganic chemistry.* **Weinheim: Wiley-VCH, 2006, ch. 9.**
The article is a comprehensive overview of enzymatic reactions relying on cobalamins ($B_{12}$). Structures of cobalamins, and the relation of cobalamins to other porphinogenic systems and model compounds, are introduced.

**Steffen A, Douglas T, Amyot M, et al. A synthesis of atmospheric mercury depletion event chemistry in the atmosphere and snow.** *Atmos. Chem. Phys.* **2008; 8: 1445–1482.**
An overview of the complex situation of the supply of mercury into the atmosphere, the ocean water and snow (in polar regions), as well as of the speciation of mercury and its depletion from the atmosphere. The article emphasizes analytical methods to reveal the distribution and speciation of mercury.

## References

1. Dance I. Ramifications of C-centering rather than N-centering of the active site FeMo-co of the enzyme nitrogenase. *Dalton Trans.* 2012; 41: 4859–4865.

2. (a) Bruska MK, Stiebritz MT, Reiher M. Regioselectivity of H cluster oxidation. *J. Am. Chem. Soc.* 2011; 133: 20588–20603; (b) Berggren B, Adamska A, Lambertz C, et al. Biomimetic assembly and activation of [FeFe]-hydrogenases. *Nature* 2013; 499, 66–69.

3. Ogata H, Lubitz W, Higuchi Y. [NiFe] hydrogenases: structural and spectroscopic studies of the reaction mechanism. *Dalton Trans.* 2009; 7577–7587.

4. Shepard EM, Duffus BR, George SJ, et al. [FeFe]-Hydrogenase maturation: HydG-catalyzed synthesis of carbon monoxide. *J. Am. Chem. Soc.* 2010; 132: 9247–9249.

5. Ragsdale SW. Metals and their scaffolds: to promote difficult enzymatic reactions. *Chem. Rev.* 2006; 106: 3317–3337.

6. (a) Cedervall PE, Dey M, Li X, et al. Structural analysis of a Ni-methyl species in methyl-coenzyme M reductase from *Methanothermobacter marburgensis. J. Am. Chem. Soc.* 2011; 133: 5626–5628; (b) Scheller S, Goenrich M, Boecher R, et al. The key nickel enzyme of methanogenesis catalyses the anaerobic oxidation of methane. *Nature* 2010; 465: 606–609.

7. (a) Matthews RG, Koutmos M, Datta S. Cobalamin- and cobamide-dependent methyltransferases. *Curr. Opin. Struct. Biol.* 2008; 18: 658–666; (b) Kozlowski PM, Kamachi T, Kumar M, et al. Reductive elimination pathway for homocysteine to methionine

conversion in cobalamin-dependent methionine synthase. *J. Biol. Inorg. Chem.* 2012; 17: 611–619.

8. (a) Buckel W. Unusual enzymes involved in five pathways of glutamate fermentation. *Appl. Microbiol. Biotechnol.* 2001; 57: 263–273; (b) Rommel JB and Kästner J. The fragmentation-recombination mechanism of the enzyme glutamate mutase studied by QM/MM simulations. *J. Am. Chem. Soc.* 2011; 133: 10195–10203.

9. (a) Streets DG, Devane MK, Lu Z, et al. All-time release of mercury to the atmosphere from human activities. *Environ. Sci. Technol.* 2011; 45: 10485–10491; (b) Mathema VB, Thakuri BC, Sillanpää M. Bacterial *mer* operon-mediated detoxification of mercurial compounds: a short review. *Arch. Microbiol.* 2011; 193: 837–844; (c) Ruiz ON and Daniell H. Genetic engineering to enhance mercury phytoremediation. *Curr. Opin. Biotechnol.* 2009; 20: 213–219.

10. (a) Lafrance-Vanasse J, Lefebvre M, Di Lello P, et al. Crystal structures of the organomercurial lyase MerB in its free and mercury-bound forms. Insights into the mechanism of methylmercury degradation. *J. Biol. Chem.* 2009; 284: 938–944; (b) Parks JM, Guo H, Momany C, et al. Mechanism of Hg—C protonolysis in the organomercurial lyase MerB. *J. Am. Chem. Soc.* 2009; 131: 13278–13285.

11. Shotyk W and Le Roux G. Biochemistry and cycling of lead. In: Sigel A, Sigel H, Sigel RKO (eds) *Biochemical cycles of elements.* Boca Raton, FL: Taylor & Francis, 2005, vol. 43, pp. 239–275.

12. Winkel LHE, Johnson CA, Lenz M, et al. Environmental selenium research: from microscopic processes to global understanding. *Environ. Sci. Technol.* 2011; 30: 571–579.

13. Itoh Y, Sekine S-i, Matsumoto E, et al. Structure of selenophosphate synthase essential for selenium incorporation into proteins and RNAs. *J. Mol. Biol.* 2009; 385: 1456–1469.

14. Chasteen TG and Bentley R. Biomethylation of selenium and tellurium: microorganisms and plants. *Chem. Rev.* 2003; 103: 1–26.

15. Stolz JF, Basu P, Oremland RS. Microbial arsenic metabolism: new twists on an old poison. *Microbe* 2010; 5: 53–59.

16. Messens J, Martins JC, Van Belle K. All intermediates of the arsenate reductase mechanism, including an intramolecular dynamic disulfide cascade. *Proc. Natl. Acad. Sci. USA* 2002; 99: 8506–8511.

17. Marakapala K, Qin J, Rosen BP. Identification of catalytic residues in the As(III) *S*-adenosylmethionine methyltransferase. *Biochemistry* 2012; 51: 944–951.

18. Ord MG and Stocken LA. A contribution to chemical defense in World War II. *Trends Biochem. Sci.* 2000; 25: 253–256.

19. Lloyd NC, Morgan HW, Nicholson BK, et al. The composition of Ehrlich's salvarsan. *Angew. Chem. Int. Ed.* 2005; 117: 963–966.

# 14 Inorganics in medicine

The Swiss medicinal alchemist and philosopher Theophrastus Bombastus von Hohenheim (1493–1541), better known as Paracelsus, once stated that *"Alle Ding' sind Gift, und nichts ohn' Gift; allein die Dosis macht, daß ein Ding kein Gift ist"* (All substances are poisons; it is the right dose that differentiates a poison and a remedy). We commonly consider arsenic, selenium, and heavy metals such as platinum and gold to be poisons. However, arsenic compounds have been used to cure several microbial diseases, trace amounts of selenium are required for the synthesis of the essential amino acid selenocysteine, and platinum and gold compounds are widely and successfully employed in the treatment of cancer and inflammations (such as arthritis), respectively. Other metals, such as iron or copper, are essential for a plethora of bodily functions, but iron and copper overloads have fatal consequences for our health.

In this chapter, we discuss consequences for our health if iron and copper homeostasis (the equilibrium concentrations of Fe and Cu) is disrupted—particularly, as far as copper is concerned, in the context of Menkes disease, Wilson's disease, and Alzheimer's disease. We also consider the mode of operation of drugs based on platinum and gold, as well as the role of bismuth in the treatment of irritations of the gastric mucosa, the therapy of bipolar disorder with lithium prescriptions, and the use of silver as a medicinal disinfectant. Furthermore, we explore the potential of the phosphate analogue vanadate as an insulin-enhancing and thus anti-diabetic agent.

We also introduce the application of short-lived radionuclei in the treatment of malignant tumours and in the palliation of patients suffering from metastasized bone cancer. In this context, the employment of $\gamma$ emitters (e.g. metastable 99-technetium) and positron emitters such as 123-iodine in $\gamma$-ray imaging of diseased tissues is also considered, as are imaging techniques based on nuclear magnetic resonance employing paramagnetic gadolinium compounds as contrast agents. These imaging methods are well established, and commonly referred to as positron emission tomography (PET) and magnetic resonance imaging (MRI).

Finally, we return, in the last section of this chapter, to Paracelsus' relativizing statement on toxic substances in conjunction with the gases carbon monoxide, nitric oxide, and hydrogen sulfide. These gases are without doubt highly toxic; but they are also produced in minor amounts in the body, where they are indispensable—for example, in the regulation of blood pressure. NO-releasing therapeutics have already been in use for decades; prodrugs for the target-specific liberation of CO and $H_2S$ are presently being developed, and future developments in this field will be outlined.

## 14.1  Metals and metalloids: an introduction

Metals and metalloids are introduced into the body via food, drinking water, breathing air, and skin re-sorption; they are also introduced 'artificially' by injection and, orally, via metallo-drugs. In the case of simple inorganic ionic metal compounds, such as water-soluble salts, the free metal ion is directly available. For most of the metallo-drugs, where the metal ion is 'hidden' in a coordination compound, the complete ligand system or part of it can be removed and/or interchanged in the body fluids, thus making the metal available for sub-sequent actions. In such a case, the drug is commonly referred to as *prodrug*. The extent to which genuine ligands can be removed, and/or interchanged for ligands provided in the physiological broth, is a matter of the complex stability, activation barriers, and kinetics.

Along with simple inorganic ions such as chloride, carbonate, and phosphate (which we discuss in Section 14.3.1) there are essentially three sets of targets for metal ions that are introduced into the body: the body's own low-molecular mass chelators; proteins; and DNA. Chelators[1] bear oligo-functional units capable of conjointly coordinating to a metal ion. Examples for self-contained low molecular mass chelators are cyclic oligopeptides (iono-phores) for alkaline metal ions (Section 3.3), and the corrin system for $Co^{1+/3+}$ (Section 13.1). In many cases, however, chelators are an integral functional part of a protein, such as the por-phinogenic systems for $Fe^{2+/3+}$ (the haem groups) in cytochromes (Sidebar 5.2), and the dithi-olene moiety of the molybdopterin cofactor for $Mo^{4+/6+}$ (Section 7.1). Proteins with metal ion chelators as an integral part can take over catalytic as well as non-catalytic functions, and they are correspondingly referred to as either metallo-enzymes or metallo-proteins.

Examples of metallo-proteins include metal ion transporters, and proteins involved in metal storage and gene expression. Replacement of the genuine metal ion in these struc-tures for a foreign metal supplied by contaminated food or in the form of a metallo-drug can disrupt the protein's original function and thus cause toxic effects. But metal ions can also induce protein misfolding by coordinating non-specifically to functional groups of any pro-tein and thus provoke abnormal protein aggregation, causing neurodegenerative diseases.

Of the 21 essential amino acids constituting peptides and proteins, there are many with side-chain functions that can bind a metal ion. Examples are histidine, arginine; tyrosine, serine, aspartate, glutamate; cysteine, selenocysteine, methionine, lysine, threonine, and tryptophane. In addition, the peptide linkage of a protein can be involved, and inorganic ligands such as water, hydroxide, sulfide and phosphate often complement the coordination sphere. For the various binding modes of metal ions in peptides and proteins, see Fig. 1.2 in Chapter 1.

As mentioned, metal ions can also target DNA. Common DNAs are Watson–Crick[2] double helices, where two right-handed helical chains entwine around the same axis. Each of the

---

[1] A chelate ligand is a ligand with two or more (commonly up to six) ligand functions. Chelate ligands exploit the chelate effect, an entropic effect contributing to the Gibbs free energy of complex formation. Here, 'entropic' refers to the increase of the number of molecules/ions per volume unit as a consequence of complex formation. An example is provided by Eq. (3.2) in Section 3.3.

[2] J.D. Watson and F. Crick co-discovered the structural features of DNA in 1953 and proposed the by now well established helical structure.

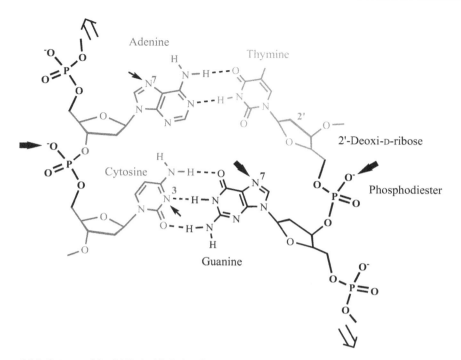

**Figure 14.1** Cut-out of the DNA double helix, showing the two base pairs that are formed between the four nucleobases (depicted in light/dark blue and grey/black). The sugar phosphates constitute the backbone of the helices. Broken lines represent hydrogen bonds. Solid arrows indicate the main binding sites for metal ions, standard arrows secondary binding sites, and open arrows the opposing sequence directions for the nucleotides in the two helical strands.

two strands is composed of four nucleobases[3]—adenine, guanine, thymine, and cytosine— arranged in the two strands in opposing sequence. As shown in Fig. 14.1, the nucleobases *within* the same strand are linked by deoxyribose and phosphate; the two strands are *interconnected* via hydrogen bonds. In order to fit into the restricted space of a nucleus, DNA is wrapped around proteins (called histones) and further packed by additional proteins to form chromatin (and thus chromosomes); such packaging serves to protect the DNA from damage. However, as soon as a DNA segment is triggered to being exposed, with the intention of synthesizing RNA or a protein, or of DNA replication, DNA becomes susceptible to attack by (metallo-)drugs, increasing the risk of mutation (whereby the DNA sequence is 'read' inaccurately). Drugs that attack DNA in this way, including bare metal ions and coordination compounds, are commonly referred to as genotoxic.

Preferential binding sites for metal ions are the imine nitrogens, N7, of adenine and, to some extent, guanine, and the negatively charged oxido groups of the phosphate linkages. The respective positions are indicated by black arrows in Fig. 14.1. Metal ions can either form crosslinks between two neighbouring bases of the same strand (*intra*strand coordination) or

---

[3] The ensemble nucleobase+ribose is referred to as *nucleoside*, the ensemble nucleobase+ribose+phosphate as *nucleotide*.

between bases of the two strands of the double helix, termed *inter*strand linkage. A further option for a metal ion is to link between a DNA base and a protein, or between two separate DNA molecules. Complexes with flat hydrophobic ligands, such as aromatic and pseudoaromatic compounds, may insert in-between base pairs, a mode of interaction which is known as *intercalation*.

Various metals and metalloids are acutely poisonous, or even lethal, when applied in a form that can enter the organism via the gastrointestinal tract. Arsenic and cadmium are examples, as are mercury and barium. The toxicity is commonly a function of the species present, and of the speciation a potential poison undergoes once incorporated. Elemental mercury is therefore less toxic than mercurous and mercuric compounds, and practically insoluble barium sulfate has been employed for decades as a radio-opaque substance to detect defects such as ulcers in the intestines—despite the fact that free $Ba^{2+}$ is highly toxic.

Other metals are not immediately toxic in low doses, but develop toxic effects over the years because they are only partially secreted and thus accumulate in specific organelles. The zinc antagonist cadmium is an illustrative example: $Cd^{2+}$ strongly coordinates to deprotonated sulfhydryl groups of cysteinyl residues in (zinc-dependent) proteins [1], disrupting the proteins' function[4] and eventually denaturing the protein and forming insoluble CdS.

Other primarily harmless metals can cause severe allergies. Nickel, often a constituent in metal jewellery, and one of the most common skin allergens, has been shown to activate a receptor in skin cells that in turn activates the immune defence system to fight alleged intruders by promoting inflammation of the skin [2].

We discuss the toxicity of the metals mercury and lead, and of the metalloids arsenic and selenium, in Section 13.2 in the context of organometallic intermediates which are formed in the course of their speciation. In the next section of this chapter, we consider dysfunctions in the metabolic pathways of iron and copper, and disease patterns related to these disturbances. Toxic levels of metal ions as a result of malfunction of homeostasis (and thus overload of otherwise essential metals), intoxication (toxic metals and metalloids), or therapy and diagnosis (heavy metals and radionuclides) are commonly dealt with by what is known as 'chelation therapy'. In this therapy, ion-selective organic ligands are injected to effectively sequester the target ion and take care of its secretion. As far as functional metals such as copper and iron are concerned, such a therapy should work, in the ideal case, without impairment of the intrinsic *in vivo* function of these metal ions. Additionally, the chelator applied in chelate therapy must be non-toxic (or at least of low toxicity only), and designed in such a way that hydro- and lipophilicity are balanced to ascertain efficient targeting of the metal in the tissue, and the metal's excretion.

## 14.2  Dysfunction of iron and copper homeostasis

Homeostasis is the ability of a living system to adjust and maintain the various processes and pathways within the parameters required for the system to function properly; it incorporates the way a living system adjusts and maintains a level of metal ions that ensures the optimal

---

[4] Cd may also substitute for Zn in such a way that the protein's function is maintained. An example is a cadmium-containing carboanhydrase in a marine diatom; for details see Section 12.2.1.

functioning of physiological processes. Both undersupply and overload of essential metals such as iron and copper can result in severe malfunction of processes and pathways essential to normal metabolism.

### 14.2.1 Iron

The adult human body contains an iron pool of ca. 4 g, about 70% of which is integrated into haemoglobin. Of the remaining 30%, the main part is bound to the transport protein transferrin and to iron storage proteins such as ferritin (Section 4.2). Most of the iron set free in the context of metabolic processes (such as the rebuilding of haems) is recycled; and an absorption rate of 1–2 mg of iron per day thus suffices to maintain the iron body pool. An undersupply of iron by defective re-adsorption in the duodenum, or by impaired haemo-globin synthesis (thalassemia; see later in section) leads to anaemia, while iron oversupply causes harmful accumulation of ferric hydroxide deposits in the tissue and oxidative stress.

Iron overload is also accompanied by comparatively high fractions of non-transferrin bound iron in the blood serum. This 'free' iron, which is mainly present in the form of ternary iron–citrate–albumin complexes [3], causes oxidative stress by generating a non-physiolog-ical excess of harmful reactive oxygen species, such as hydroxyl and hydroperoxyl radicals as represented by the Fenton reaction, Eq. (14.1), and hydroxyl radicals from superoxide as formed according to the Haber–Weiss reaction, Eq. (14.2).

$$Fe^{2+} + H_2O_2 \rightarrow Fe^{3+} + HO^\bullet + OH^-; Fe^{3+} + H_2O_2 \rightarrow Fe^{2+} + HOO^\bullet + H^+ \qquad (14.1)$$

$$^\bullet O_2^- + H_2O_2 \rightarrow {}^\bullet OH + OH^- + O_2 (\text{catalysed by } Fe^{3+}) \qquad (14.2)$$

Several of the key steps in iron homeostasis [4] are sketched in Fig. 14.2: *dietary* iron is taken up in its ferrous ($Fe^{2+}$) form in the intestinal lumen by absorptive cells called enterocytes, and released into the plasma by the trans-membrane iron transport protein ferroportin, again as $Fe^{2+}$. $Fe^{2+}$ is then oxidized to ferric ($Fe^{3+}$) iron, a reaction catalysed by a copper-based oxidase, followed by firm coordination of $Fe^{3+}$ to the transport protein transferrin (Fig. 4.3 in Section 4.2). The lion's share of *recycled* body iron ends up in macrophages that devour senescent red blood cells. $Fe^{2+}$ thus recovered from haemoglobin is again released via ferroportin into the plasma and coupled to transferrin after oxidation to $Fe^{3+}$.

These two processes—the release of iron into the plasma by macrophages and entero-cytes—are regulated by the peptide hormone hepcidin, produced in the liver. In the case of excessive iron supply, hepcidin down-regulates the release of $Fe^{2+}$ into the plasma by initiat-ing degradation of ferroportin. On the other hand, in the case of iron deficiency, expression of hepcidin is suppressed and release of iron into the plasma is thus facilitated. A specific iron-sensing protein called haemochromatosis protein is responsible for the regulation of hepcidin expression.

Hereditary haemochromatosis is a disease induced by iron overload in several organs, such as the liver, heart and pancreas, where excess iron induces the severe dysfunction of these organs, and diseases such as liver cirrhosis, cardiomyopathy (deterioration of the function of the heart muscle) and diabetes mellitus result. Hereditary haemochromatosis is caused by abnormally elevated iron absorption in the intestines: 8–10 mg instead of the normal 1–2 mg per day; this heightened absorption causes progressive accumulation of iron in the tissues.

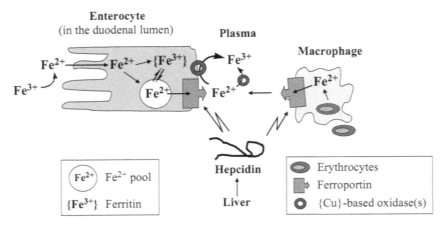

Figure 14.2 Dietary supply of iron (uptake by the intestinal lumen; left) and body's own regulation of iron homeostasis (recycling of iron from haemoglobin; right). The peptide hormone *hepcidin*, produced in the liver, targets the ferrous ion transporter *ferroportin*. In a healthy individual, hepcidin levels and iron efflux increase in the case of low iron supply and hypoanaemia, and decreases in the case of excessive iron supply and hyperanaemia. See also [4]. **Also reproduced in full colour in Plate 9.**

This hyperabsorption of dietary iron is caused by a (recessive) one-point mutation in the gene expressing a protein that regulates cellular iron homeostasis. In this mutated protein, a cysteine (Cys260) is replaced by tyrosine, causing disruption of the secondary and tertiary structural features of this specific protein, whose loss of function leads to the deregulation of hepcidin expression and, hence, iron homeostasis.

Hepatitic and cardiac iron loading is also a consequence of the treatment of thalassemia. Thalassemia is a genetic disease, which is widespread mainly in areas affected by malaria, because people suffering from thalassemia develop some resistance against malaria. In other words, thalassemia has been selected for in evolutionary terms because of the protection against malaria that it confers upon those with the condition. In individuals suffering from thalassemia, the synthesis of haemoglobin is partially disrupted, and these patients are therefore commonly and constantly treated with blood transfusions, causing harmful accumulation of iron within a couple of years. To counteract iron loading, chelation therapy is applied. The three common chelators employed to counteract iron overloading are Desferal, Deferriprone, and Deferasirox, whose structures are illustrated in Fig. 14.3.

All three chelators effectively coordinate ferric ions present in the plasma in the form of non-transferrin bound iron such as the ternary iron-citrate-albumin complexes. Desferal, also known as deferoxamine, is a naturally-occurring siderophore, produced by the bacterium *Streptomyces pilosus*. It contains three hydroxamic acid functions (the main coordination sites for $Fe^{3+}$), two peptide groups, and a terminal amino group. Deferriprone is a hydroxypyridinone, and Deferasirox a derivative of 1,2,4-triazole.

### 14.2.2 Copper

Copper is an indispensable cofactor in a variety of enzymes involved in redox reactions, which exploit the oscillation between $Cu^+$ and $Cu^{2+}$. Examples are ceruloplasmin and hephaestin

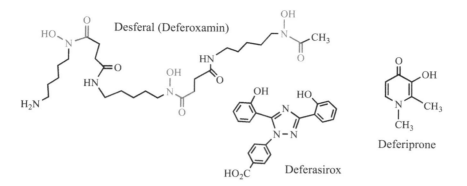

**Figure 14.3** Ferric ion chelators used in the chelation therapy of iron overload diseases. The iron binding hydroxamic acid groups in Desferal are highlighted in blue.

(indicated by rings in Fig. 14.2), enzymes that catalyse the oxidation of $Fe^{2+}$ to $Fe^{3+}$. Several neurodegenerative diseases are linked to either copper overload (Wilson's disease), or copper deficiency (Menkes disease), or copper imbalance (Alzheimer's disease). An insufficient supply of copper can also arise from molybdenum-induced copper deficiency, a potentially fatal problem for ruminants grazing on grounds bearing high amounts of molybdenum in the form of water soluble molybdate $MoO_4^{2-}$: Molybdate is converted to tetrathiomolybdate $MoS_4^{2-}$ in the digestive tract of ruminants; $MoS_4^{2-}$ forms insoluble cluster compounds with $Cu^+$ and $Cu^{2+}$.

Potential Cu–Mo clusters based on $MoS_4^{2-}$ were first modelled more than 30 years ago [5a]; the pseudo-cubane in Fig. 14.4a is an example. More recently, a cluster with the composition {Cu[(CuAtx1)$_3$MoS$_4$]} has been structurally characterized [5b]; Fig. 14.4b. Atx1 is a metallochaperone[5] from baker's yeast, and the structural characterization of this cluster containing four copper(I) centres, thiomolybdate(VI), and three Atx1 protein molecules (linked to three Cu centres via six cysteinyl residues) demonstrates that thiomolybdate does not necessarily remove copper ions from a protein, but can also effectively disturb Cu trafficking by directly coordinating to the protein-bound copper.

The overall amount of copper in an adult body is just 8 mg (iron: 4 g!); the daily dietary intake and discharge amounts to 0.8 mg [6]. Dietary copper is commonly in the +II state, and reduced to the +I state once in the body circulation by reductants such as glutathione. Cu enters the body via the enterocytes, the absorptive cells for nutritional uptake in the small intestine. It is further taken up, distributed to the body circulation, and eventually secreted into the bile via liver cells (hepatocytes). Free copper ions induce Fenton-like reactions, i.e. reactions analogous to Eq. (14.1) in the preceding section, and thus are potentially toxic. Consequently, free copper ions are acquired by metal ion transporters (chaperones) for delivery to their enzymatic sites of action, among these cellular superoxide dismutase SOD,

---

[5] Metallochaperones are proteins that deliver metal ions to specific targets in the cytoplasm. A typical target for a chaperone is a metallo-apoenzyme, i.e. a metal-dependent enzyme depleted of its metal centre(s). Yeast Atx1 is closely related to the human chaperone Atox1 which delivers copper ions to the oxidase ceruloplasmin. Atox1 binds $Cu^I$ at a site bearing a Met and two Cys.

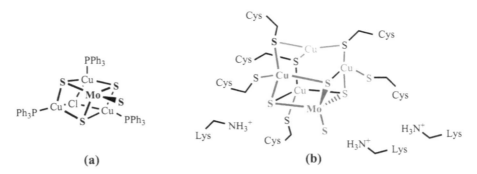

Figure 14.4 Examples for structurally characterized clusters formed between copper ions and $MoS_4^{2-}$. The pseudo-cubane cluster (a) with $Cu^+$ is an early model system [5a] generated in the reaction between $CuCl_2$, $MoS_4^{2-}$, and triphenylphosphine $PPh_3$. The nest-shaped cluster (b), $[Cu(CuAtx1)_3MoS_4]^{4-} \equiv [Cu_4(Cys)_6\{MoS_4\}]^{4-}$ [5b], forms by direct interaction of $MoS_4^{2-}$ with the copper chaperone Atx1. The three Atx1 units, linked to the four $Cu^+$ via the side-chain thiolates of six cysteinyl (Cys) residues, are arranged around the central cluster in a $C_3$-symmetrical manner.[6] Colour code: grey: $MoS_4^{2-}$; dark blue: $Cu^+$ ions, linking to thiomolybdate, in a distorted tetrahedral environment; light blue: $Cu^+$ in a trigonal-planar environment. In addition, there are electrostatic and hydrogen bonding interactions between the negatively charged cluster and positively charged side-chain ammonium groups of three lysines.

a $Cu^{2+}$–$Zn^{2+}$ based enzyme that efficiently detoxifies superoxide according to Eq. (14.3). For details with respect to SOD, see Section 6.2.

$$2O_2^- + 2H^+ \rightarrow H_2O_2 + O_2 \tag{14.3}$$

Crucial for the delivery of $Cu^+$ from the cytoplasm to the intracellular Golgi apparatus[7] is a copper-ATPase, a pump for $Cu^+$, energized by the hydrolytic cleavage of the terminal phosphate (the γ-phosphate) of the adenosine triphosphate (ATP, Section 3.4) moiety. In mammalian cells, there are two different Cu-ATPases, located in the intestinal enterocytes (ATP7A) and the liver hepatocytes (ATP7B), respectively. Mutations in the genes encoding the intestinal ATP7A lead to defects in dietary Cu absorption and successive Cu distribution in tissues (case 1), while mutations encoding the hepatic ATPase, ATP7B, result in defective distribution and detoxification of Cu mediated by the liver cells (case 2). The consequence is either copper undersupply—Menkes disease (case 1), or copper overload—Wilson's disease (case 2) [6].

Copper overload in the case of Wilson's disease induces neurological and hepatic syndromes. Neurological manifestations include movement disorder and psychoses; hepatic syndromes include severe hepatitis. The disease is diagnosed through greenish to brownish rings encircling the iris of the eyes (so-called Kayser–Fleischer rings) due to copper depositions.

As a consequence of the reducing conditions in the cytosol, copper is essentially present here in its cuprous form, and chelation therapies targeting $Cu^+$ preferentially are

---

[6] For 'symmetry', see Sidebar 4.4.

[7] The intracellular Golgi apparatus packages and labels items [here: copper-dependent enzymes such as ceruloplasmin and hephaestin] which it then sends to different parts of the cell and, eventually, across the cell membrane.

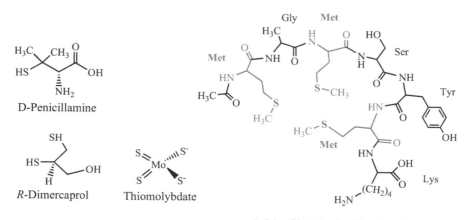

Figure 14.5 Drugs (penicillamine, dimercaprol, thiomolybdate) and a potential (pro-)drug for chelate therapy of copper overload. The peptide models a copper binding site of the membrane-bound Cu transporter Ctr1. The three methionine moieties, coordinating Cu$^+$ through the thioether functions, are highlighted in blue.

carried out with drug formulae that have soft ligand functions such as those provided by sulfur-based ligands. Examples are D-penicillamine, $R,S$-2,3-dimercapto-1-propanol (dimercaprol) and thiomolybdate $MoS_4^{2-}$, all of which strongly bind Cu$^+$ and thus stimulate excretion of Cu with the urine. The ligands are depicted in Fig. 14.5; for thiomolybdate see also Fig. 14.4. Future development of Cu purging drugs include small peptides with motifs present in copper transporters, such as the cross-membrane Cu transporter Ctr1, which delivers extracellular Cu$^+$ to intracellular copper-chaperones. The copper binding domains in Ctr1 contain methionine at the extracellular site, and cysteine plus histidine at the intracellular site of the transporter [7]. A model peptide for the extracellular domain is included in Fig. 14.5.

Menkes disease, a genetic disorder of intestinal copper absorption, which results in copper deficiency, is due to a recessive mutation of the copper transport gene responsible for the respective copper transport protein. This gene is located on the X chromosome and so the condition affects males almost exclusively. Phenotypic characteristics include fuzzy hair and fair complexation. Mental and growth retardation and neurodegeneration are typical symptoms. If not treated, children die during their early childhood. Treatment with copper-histidine formulations within the first two months after birth, or even prenatal parental treatment, can effectively ameliorate the symptoms.

While Wilson's and Menkes diseases are comparatively rare (1–4 in 100,000 people in the case of Wilson's disease, 1 in ca. 300,000 in Menkes disease), Alzheimer's disease affects 1 person in 20 over the age of 65, and 1 in 5 over the age of 80; it causes progressive cognitive impairment, and hence loss of independence. The disease appears to stem from an imbalance of copper supply in the brain [8,9] (dyshomeostasis) in the sense that copper speciation is disrupted, where 'copper speciation' refers to the quality of Cu coordination (strong vs. labile) and the Cu oxidation state (cupric vs. cuprous). Accordingly, there is an increased pool

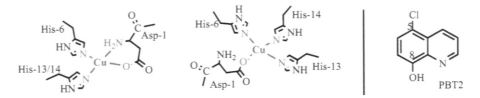

Figure 14.6 Models of coordination sites of Cu²⁺ in amyloid-β [9a], and a potential drug (5-chloro-8-hydroxyquinoline, PBT2) for copper re-equilibration in the brain. PBT2 chelates Cu²⁺ via the pyridine-N and the phenolate-O.

of *labile* copper in brain regions affected in Alzheimer's, along with elevated levels of the peptide amyloid-β[8], Aβ.

Aβ, a peptide typically comprising 40 or 42 amino acids, is cleaved from the 'amyloid pre-cursor protein', which is a trans-membrane glycoprotein. Monomers of Aβ are present in healthy brains and biological fluids and possibly function as antioxidants. Cu²⁺ ions in the labile copper pool promote the formation of soluble Aβ *oligomers*, the toxic forms of the peptide, which are responsible for brain tissue damage caused by oxidative stress—the destruction of functional brain structures by highly toxic reactive oxygen species such as superoxide and hydroxyl radicals. Eventually, insoluble amyloid plaques rich in Cu, Fe and Zn are deposited in affected brain regions, and are diagnostic for Alzheimer's patients.

Copper(II) most likely binds to amyloid-β in a distorted square-planar coordination environment provided by three nitrogens of 3 histidines, or 2 histidines +1 backbone amide, and a carboxylate of aspartate, Fig. 14.6 [9a]. In the light of a likely interconnection between Alzheimer's disease and the maldistribution of copper in the brain, it is suggested that a therapy be directed towards re-equilibration of the copper pools in the brain, rather than towards Cu depletion or Cu supplementation. The compound 5-chloro-8-hydroxiquinoline (PBT2), Fig. 14.6, has shown promising effects in this respect in clinical tests. The compound coordinates to Cu²⁺, and readily transmits Cu²⁺ to binding sites of proteins.

## 14.3  Metals and metalloids in therapy

### 14.3.1  Historical and general notes

The metal-based treatment of diseases is possibly as old as mankind. Traditional medications include recipes based on mercury, gold and antimony. Arabian quicksilver ointments for the treatment of skin diseases such as scabies have possibly motivated Paracelsus, in the early decades of the sixteenth century, to use mercury to cure syphilis and other diseases, including leprosy and lupus. For the medieval alchemists, mercury epitomized the soul. Paracelsus' mercury medications contained this 'transformative agent' in the form of 'red mercury' (cinnabar, HgS), calomel ($Hg_2Cl_2$), or sublimate ($HgCl_2$).

---

[8] Amyloid='akin to starch' derives from *amylum*, the Latin word for starch. Just as starch, amyloids show a bluish coloration when treated with iodine.

(1) $IO_3^- + 5Fe^{2+} + 6H^+ \longrightarrow 1/2I_2 + 5Fe^{3+} + 3H_2O$

    ($KIO_3$ is present as a minor component in $KNO_3$)

(2) $I_2 + 2e^- \longrightarrow 2I^-$

    (The reducing equivalents are delivered by organics
    extracted from bamboo)

(3) $2Au + 1/2O_2 + 4I^- + 2H^+ \longrightarrow 2[AuI_2]^- + H_2O$

Scheme 14.1 Conversion of gold to the physiologically potent antioxidant diiodoaurate(I) in a suspension of gold tinsels in an acidified aqueous solution of potassium nitrate plus ferrous sulfate contained in a bamboo cane.

Elixirs based on gold, a potent constituent in medications of rheumatoid arthritis, have been used in many cultures. A Chinese formulation from 600 BC describes an elixir for acquiring immortality. According to this formulation, gold foil is suspended in an aqueous solution of acetic acid, potassium nitrate and ferrous sulfate, contained in a bamboo cane. Such a system generates, surprisingly on first sight, the complex anion diiodoaurate $[AuI_2]^-$ with $Au^+$. The potassium nitrate employed here apparently contained impurities of potassium iodate, which acted both as a powerful oxidizing agent and as the source for iodine. In acidic solutions, iodate(V) is reduced by $Fe^{2+}$ to iodine and further to iodide, mediated by organic impurities provided by the bamboo. In the presence of iodide, atmospheric oxygen oxidizes elemental gold to $Au^+$. The reaction sequence is compiled in Scheme 14.1. We discuss today's medicinal use of gold compounds in the next section.

Antimony, in the form of finely powdered stibnite ($Sb_2S_3$), was another medieval formulation applied in the treatment of infectious skin impurities. Potassium antimonyl tartrate ('emetic tartar') $K_2Sb_2(C_4H_2O_6)_2$ was one of Paracelsus' favourite medicines. The compound forms as tartaric acid (a constituent of red wine) or tartar (sparingly soluble potassium tartrate) is in contact with a vessel containing antimony. Emetic tartar causes vomiting, and has thus been employed to 'cure' all kinds of stomach problems. Since antimony is poisonous, with symptoms resembling those of arsenic poisoning[9], it fell into disrepute in the nineteenth century, but became reintroduced for the treatment of tropical diseases such as leishmaniasis and schistosomiasis.

As has been noted in Section 14.1, a metallo-drug that enters the body usually undergoes speciation prior to reaching its final target—commonly a tissue protein or DNA. In the course of this speciation, the original ligand system in the drug (designed to optimize resorption in the intestines and transport through the body, and to minimize acute toxicity) is partially or completely replaced by constituents present in the blood serum, in other extracellular body fluids, and in the intracellular space, the cytosol. The main constituents of blood serum are listed in Table 14.1. The predominant *inorganic* serum anions capable of replacing ligands in the original metallo-drug are chloride, hydrogenphosphate, and hydrogencarbonate, along

---

[9] For arsenic poisoning, see Section 13.2.

Table 14.1  Mean concentrations $c$ of the main (potential) ligands for metal ions in human blood.

|  | $c$/mM |  | $c$/mM |
| --- | --- | --- | --- |
| Phosphate $H_2PO_4^-$/$HPO_4^{2-}$ | 1.2 | Glycine | 2.3 |
| Carbonate $HCO_3^-$ | 25 | Histidine | 0.08 |
| Chloride | 104 | Cysteine | 0.03 |
| Sulfate | 0.3 | Glutamate | 0.06 |
| Lactate | 1.5 | Glutathione (GSH) | 0.003 |
| Citrate | 0.1 | Albumin ($M \cong 70$ kDa) | 0.6 |
| Oxalate | 0.015 | Transferrin ($M \cong 80$ kDa) | 0.035 |
| Ascorbate | 0.06 | Immunoglobulin G ($M \cong 150$ kDa) | 0.084 |

with water/hydroxide. As shown by the $pK_a$ for the equilibrium represented by Eq. (14.4), mono- and dihydrogenphosphate are present in about equal amounts at pH 7.

$$H_2PO_4^- + H_2O \rightleftharpoons HPO_4^{2-} + H_3O^+ \quad pK_a = 7.21 \tag{14.4}$$

Hydrogencarbonate, $HCO_3^-$, makes up about 90% of the overall dissolved carbon dioxide, hydrated carbon dioxide, $CO_2 \cdot aq$, for the main part of the remaining $CO_2$. The kinetics for the formation of hydrogencarbonate from $CO_2$ and $H_2O$ are slow, but the reaction is effectively sped up, by a factor of $10^7$, with the zinc-based enzyme carbonic anhydrase, which we discuss in Section 12.2.1. Along with inorganics, organic low molecular mass ligands such as lactate and glycine, and high-molecular mass ligands, in particular albumin and transferrin, are prone to act as secondary ligands for metal ions.

The cytosolic glutathione concentration is larger by three orders of magnitude ($\approx 3$ mM) than the blood serum concentration, making GSH an important secondary ligand, as well as a protective agent for reactive oxygen species (including hydrogen peroxide, OH radicals, superoxide) generated in the cytosol by redox-active free metal ions. GSH is in redox equilibrium with its oxidized form, the disulfide GSSG; Eq. (14.5).

$$2GSH \rightleftharpoons GSSG + 2H^+ + 2e^- \tag{14.5}$$

Due to the reducing conditions in the intracellular medium, the ratio GSH:GSSG is $\approx 500$. Once oxidized, GSSH is reduced back to GSH by NADH, a reaction that is catalysed by the enzyme glutathione reductase.

Yet another potential detoxifier for metal ions forming sparingly soluble sulfides are the metallothioneins, cysteine-rich small proteins which otherwise store zinc ions (Section 12.4). The cytosolic concentration of metallothioneins amounts to $\approx 2$ mM, but can increase when triggered by a non-physiologically high concentration of (foreign) metal ions.

### 14.3.2  Treatment of arthritis with gold compounds

Rheumatoid arthritis, or arthritis for short, is a chronic systemic autoimmune disease associated with inflammatory disorder, which causes the immune system to attack mainly flexible joints. The disorder results in an inadequate supply of lubricating fluid in the capsules around the joints, accompanied by swelling, joint pain, destruction and deformation of the bone structure, and finally loss of flexibility. The disease is triggered and maintained by the migration of white blood cells (phagocytes and leucocytes) into the tissues of joints, the sinovial tissue.

The first gold compound to be introduced in modern chemistry was the sodium salt of dicyanoaurate(I), Na[Au(CN)$_2$]. In 1890, Robert Koch discovered that this compound inhibits the growth of tubercle bacilli. Erroneously assuming that arthritis is a variant of tuberculosis, successful treatments of arthritis with the less toxic *thiolato* complexes of gold were carried out in the 1920s by K. Landé and E. Pick (1927) and, more systematically, by F. Forrestier, starting in 1929 [10]. The gold compounds employed in these early treatments, solganol and allochrysine, are still in use and are pictured, together with a more recently developed therapeutic agent, myochrisine, in Fig. 14.7. In all of these polymeric complexes, Au$^+$ is linearly coordinated to two thiolate groups of organic sulfides [11]. Polymerization occurs via the thiolate sulfur, linked to two Au$^+$ and the organic residue. A fourth compound, auranofin, also shown in Fig. 14.7, is monomeric and contains the ligands acetyl-thioglucose and triethylphosphane, again in a linear arrangement. Thiolates and phosphanes are typical ligands for the soft d$^{10}$ ion Au$^+$.

While most of the gold drugs are applied intravenously, auranofin (AcGluS-Au-PEt$_3$) is given orally. A possible speciation sequence after administration of auranofin is provided in Scheme 14.2: After absorption, the thiolate ligand is rapidly released, allowing for binding of the {Et$_3$PAu$^+$} fragment to the cysteinyl thiolate of serum albumin (Alb), Cys34. Alb thus acts a transporter for the phosphane–gold fragment. Further speciation includes the formation of cyanoaurate(I) and –(III), and re-reduction of Au$^{3+}$ to Au$^+$ by thiols.

Cyanide is present in the body in minor, non-toxic concentrations (see, e.g. Section 13.1). It can be supplied through the oxidation of thiocyanate SCN$^-$ by reactive oxygen species, ROS. Since outbursts of elevated ROS concentrations are typical for arthritic inflammation, one of the functions of gold therapy is the effective capture of CN$^-$ by Au$^+$, and

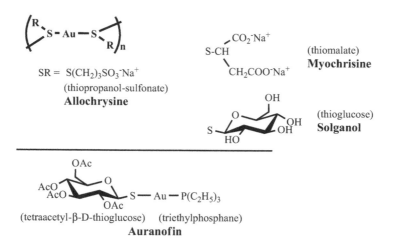

Figure 14.7 Gold-based drugs employed in the treatment of rheumatoid arthritis.

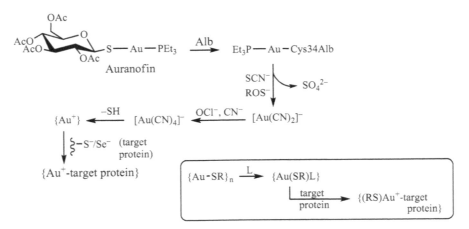

thus a shift of the redox equilibrium ROS+thiocyanate ⇋ sulfate+cyanide in favour of sulfate+cyanide. Oxidative conversion of dicyanoaurate(I) by, for example, hypochlorite yields teracyanoaurate(III), which is than reduced back to $Au^I$ by cysteinyl residues in peptides and proteins. The $Au^+$ ions finally can target cysteinyl and/or selenocysteinyl sites in proteins involved in the activation and maintenance of arthritis.

One of the central target proteins for $Au^+$ at the effector level are the cathepsins, in particular cathepsin-K. Cathepsin-K is a cysteine protease, which is expressed in bone cells (osteoblasts). Proteases degrade other proteins by hydrolytic breakage of peptide bonds. Cathepsin-K in particular is involved in bone remodelling, and thus also in the intermittent degradation of bone structures to be rebuilt. Where regeneration of the genuine bone structure is hampered, such as in the case of osteoporosis or arthritis, cathepsin-K acts solely destructively. By firm coordination to gold(I), this destructive potential of cathepsin-K is annulled. The complex formed between cathepsin-K and monomeric units of the drug myochrisine (Fig. 14.7), has been structurally characterized; the relevant section of the structure is shown in Fig. 14.8.

### 14.3.3 Cancer treatment

Thousands of coordination compounds of various metals have so far been tested—including in clinical trials—for their anticancer properties. Among these are (potential) agents based on early transition metals such as titanium and vanadium, and late transition metals such as ruthenium, rhodium, palladium, platinum, and gold. However, only platinum complexes are actually being clinically employed in cancer chemotherapy. In this section, emphasis will thus be laid on anticancer drugs based on platinum. A selection of promising clinical trials with other metals will also briefly be addressed at the end of this section. We discuss the use of radio-pharmaceuticals in cancer treatment in Section 14.3.5.

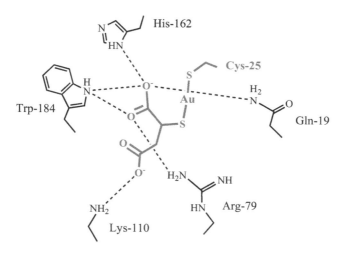

**Figure 14.8** Blockade of the active site Cys25 in the protease cathepsin-K by monomers of the drug myochrisine [12]. Gold(I) is almost linearly positioned between the thiolate sulfurs of the original thiomalate ligand and the cysteine-25 of the protein. The structure is further stabilized by hydrogen bonds (dashed lines) between the carboxylate oxygens of the ligand, and amino acid side-chain functions of the protein. See also the inset in Scheme 14.2.

## Platinum

The discovery of the anticancer potential of platinum compounds dates back to the mid-1960s, when it was found[10] that cell division, but not growth, of the enterobacterium *Escherichia coli* was inhibited in an electrical field applied to a culture medium containing these bacteria. The electric field was generated between platinum electrodes, and the growth medium contained chloride and ammonium ions, an experimental set-up that led to the generation of small amounts of simple inorganic platinum complexes, among them *cis*-diamminedichloridoplatinum(II), *cis*-[Pt(NH$_3$)$_2$Cl$_2$], or cisplatin. As shown in Fig. 14.9, this Pt$^{2+}$ complex (d$^8$ electron configuration) has a square-planar geometry. The compound later turned out to suppress the growth of sarcoma[11] tumours implanted into mice. Clinical tests commenced in the early 1970s, and cisplatin was approved for the treatment of testicular and ovarian cancer in 1978, reducing the death rates for testicular cancer from more than 90% to less than 5%.

A couple of tumours other than testicular and ovarian cancer are also susceptible to treatment with platinum-based drugs, including cervical cancer and neck cancer. New developments following the clinical introduction of cisplatin have been directed to (a) broaden the spectrum of application, (b) enhance the physiological stability of the drug, (c) minimize toxicity, and (d) counteract gradually increasing drug resistance. Two additional Pt-based

[10] The initial work goes back to the group of Barnett Rosenberg from the Michigan State University, East Lansing.

[11] Sarcomas belong to a cancer type derived from cells of the connective tissue, and thus are distinct from carcinomas, which derive from epithelial cells.

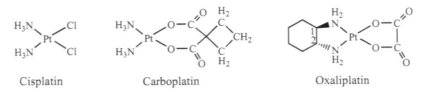

| Cisplatin | Carboplatin | Oxaliplatin |

Figure 14.9 Platinum compounds that have found worldwide approval in clinical treatments of cancer. The bidentate ligand in carboplatin is cyclobutane-1,1-dicarboxylate, the bidentate ligands in oxaliplatin are oxalate and cyclohexane-1,2-diamine (with the carbons C1 and C2 in the R configuration).

anticancer pharmaceuticals are presently in clinical use worldwide, carboplatin and oxaliplatin, whose structures are shown in Fig. 14.9. All three drugs have the same basic requirements: $Pt^{II}$ is in a four-coordinate, cis-configurated planar environment, i.e. cis-$[PtL_2L'_2]$, where L is a neutral NH-functional ligand, and L' represents a monoanionic ligand. Both carboplatin and oxaliplatin contain bidentate ligands and are therefore more stable under physiological conditions than cisplatin as a consequence of the chelate effect (see Sidebar 4.1).

Cisplatin is commonly applied via intravenous infusions in the form of a saline solution (154 mM NaCl), adjusted to ~pH 4 in order to suppress the formation of μ-hydroxido oligomers. In the physiological medium, with an extracellular chloride concentration of 104 mM and a pH of 7.4, speciation occurs, as summarized in Scheme 14.3. The main conversions affect the {Pt-Cl} moiety; the {cis-Pt(NH_3)_2} moiety is comparably stable [13]. Exchange of the chlorido by the aqua ligand is slow; the pseudo-first order[12] rate constants are in the order

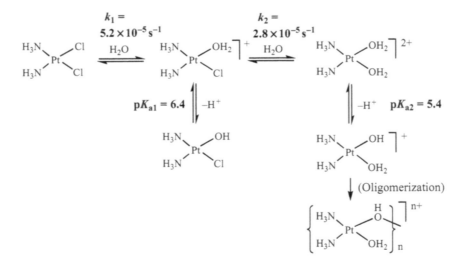

Scheme 14.3 Selection of speciation steps for cisplatin in a physiological environment [13]. Similar speciation steps apply to carboplatin. The likely additional involvement of hydrogencarbonate in speciation is not considered here.

[12] The reactions are pseudo-first order because of the practically constant concentration of water (55.5 mol L⁻¹) in the aqueous medium.

of magnitude of $10^{-6}$ to $10^{-5}$ s$^{-1}$. By contrast, the deprotonation of the aqua to the hydroxido ligand is very fast. The p$K_a$ values indicate that, in blood serum, the predominant species by far are the monohydroxido complexes. The substitution kinetics of carboplatin is slower than that of cisplatin because, in aqueous media, carboplatin exists in a monomer–dimer equilibrium, a fact which accounts for its more pronounced long-term stability.

At a later stage, the $\{(NH_3)_2Pt^{2+}\}$ moiety, derived from cisplatin or carboplatin, becomes coordinated to histidine, methionine and the cysteine-34 residue of serum albumin, and the NH$_3$ ligands are also now subjected to substitution. Since the original drug is neutral and thus exhibits roughly balanced hydro-/lipophilicity, passive transport by diffusion across the cell membrane accounts for the transport of some of the platinum into the cell. Due to the lower intracellular chloride concentration (ca. 4 mM), the equilibria in Scheme 14.3 will shift towards the aqua/hydroxido species.

For Pt$^{2+}$ devoid of its original ligand system, membrane-bound proteins that otherwise transport copper ions also play a role [14]. An example is the membrane-bound copper transporter Ctr1, shown in Fig. 14.5. Further transport in the cytoplasm is possibly achieved by copper chaperones. Binding to sulfur donors such as those provided by cysteine and methionine residues, {-SR}, is kinetically favoured (though thermodynamically disfavoured with respect to nitrogen donors). Intermediate cytosolic {Pt-SR} species should therefore be involved in the delivery of Pt$^{2+}$ to its targets.

The main final target for Pt-based anticancer drugs and their speciation products is DNA [15]. Platinum compounds enter both healthy and tumour cells, but specifically attack the DNA in testicular and ovarian cancer cells. Healthy cells, as well as tumour cells that are—or progressively become—(partially) resistant to platinum chemotherapy, can get rid of the drug and/or its speciation products. Copper transport pathways are possibly also responsible for this cellular *export* of Pt$^{2+}$.

The N7 of guanosine in the DNA double helix is a preferential target for the soft Pt$^{2+}$ in the $\{Pt(NH_3)_2^{2+}\}$ or a related moiety (Fig. 14.1), where platinum predominantly coordinates to two adjacent guanosine bases in the same strand, Fig. 14.10; this interaction generates an

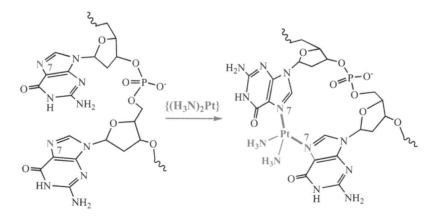

Figure 14.10 Placement of the *cis*-{Pt(NH$_3$)$_2$} unit between two adjacent (intrastrand) guanines; redrawn from [16]. The coordination of Pt$^{2+}$ to the N7 of guanine induces a bend in the DNA backbone. For a cut-out of the DNA double helix, cf. Fig. 14.1.

*intra*strand cross-link, which results in a kink in the DNA of 20–45°. The DNA sequence to be replicated, or to be transcribed for RNA or protein synthesis, is thus no longer recognized by proteins involved in reading DNA. Various cellular functions are thus disabled, including cell division. If this damage caused by platination of the DNA is not repaired, apoptosis and finally cell death result. Tumour cells are particularly active with respect to metabolic processes and cell division, which reduces the time span available for potential repair. This makes tumour cells generally more susceptible to damage than intact cells—along with the possibly less efficient detoxification paths such as the cellular export of platinum mentioned previously.

## Ruthenium

Along with platinum, ruthenium has been in the focus of anticancer research in recent years [17]. However, none of the newly developed cytotoxic ruthenium coordination compounds has so far passed phase II of clinical tests. Promising results from phase I clinical tests have been obtained for the $Ru^{3+}$ complexes NAMI-A and KP1019, Fig. 14.11a. Both compounds are octahedrally coordinated anionic complexes, with four chlorido ligands in the equatorial plane. In NAMI-A, the axial positions are occupied by imidazole and dimethyl sulfoxide, in KP1019 by two benzpyrazoles. The counter-ions are imidazolium and benzpyrazolium, respectively.

The presence of chlorido ligands suggests a mechanism comparable to that of cisplatin, i.e. (i) replacement of $Cl^-$ by $H_2O/OH$, followed by (ii) preferential binding to His and/or Cys sites of serum proteins such as albumin, (iii) delivery into the target tumour cells, and (iv) linkage to DNA. However, in contrast to $Pt^{2+}$, $Ru^{3+}$ ($d^5$) is redox-active in as far as it is readily bio-reduced to $Ru^{2+}$ and may thus additionally act by modulating the level of ROS.

Another mechanism of action is also suggested by the differing type of tumours targeted by Pt- and Ru-based drugs: While cisplatin is particularly effective in the treatment of testicular

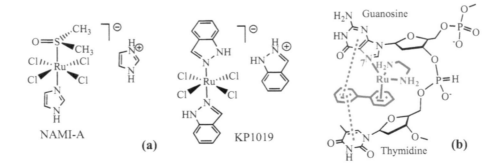

Figure 14.11 (**a**) NAMI-A and KP1019 are ruthenium(III) complexes which have successfully been tested clinically (phase I). NAMI is the acronym for new anti-tumour metastasis inhibitor, and KP [17a] relates to the research group of B.K. Keppler, University of Vienna, where this compound (among many others) has been developed and investigated. (**b**) The ruthenium(II) complex [(bp)Ru(en)Cl][PF$_6$] (bp=biphenyl, en=ethylenediamine) [18b] provides the {bpRu(en)} moiety that interacts with DNA in two ways: coordination of $Ru^{2+}$ to N7 of guanine on the one hand, and intercalation (indicated by dashed lines) of bp in between guanine and thymine on the other hand.

and ovarian cancer, NAMI-A stops tumours from spreading to other parts of the body, known as metastasis, and KP1019 preferentially targets colon carcinomas.

Various additional ruthenium coordination compounds do have anti-tumour activity *in vitro* and *in vivo* [18a]. An example is the cationic species [(bp)Ru(en)Cl]$^+$, a ruthenium(II) complex containing side-on coordinated biphenyl (bp) along with the chlorido and the ethylenediamine (en) ligand. The chloride is exchangeable, and the {(bp)Ru(en)}$^{2+}$ moiety binds to the N7 of guanine, hence a binding mode comparable to that of cisplatin (Fig. 14.10). In contrast to cisplatin, however, there is additional interaction, namely π stacking between the side-on coordinated biphenyl and the adjacent DNA bases guanine and thymine [18b], a mode of interplay also referred to as *intercalation*. This dual binding to DNA is illustrated in Fig. 14.11b.

## Other metals

Metallo-intercalators as potential anticancer drugs are of interest because their mode of action broadens, and thus amplifies, the effectiveness of drugs such as cisplatin that 'simply' deform DNA by coordination of the metal centre to the N-functions of the nucleo-bases and, eventually, the O-functions of the phosphates. Additional modes of operation of a metal-based drug or prodrug include modifications of DNA through redox interaction, hydrolytic cleavage, and the formation of radicals by abstraction of H. Fig. 14.12 illustrates assorted metal–organic (**1**, **2**) and organometallic (**3**, **4**) coordination compounds that have proven anti-tumour activity *in vitro* and *in vivo*, and have passed phase I clinical tests.

All of the complexes carry aromatic residues in the coordination periphery and are therefore potential intercalators for DNA. Compounds **2**, **3**, and **4** also contain leaving groups (ethanolate, chloride, pseudohalide), allowing for a direct attack of the metal centre on a suitable ligand function in DNA. For the hard early transition metal ions Ti$^{IV}$ (d$^0$) and V$^{IV}$ (d$^1$), phosphate-O$^-$ is the most likely target.

Organometallic compounds with the metal centre 'sandwiched' by two cyclopentadienide(1–) ligands bonded in the η$^5$ mode (cf. Sidebar 13.1) are referred to as metallocenes. Vanadocenes and titanocenes have well-known cancerostatic potential, and investigations based on cytotoxic titanium complexes have recently been revived [19]. But so far, none of these compounds has come into clinical use.

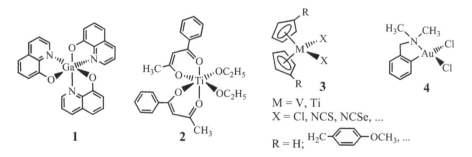

M = V, Ti
X = Cl, NCS, NCSe, ...
R = H; H$_2$C—⟨  ⟩—OCH$_3$, ...

Figure 14.12 Selected anti-tumour metal complexes that carry aromatic residues in the coordination periphery. Complexes with the following ligands are shown: **1**, 8-oxichinolinate(1–); **2**, 1-phenyl-2-methyl-β-butane diketonate(1–); **3**, derivatives of η$^5$-cyclopentadienide(1–); **4**, dimethylaminoethylphenyl(1–). Compound **2** is also known as budotitane.

The square-planar complex **4** contains gold in the oxidation state +III. $Au^{3+}$ is isoelectronic with $Pt^{2+}$—both ions are in the $d^8$ electronic configuration—and the gold complexes may thus be considered to act on DNA in a way comparable to platinum-based drugs. However, $Au^{3+}$ is redox-labile and hence normally easily reduced to $Au^+$ ($d^{10}$). Although the presence of a $Au-C(\sigma)$ bond in **4** substantially stabilizes the +III oxidation state, $Au^{3+}$ is not likely to survive the reductive potential of thiol residues present in peptides and proteins.

### 14.3.4 Further metal-based medications

In the previous sections, we discussed major health problems involving metals in some detail: diseases linked to copper and iron homeostasis; therapy of rheumatoid arthritis with gold compounds; platinum prescriptions in clinical use for the treatment of cancer; and a few other metal-based anti-tumour drugs that have been introduced into clinical trials. In this section we focus on additional metals of traditional and/or current medicinal interest: the potentiality of vanadium in the amelioration of diabetes mellitus, lithium in the medication of mood disorder; bismuth in the treatment of gastrointestinal disorders; and silver in wound healing. For the metalloid arsenic in the (historical) treatment of syphilis see Section 13.2.

### Vanadium

Vanadium is very likely an essential element. Its average concentration in the human body is ca. $0.3\,\mu M$, corresponding to ca. 1 mg V in an individual of 70 kg body mass. The mean vanadium content in food and drinking water secures the daily supply of this omnipresent element. In a physiological environment, the primary oxidation states of vanadium are +V ($VO^{3+}$, $VO_2^+$, $H_2VO_4^-$) and +IV ($VO^{2+}$). The main complexing agent in blood serum is the ferric ion transporter transferrin. At oxic conditions, micromolar concentrations, and a pH of 7.4, the predominant vanadate(V) species present in physiological liquids is dihydrogenvanadate $H_2VO_4^-$. Scheme 14.4 shows how vanadate is a phosphate analogue; the (likely) essential nature of vanadate in all probability goes back to its regulatory function in metabolic processes involving phosphate.

The vanadate/phosphate analogy (see also Sidebar 12.1) is also most likely the key to vanadium's insulin-mimetic (or insulin-enhancing) potentiality [20]. Diabetes is caused by insufficient insulin supply (type 1) or insulin response (type 2). In type 1 diabetes, insulin production by the β cells in the islets of Langerhans in the pancreas is drastically reduced; in type 2 diabetes, the cellular insulin receptors do not respond properly to insulin. Type 1 diabetes, an auto-immune disease, is caused by the destruction of the β cells and often appears during childhood or adolescence. Type 2 diabetes, the frequency of which is about ten times that of type 1, typically affects elderly individuals, but is also increasingly becoming a problem with

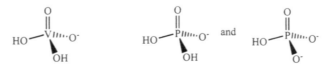

Scheme 14.4 Structural similarity between vanadate and phosphate. Shown are the main species that are present at pH ≈ 7.

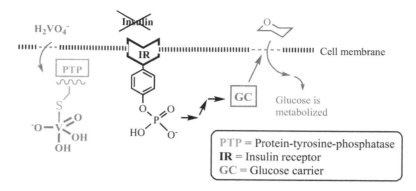

**Figure 14.13** Restoring the signalling cascade for glucose import in diabetic individuals: In the case of impairment of insulin supply or response, vanadate inhibits PTP by coordinating to a cysteinate residue (S) at the enzyme's active site, thus preventing *de*phosphorylation of the IR by PTP and keeping the signalling cascade (black arrows) for the activation of the GC operative.

*obese* young people. In either case, glucose uptake by the cells and thus glucose metabolism within the cells is hampered, resulting in glucose overload in the blood plasma (hyperglycemia). Concomitantly, lipolysis is stimulated and lipogenesis is restrained, resulting in an accumulation of harmful keto acids such as acetoacetic acid in blood and tissues—factors that are responsible for the more serious diabetes symptoms such as vision disturbance and tissue damage.

The likely mode of action of vanadate at the cellular level is demonstrated in Fig. 14.13: In the case of an *intact* glucose metabolism, insulin docks to the extracellular site of the insulin receptor (IR), triggering phosphorylation of tyrosine at the intracellular site of the IR, thus initiating a signal cascade by which a glucose carrier (GC) is activated. Glucose thus becomes transported into, and metabolized within, the cell. In the case of insufficient insulin supply, or severely diminished insulin response, phosphorylation of the IR is annulled by the action of a phosphatase (a protein tyrosine phosphatase, PTP)—an enzyme that catalyses the hydrolytic dephosphorylation of the insulin receptor. The signal cascade for glucose import thus becomes blocked. This is where vanadate comes in. Vanadate can enter cells via phosphate channels and, once in the cytosol, can inhibit PTP by coordinating to a cysteine residue at its active site. The signalling path is thus restored.

Among the numerous vanadium complexes that have so far been tested *in vitro* and *in vivo*, just one compound, [VO(ethylmaltol)$_2$] (BEOV), has entered phase I and II clinical tests [21]. BEOV is applied orally, and is readily absorbed in the gastrointestinal tract. As shown in Fig. 14.14, the compound undergoes speciation, including hydrolysis, oxidation, and, once in the blood stream, coordination of the VO$^{2+}$ moiety to transferrin. Vanadium is then distributed to tissues either by transferrin via endocytosis, and/or in the form of vanadate across phosphate channels. A substantial quantity of vanadium is temporarily stored in bone, underlining the similarity between vanadate and phosphate.

## Lithium

Lithium therapy is being successfully applied in the treatment of bipolar disorder, also known as mood disorder or manic depression. Both manic and depressive episodes are ameliorated

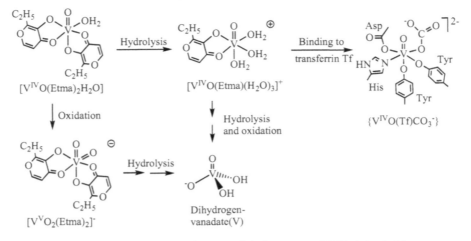

Figure 14.14 Selection of transformations of the anti-diabetic compound [VO(ethylmaltol)$_2$] ([VO(Etma)$_2$H$_2$O], BEOV) in body fluids.

by the treatment of patients with Li$^+$. As already outlined in Section 3.1, the ions Li$^+$ and Mg$^{2+}$ are of almost the same size: The ionic radii are 76 pm for Li$^+$ and 72 pm for Mg$^{2+}$, suggesting a competitive mode of action for Li$^+$ in key enzymes depending on the presence of Mg$^{2+}$, and also indicating that lithium has a narrow therapeutic window and hence is potentially toxic. On the other hand, the largely differing charge densities[13] of Li$^+$ (1.32) and Mg$^{2+}$ (2.78) also protect many vital cellular Mg$^{2+}$ proteins from being 'attacked' by Li$^+$ as long as lithium concentrations remain low [22].

Disabling mood swings in patients suffering from bipolar disorder are believed to be initiated by comparatively high levels of inositol, generated by an elevated activity of inositol monophosphatase, IMPase. IMPase is a Mg$^{2+}$ dependent enzyme that catalyses the hydrolysis of inositol phosphate; Fig. 14.15a provides an overview of the reaction path inositol phosphate → inositol → phosphatidylinositol. The free inositol thus formed can be transformed into phosphatidylinositol, a phospholipid responsible for cell signalling, including pathologic signalling as in the case of dipolar disorder.

IMPase contains three Mg$^{2+}$ binding sites, with the magnesium ions coordinated to side-chain carboxylates of aspartate and glutamate, and to water. Replacement of Mg$^{2+}$ by Li$^+$ in one of these sites [23] (Fig. 14.15b) inactivates the enzyme and thus prevents hydrolytic cleavage of the inositol-phosphate bond and hence the availability of inositol for signalling.

## Bismuth

Bismuth compounds have a long-held role in the treatment of diseases caused by the bacterium *Helicobacter pylori*. The bacterium colonizes the gastric mucosa [24] and is occasionally considered a commensal microorganism that is involved in the regulation of the stomach pH and the appetite. Infectious malfunction can cause gastritis, gastric and duodenal ulcers

---

[13] Charge density is defined as the quotient *ionic charge/ionic radius* (in Å); Section 3.1.

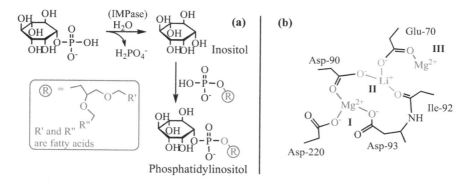

Figure 14.15 (a) Dephosphorylation of inositol phosphate catalysed by the $Mg^{2+}$-dependent inositol monophosphatase, IMPase, and subsequent build-in of inositol into phosphatidylinositol. (b) $Li^+$-inhibited IMPase [23]; aqua ligands supplementing the coordination sphere of the ions (octahedral for $Mg^{2+}$, tetrahedral for $Li^+$) are not shown. I, II and III indicate the three $Mg^{2+}$ binding sites in genuine IMPase.

and, in the case of chronic infections, stomach cancer. The currently available bismuth drugs are based on the subsalicylates and subcitrates of bismuth(III). The prefix 'sub' refers to partial hydrolysis of the idealized formulations $Bi(Hsal)_3$ and $[N]^+[Bi(cit)]^-$, where Hsal is salicylate(1−), cit is citrate(4−), and $[N]^+$ represents a protonated nitrogen base such as ethylenediamine. Partial hydrolysis of these precursor compounds results in the formation of colloids with basic structural units related to those shown in Fig. 14.16.

The mode of action of bismuth drugs is multifaceted. In the strongly acidic medium of the stomach—0.5 M HCl (pH around 2)—the drugs undergo speciation to {BiOCl} and basic bismuth citrates/salicylate. These polymeric species form a protective coating, preventing adherence of *Helicobacter* to the mucosa [24]. But bismuth ions can also interact with lactoferrin, a ferric iron transporter akin to transferrin (Section 4.2), thus depriving *Helicobacter* of iron acquisition. Finally, once bismuth is internalized by *Helicobacter*, the activity of bacterial enzymes such as urease (a $Ni^{2+}$ dependent enzyme; Section 10.4) is hampered either by direct interaction of $Bi^{3+}$ with the protein, and/or by the interference with $Ni^{2+}$ homeostasis. Urease catalyses the breakdown of urea to $NH_3$ and thus produces a less acidic to neutral

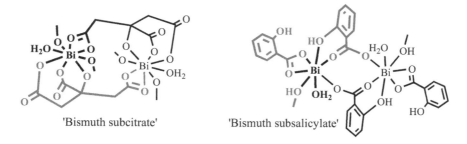

Figure 14.16 The dinuclear units in structurally characterized polymers/oligomers present in colloidal solutions of bismuth-based drugs that are applied in the treatment of gastric and duodenal ulcers. Examples for licensed drugs are Pepto-Bismol® (Bi subsalicylate) and De-Nol® (Bi subcitrate). Citrate and salicylate in the monomeric subunits are highlighted in blue.

environment in the immediate surroundings of *Helicobacter*—namely one that is life-sustaining from the point of view of this bacterium.

## Silver

Silver is an efficient sterilizing agent. Its antibacterial potential has been utilized for centuries, and exploited—unwittingly—by using silver vessels for the storage of drinking water, but also in wound disinfection and thus in the promotion of wound healing. The formulations of silver have changed during the past decades, from bulk silver in the form of silver foils or fabric impregnated with metallic silver, to ionic silver in the form of silver nitrate $Ag^+NO_3^-$ or $Ag^+$ absorbed in the micropores of carrier materials such as zeolites, and lately to silver nanoparticles. The latter can either be produced biogenically, or by physical and chemical procedures. Biogenic production of Ag nanoparticles resorts to the lactic acid bacterium *Lactobacillus fermentum*. This bacterium can employ external $Ag^+$ as the final electron acceptor in bacterial respiration, depositing nano-particulate $Ag^0$ at its outer cellular membrane. Due to the larger specific surface area, biogenically generated nanoparticulate silver tends to be clearly more efficient than chemically or physically produced nanosilver [25a].

When applied externally in wound disinfection, the antibacterial potency of silver ions is achieved either directly (silver nitrate, zeolites loaded with $Ag^+$) or by oxidative conversion of nanosilver to $Ag^+$. Oxidation of $Ag^0$ to $Ag^+$ is effected by direct oxidation with atmospheric $O_2$ (Eq. (14.6)), or by oxidation with $H_2O_2$ present in the inflammatory regions of a wound; Eq. (14.7).

$$2\,Ag + \tfrac{1}{2}O_2 + H_2O \rightarrow 2\,Ag^+ + 2\,OH^- \tag{14.6}$$

$$2\,Ag + H_2O_2 + 2\,H^+ \rightarrow 2\,Ag^+ + 2\,H_2O \tag{14.7}$$

The toxicity of silver ions is similar to that of metal ions such as $Au^+$, $Pt^{2+}$, $Cu^{+/2+}$, $Cd^{2+}$, and $Hg_2^{2+}/Hg^{2+}$. Common mechanisms for cell toxicity [25b] thus are (i) disruption of the function of enzymes and other proteins through coordination of $Ag^+$ to thiolates of cysteinyl residues, (ii) disruption of DNA replication by direct interaction with the DNA bases, and (iii) generation of ROS. In addition, direct damage of the bacterial cell membrane plays a part, allowing leakage of intracellular contents, and resulting in the suppression of growth and replication, and also increasing the membrane permeability for antibiotics [25c].

### 14.3.5 **Radiopharmaceuticals**

Radiotherapy, as well as radio-imaging (Section 14.4), exploits short-lived radionuclei, i.e. nuclei which—in contrast with stable nuclei—decay to form a daughter nucleus, concomitantly emitting radiation. Radionuclei employed in medicine emit $\alpha$, $\beta^-$, or $\beta^+$ particles along with neutrinos and, in most cases, $\gamma$ radiation. An overview of radioactivity is provided in Sidebar 14.1.

---

**Sidebar 14.1    Radioactivity**

The nucleus of an isotope of a chemical element is characterized by its nuclear charge $z$ and its mass number $m$. The nuclear charge $z$ is equal to the number of protons in the nucleus; $z$ is indicated as a lower index at the left of the element symbol. The mass number $m$ is the sum of neutrons and

Compilation of some elemental particles

| Name | Symbol | Location/origin | Rest mass (g mol$^{-1}$) | Charge |
|---|---|---|---|---|
| proton | p, H$^+$ | nucleus | 1.00728 | +1 |
| neutron | n | nucleus | 1.00867 | 0 |
| electron | e$^-$, β$^-$ | atomic shell/decay of a neutron | 0.00055 | −1 |
| positron | e$^+$, β$^+$ | decay of a proton | 0.00055 | +1 |
| neutrino | $\nu$ | decay of proton/neutron[a] | 0 | 0 |

[a] The decay of a neutron (to an electron and a proton) delivers an antineutrino, the decay of a proton (to a neutron and a positron) a neutrino.

protons; this information is provided by the upper left index of the element symbol. Examples include (radioactive isotopes in blue):

$^{1}_{1}H$ $\quad$ $^{2}_{1}H$ $\quad$ $^{3}_{1}H$ $\qquad$ The three hydrogen isotopes protium, deuterium and tritium

$^{12}_{6}C$ $\quad$ $^{13}_{6}C$ $\quad$ $^{14}_{6}C$ $\qquad$ The three main carbon isotopes

$^{235}_{92}U$ $\quad$ $^{238}_{92}U$ $\qquad$ Two of the more common uranium isotopes

Isotopes with an unbalanced number of protons and neutrons (see the table above for properties of elemental particles) are unstable, i.e. they decay, commonly forming a new element. For the main decay modes see the medicinally relevant examples (adjoining equations). $^{99m}$Tc is a *metastable* state of the isotope $^{99}$Tc.

$$\alpha \text{ decay:} \quad ^{212}_{83}Bi \longrightarrow ^{208}_{81}Tl + ^{4}_{2}He \ (\equiv \alpha)$$

$$\beta^- \text{ decay:} \quad ^{89}_{38}Sr \longrightarrow ^{89}_{39}Y + ^{0}_{-1}e^- \ (\equiv \beta^-)$$

$$\beta^+ \text{ decay:} \quad ^{123}_{53}I \longrightarrow ^{123}_{52}Te + ^{0}_{1}e^+ \ (\equiv \beta^+)$$

$$\gamma \text{ decay:} \quad ^{99m}_{43}Tc \longrightarrow ^{99}_{43}Tc + \gamma$$

The radioactive decay is described by the following differential equation: $-dN/dt = \lambda N$, where $N$ is the number of nuclei, $t$ the time, and $\lambda$ the decay constant (characteristic for the activity of a specific nucleus). Integration yields $\ln(N/N_0) = -\lambda t$, or $N = N_0 e^{-\lambda t}$, with $N_0$ the number of nuclei when $t = 0$. More descriptive is the half-life $t_{1/2} = \ln2/\lambda$. The half-life indicates the time interval in which half of a given number of radioactive nuclei decays. The shorter $t_{1/2}$, the more active is the nucleus.

The *activity* of a radionuclide is quantified by the unit Bq (derived from Bequerel): 1 Bq = 1 decay per second. For judging the *biological* effect of exposure to radiation, the *absorbed dose* (unit: Gray, Gy) and the *dose equivalent* (unit: Sievert, Sv) are used. The units of Gy and Sv are J kg$^{-1}$. Absorbed dose and dose equivalent are related to each other through a dimensionless factor $Q$: dose equivalent = $Q \times$ (*absorbed dose*), where $Q$ depends on the nature of the radiation: $Q = 20$ for α, 1 for β, γ, and X-rays, and 3–10 for neutrons (depending on the neutrons' velocity).

In conventional radiotherapy—for the treatment of cancer, for example—a narrow beam of radiation is moved around the tumour in such a way that the tumour is constantly in the beam's focus, while the surrounding tissue is subjected to minor or negligible damage only. This therapeutic method is, however, restricted to localized tumours, i.e. it cannot be

Table 14.2 Some short-lived radioactive nuclei (including the non-metals fluorine and iodine) employed in diagnostics (Section 14.4) and therapy (this section). For details concerning radioactive decay, see Sidebar 14.1.

| Nuclide; $q^a$ | Application in | Main radiation/ Progeny | Maximal energy (MeV) | Half-life |
|---|---|---|---|---|
| $^{18}F$ $q=9$ | Imaging | $\beta^+$, and $\gamma^b$/$^{18}O$ | 0.63 | 110 minutes |
| $^{68}Ga$ $q=31$ | Imaging | $\beta^+$, and $\gamma^b$/$^{68}Zn$ | 1.9 | 67.6 minutes |
| $^{89}Sr$ $q=38$ | Therapy | $\beta^-$/$^{89}Y$ | 1.5 | 50.6 days |
| $^{90}Y$ $q=39$ | Therapy | $\beta^-+\gamma$/$^{90}Zr$ | 2.3 | 64.1 hours |
| $^{99m}Tc$ $q=43$ | Imaging | $\gamma$/$^{99}Tc$ | 0.14 | 6 hours |
| $^{123}I$ $q=53$ | Imaging | $\gamma$(e$^-$ capture)/$^{123}Te$ | 0.13 | 13 hours |
| $^{153}Sm$ $q=62$ | Therapy | $\beta^-+\gamma$/$^{153}Eu$ | 0.7 | 46.3 hours |
| $^{186}Re$ $q=75$ | Therapy | $\beta^-+\gamma$/$^{186}Os$ | 1.08 | 3.8 days |
| $^{188}Re$ $q=75$ | Therapy | $\beta^-+\gamma$/$^{188}Os$ | 2.12 | 17 hours |

$^a$ Nuclear charge (number of protons).
$^b$ Generated by recombination of the positron with an electron.

applied to metastasized cancers. To cope with this latter problem, radiopharmaceuticals have been designed that specifically target, and are absorbed by, the malignant tissue, where they accumulate and deliver a destructive radiation dose. For this purpose, the design of the coordination sphere of the radioactive nuclide has to take into account, to ascertain that the pharmaceutical survives its transport to and uptake by the target. In addition, the radioactive nuclide in the pharmaceutical has to be chosen in such a way that it decays sufficiently slowly to secure the complete destruction of the tumour cells, but sufficiently fast to minimize damage of surrounding and to rebuild healthy tissue. In addition, the decay product(s) should be non-toxic and/or rapidly cleared from the body.

A crucial point in choosing the adequate radionuclide is therefore its half-life, and its mode of decay: Given the very small operating distance of $\alpha$ particles (less than 0.1 mm), $\beta$ emitters are used exclusively. The range of operation of a $\beta$ emitter depends on the energy of the $\beta$ radiation; the range typically encompasses a few millimetres up to around 1 cm. Table 14.2 lists sources for and properties of selected radionuclei used in radiotherapy and radio-imaging. It should also be noted that tissue damage is not only caused by direct targeting of cells with $\beta$ radiation, but also by radicals such as $^{\cdot}OH$ and others, produced from the radiolysis of water.

$^{89}Sr$ is one of the rare nuclides that decay without accompanying $\gamma$ radiation. This is the reason why it is not easily detected as a component of radioactive contamination in the aftermath of a nuclear accident. Strontium is chemically related to calcium and is therefore built into the hydroxyapatite[14] of the bone structure. $^{89}Sr$, in the form of aqueous solutions of $SrCl_2$, has therefore been applied in the palliation of pain caused by bone metastases in advanced cases of prostate, breast, and lung cancers. Other prescriptions which preferentially target osteoblasts (cells associated with bone production) contain $^{90}Y$, $^{186}Re$, or $^{188}Re$. Yttrium and rhenium compounds are supplied to clinics in the form of inorganic salts ($YCl_3$; $Na[ReO_4]$) that, prior to application, are processed into conjugates with organic carriers. Examples are displayed in Fig. 14.17a and b [26]. In contrast with $^{89}Sr$, the $\beta^-$ decay of $^{90}Y$ and $^{186/188}Re$ is accompanied by $\gamma$ radiation, allowing for concomitant localization of the bone metastases by imaging.

[14] See Section 3.5.

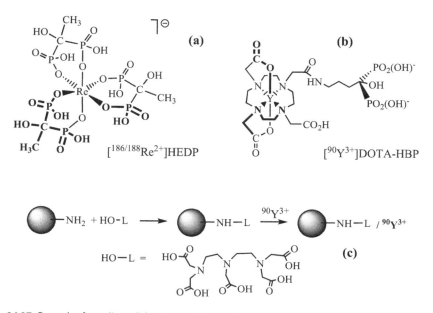

**Figure 14.17** Examples for radionuclides in organic ligands/linkers used in the palliation of cancers metastasized to bone. (**a**) Structural unit of oligomers formed between hydroxyethane-diphosphonate (HEDP) and $^{186/188}Re^{2+}$ (generated from $ReO_4^-$ by reduction with $SnCl_2$+ascorbic acid). (**b**) $^{90}Y^{3+}$ coordinated to DOTA-HBP. DOTA=tetraazacyclododecane-tetraacetic acid, HTB=4-amino-1-hydroxybutane-1,1-*bis*(phosphonate). (**c**) Coordination of $^{90}Y^{3+}$ to the conjugate formed between a monoclonal antibody (symbolized by a shaded circle) and ethylenetriamine-pentaacetate.

A more recent approach towards specific targeting of tissues by radiopharmaceuticals is radioimmunotherapy: Here, a radioactive species such as the ion $^{90}Y^{3+}$ is linked, via an auxiliary multidentate ligand, to a monoclonal antibody (Fig. 14.17c) [27], carrying the radionuclide and delivering it to the target tissue. A monoclonal antibody is an immunologically active protein that binds to a specific site of an antigen.

## 14.4 Metals and metalloids in diagnostic imaging

There are three main clinical applications of metals and metalloids in the context of the non-invasive imaging of structures inside the body: X-ray enhancement of soft tissue; nuclear magnetic resonance (NMR) imaging; and γ ray imaging. X-ray enhancement employs strong X-ray absorbers to visualize soft tissue structures otherwise invisible or only barely visible by radiography. Classical examples are insoluble barium sulfate $BaSO_4$, and derivatives of 1,3,5-triiodobenzene. $BaSO_4$ is applied orally, and its deposits in the gastrointestinal tract allow for the X-ray detection of, e.g. ulcers in the small intestines. Triiodo derivatives of benzene are used for coronary angiography, i.e. visualization of the inside of blood vessels of the heart muscle.

NMR (Sidebar 14.2) is exploited anatomically in magnetic resonance imaging (MRI). The individual to be studied, or a body part such as a sprained ankle, is moved into the cylindrical cavity of a magnet, typically operating at a magnetic field strength of 1.5 Tesla, and is exposed to radio frequency h$\nu$ in the megahertz range. MRI basically targets the protons of water which, depending on the surroundings, mobility and concentration in different tissues, experience differing relaxation times $T$. The different $T$ values of the re-emitted h$\nu$ are registered by a detector and transformed into images, allowing for a detailed judgment of tissue damage such as that caused by a sprain. In order to enhance the contrast of the NMR signals of the protons, contrast agents in the form of high-spin transition metal ions are applied. The high-spin state ascertains a high paramagnetic moment (see Sidebar 4.3), and thus enhances relaxation (decreases $T$) and improves the sensitivity of the imaging process and the contrast in the image.

Widely employed metal ions in contrast agents are gadolinium(3+) and ferric ions. The electronic configurations of high-spin $Gd^{3+}$ and $Fe^{3+}$ are $4f^7$ and $3d^5$, respectively.[15] The probes are applied intravenously, $Fe^{3+}$ commonly in the form of nano-particulate ferric oxide, and

---

**Sidebar 14.2   Nuclear magnetic resonance spectroscopy**

Nuclei with an odd number of protons and/or neutrons possess a permanent nuclear spin $I$, i.e. they rotate about an internal axis. Nuclei with an even number of protons and an even number of neutrons do not have a spin; they are non-magnetic and hence inaccessible to NMR. Examples are $^{12}C$, $^{16}O$, and $^{32}S$. For magnetic nuclei, $I$ can take values of $\frac{1}{2}$, or $> \frac{1}{2}$. Examples for $I = \frac{1}{2}$ nuclei are $^{1}H$, $^{13}C$, $^{31}P$, and $^{195}Pt$, examples for $I > \frac{1}{2}$ nuclei $^{2}H$ ($I=1$), $^{14}N$ ($I=1$), $^{17}O$ ($I=5/2$), $^{99}Tc$ ($I=9/2$).

The nuclear spin, or angular momentum, of the positively charged nucleus gives rise to a magnetic moment $\mu$. Placed into an external magnetic field $B_0$, a spin $\frac{1}{2}$ nucleus can go into an antiparallel (excited), or a parallel (ground state) orientation with respect to $B_0$, the parallel orientation being the energetically favoured one. The energy difference $\Delta E$ between the two orientations, or spin states, depends on the strength of $B_0$. The energetically less stable orientation becomes populated when irradiating microsecond radiofrequency (h$\nu$) pulses, fulfilling the condition for resonance h$\nu = 2\mu B_{loc}$.

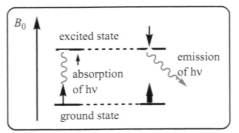

When returning from the excited to the ground state, a radio pulse is emitted (an echo signal) which can be analysed externally. $B_{loc}$ (loc for local) is the magnetic field the nucleus experiences in its electronic environment as modulated by its *chemical* environment. Differing chemical environments of the protons in a more complex (for example, organic) compound give rise to distinct local fields for each chemically distinct hydrogen and thus a distinct echo signal, or NMR signal, characterized by its *chemical shift*. The magnetic moments of protons in non-equivalent positions in a molecule further interact, or couple, giving rise to splitting of the NMR signal, quantified by *coupling constants*.

In magnetic resonance imaging (MRI), the dominant targets are the protons of tissue water. Here, the primary chemical environment is provided by the temporarily present, rapidly rearranging water

---

[15] In addition to gadolinium, the lanthanoids dysprosium and holmium can be efficient MRI probes [28]. Ho has a particularly high magnetic moment. Its receptivity ($1.16 \times 10^3$) is thus four orders of magnitude higher than that of Gd (0.124).

clusters. Secondary environmental influences arise from hydrogen bonding interactions with tissue constituents, in particular the proteins in the tissue structures; see the graph below. Due to fast fluctuations of the protons within all of these structural arrangements involving water, differences in chemical environment are equilibrated, and there is commonly just one somewhat broadened proton NMR signal. Here, an additional parameter in NMR spectroscopy gains importance: the relaxation time $T$, the time it takes for a nucleus in its excited state (excited by absorption of the radiation $h\nu$) to 'relax', i.e. to return into its ground state. Different tissue concentrations of water, as well as different environments, give rise to modulations of $T$. These modulations in $T$ are 'translated' into modifications in intensities (and *widths*) of the NMR signal, and this is what becomes processed into different levels of contrast in an MRI image.

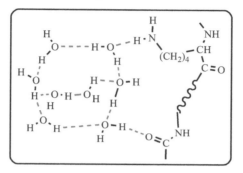

In addition, the magnetic field applied is not constant throughout the body volume. Rather, there is a field gradient, allowing for modulation of the resonance conditions as a function of the (infinitesimal) volume elements where the water molecules are located, thus refining the images.

Contrast agents in MRI such as magnetic ferric oxide nanoparticles, or paramagnetic gadolinium(III) ($f^7$) compounds with an 'open' site accessible to water molecules, enhance the relaxation (shorten $T$) of excited magnetic states of the protons in water molecules. As the contrast agent enters one specific tissue type in preference to another one, contrast enhancement results. The figure below shows MRI images of a foot scanned without (left) and with (right) a gadolinium-based contrast agent. The images show a cystic lesion (with hyperemia), indicated by an arrow, at the interface of the shinbone (tibia) and the anklebone (talus). Exchangeable water molecules $H_2O^*$, in particular in the cystic area and the blood vessels, appear bright (contrast enhanced) in the image to the right, due to direct contact with $Gd^{3+}$ through the fast exchange $LGd(H_2O) + H_2O^* \leftrightarrows LGd(H_2O^*) + H_2O$; for L, see Fig. 14.18.

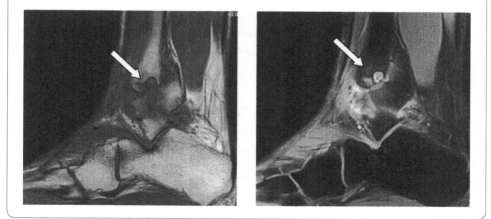

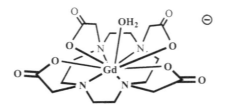

Figure 14.18 The anion [Gd(dota)H$_2$O]$^-$ of Dotarem®, used in magnetic resonance imaging. Dota=1,4,7,10-tetraazacyclododecane-*N,N′,N″,N‴*-tetraacetate. The counter ion of the complex anion is *N*-methylammonium-glucamin. The eight ligand functions (4 amine-N+4 carboxylate-O) form a square antiprism encasing Gd$^{3+}$ in a bowl-like fashion, and thus allowing exchange of internal with external water, Eq. (14.8).

Gd$^{3+}$ in the form of coordination compounds. Appropriate gadolinium complexes have to be designed so as to provide water solubility, high stability (and thus low Gd$^{3+}$-induced toxicity), and a site available for water exchange. High stability is guaranteed by a multidentate ligand L$_n$ possessing hard, i.e. *O*- or *ON*-functions. The accessibility for external water to the Gd$^{3+}$ centre is secured by an appropriate complex geometry, providing an open site for a water ligand to be exchanged, thus ensuring that (magnetically excited) tissue water 'senses' the paramagnetic Gd$^{3+}$ centre and consequently relaxes rapidly. The exchange situation is represented by Eq. (14.8), with tissue water labelled by an asterisk; an example for a gadolinium complex used in MRI is shown in Fig. 14.18.

$$[GdL_n(H_2O)]^- + H_2O^* \rightleftharpoons [GdL_n(H_2O^*)]^- + H_2O \tag{14.8}$$

The second medicinal domain of non-invasive imaging is based on γ radiation originating from radioactive nuclei. Gamma rays have a low effective cross section and thus a large range, allowing detection of a γ source, which has accumulated in a specific body tissue after injection, by a γ counter positioned outside the body. The metastable technetium isotope $^{99m}$Tc is typically employed as a direct γ source. In addition, positron (β$^+$) emitters are widely in use as diagnostic tools. Clinically well-established examples are $^{18}$F and $^{123}$I, while $^{68}$Ga-labelled tracers are still in trial. For the properties of these nuclei, see Table 14.2 in the preceding section. An example for β$^+$ decay ($^{123}$I) is provided in Sidebar 14.1.

Positrons are the antiparticles of electrons; in the case of β$^+$ decay, γ rays are produced as annihilation radiation (β$^+$+β$^-$→2γ), immediately succeeding the β$^+$ decay. The γ rays emitted by direct γ sources or in the context of positron emission are detected, and processed into an image, by a specific system known as single photon emission computed tomography (SPECT). In the case where the γ rays originate from β$^+$ emitters, the method is referred to as positron emission tomography (PET).

The metastable technetium isotope $^{99m}$Tc embedded in a coordination compound is widely used in imaging of malfunctioning organs such as kidney, liver, heart, brain, and, in particular, bone [29]. In order to channel a γ source such as $^{99m}$Tc into a specific tissue to be imaged—typically occult bone metastases in cancer patients or other bone diseases—bifunctional agents are in use. These are complexes that provide a coordination site for the technetium ion, and a biologically active functional part for recognition by a specific receptor of the target cell. Osteoblasts-associated hydroxyapatite Ca$_5$(PO$_4$)$_3$(OH) is the preferential target in

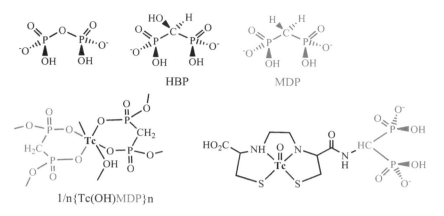

**Figure 14.19** Biphosphate (upper left) and *bis*(phosphonates) (centre and right) used for direct bone targeting formulations containing radioactive nuclei such as $^{99m}$Tc (lower left), and as part of bifunctional ligands (lower right). The *bis*(phosphonate) moiety is highlighted in blue. MDP stands for *methylene-diphosphonate*. The oxidation states of Tc are +IV in the oligomeric Tc(OH)MDP, and +V in the bifunctional system. The ligand motif coordinating to Tc$^{5+}$ contains two cysteinyl moieties. For a gallium complex containing *hydroxy*bisphosphonate (HBP) see Fig. 14.20.

affected bone structures, and biphosphonates as functional units in technetium complexes have been shown to display a strong affinity to apatite.

*Bis*(phosphonates) (Fig. 14.19) are structural analogues of inorganic biphosphate $H_2P_2O_7^{2-}$ (also known as pyrophosphate), but have the advantage that the labile P-O-P motif in $H_2P_2O_7^{2-}$ is replaced by the stable P-CRR'-P or P-C(OH)R-P unit. Direct coordination of Tc$^{4+}$ to *bis*(phosphonates), such as oligomeric {Tc(OH)MDP}$_n$ (where R=R'=H), provide efficient imaging agents for pathological bone structures. In view of a *bifunctional* approach, oxidotechnetium(V) complexes have been introduced that contain a tetradentate $N_2S_2$ ligand, thus stabilizing the Tc$^{5+}$ ion during transport, and *bis*(phosphonate) in the periphery of the coordination sphere for targeting. An example is displayed in Fig. 14.19.

The bifunctional approach for the design of a bone-targeting drug is also applied in diagnostic agents for positron emission tomography. An example is the $^{68}$Ga complex of the ligand NOTA-HBP, where NOTA is 1,4,7-triaza-cyclononane-triacetic acid, and HBP the hydroxy*bis*(phosphonate) moiety [29]. The corresponding Ga$^{3+}$ complex is depicted in Fig. 14.20. Fig. 14.20 also shows an iodine-labelled organic compound, the fatty acid analogue *p*-iodophenyl-3-methylpentadecanoic acid, $^{123}$I-BMIPP. The β$^+$ emitter $^{123}$I is used for imaging both the iodine-metabolizing thyroid gland, and organs with malfunctions due to restricted blood circulation (and thus insufficient oxygen supply to tissues) as a consequence of vasoconstriction. An example for this mode of application is the imaging of the muscle tissue of the heart (myocardial imaging) in the case of angina pectoris.

## 14.5 The toxic and therapeutic potential of CO, NO, and H$_2$S

Carbon monoxide (CO), nitric oxide (NO), and hydrogen sulfide (H$_2$S) are highly toxic gases, the toxicity of which is mainly due to the inhibition of cellular respiration by blockage of

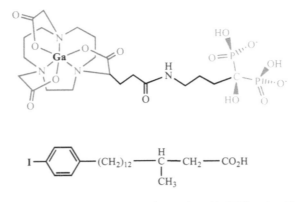

Figure 14.20 Examples for compounds containing nuclei employed in PET imaging: The bifunctional $^{68}$Ga complex [Ga$^{3+}$(NOTA-HBP)] (NOTA in dark blue, HBP in light blue) emerged as promising in the visualization of bone structures. The $^{123}$I derivative of phenyl-3-methylpentadecanoic acid is an efficient myocardial imaging agent.

the iron site in cytochromes. NO additionally acts through its radical character and thus can cause defects similar to those provoked by reactive oxygen species, ROS. While we have available efficient sensors for the detection of even very low concentrations of $H_2S$, with its characteristic foul odour of rotten eggs, no such warning systems exist for CO and NO. In trace amounts, however, all three gases are essential for a variety of body functions. Consequently, our bodies can synthesize the necessary quantities of these physiologically relevant gases. In fact, the three gases do have a therapeutic potential. An example is their ability to widen the blood vessels and thus to regulate blood pressure in the case of hypertension.

Nitric oxide is a by-product of combustion processes, and is naturally produced in the atmosphere by electric discharge (lightning). But NO is also generated by biological processes, one of which, denitrification, starts from nitrate $NO_3^-$ or nitrite $NO_2^-$, as described in detail by Eq. (9.11) in Section 9.3. Nitrate is a common nutritional ingredient, in particular in leafy vegetables such as spinach and salads; nitrite enters the food chain mainly as a preservative of meat. Another source for nitrate is the oxidative conversion of NO by oxyhaemoglobin (with ferrous iron) to methaemoglobin (MetHb, containing ferric iron), Eq. (14.9). An alternative source for nitrite is the oxidation of NO by $O_2$ as catalysed by the copper-dependent enzyme ceruloplasmin, Eq. (14.10).

$$NO + Hb(Fe^{2+} - O_2) \rightarrow NO_3^- + MetHb(Fe^{3+}) \qquad (14.9)$$

$$NO + \tfrac{1}{2}O_2 + [H] \rightarrow NO_2^- + H^+ \text{ (catalysed by ceruloplasmin)} \qquad (14.10)$$

Both $NO_3^-$ and $NO_2^-$ can be viewed as mammalian storage pools for NO, complementing the oxygen-dependent production of NO from arginine as catalysed by NO-synthase (Fig. 9.6 in Section 9.4), and backing up NO production from arginine in the case of hypoxia ($O_2$ undersupply) [30]. In general, the conversion of nitrate to nitrite in mammals is carried out by bacteria that are thriving in the saliva and in the gastrointestinal tract. In the acidic conditions of the stomach, salivary nitrite is protonated to nitrous acid, which disproportionates to form NO and higher-valent nitrogen oxides, such as $NO_2$, Eq. (14.11). Additionally, dietary reductants such as vitamin C (ascorbic acid, $AscH_2$) can reduce nitrous acid to nitric oxide, Eq. (14.12). In this redox reaction, $AscH_2$ is oxidized to dehydroascorbic acid, Asc.

$$NO_2^- + H^+ \rightarrow HNO_2; \; 2HNO_2 \rightarrow NO + NO_2 + H_2O \tag{14.11}$$

$$2HNO_2 + AscH_2 \rightarrow 2NO + Asc + 2H_2O \tag{14.12}$$

The gastrointestinal presence of $NO_2^-/NO$ is believed to have a protective effect on ulcer development through the inhibition of *Helicobacter pylori*, a bacterium that can be responsible for inflammations and occasional tumours in the stomach mucosa.[16] This effect is very much reminiscent of the antimicrobial effect of nitrite used as a preservative in raw meat products such as ground meat. Nitrite that ends up in the blood circulation is reduced to NO by deoxyhaemoglobin, which also binds and transports NO; Eq. (14.13). Even more effective in this respect is myoglobin.

$$NO_2^- + Hb(Fe^{2+}) + H^+ \rightarrow NO + MetHb(Fe^{3+}) + OH^- \tag{14.13a}$$

$$NO + Hb(Fe^{2+}) \rightarrow Hb(Fe^{2+}-NO) \tag{14.13b}$$

Traditional organic sources for nitrite such as amyl nitrite, which is metabolized *in vivo* to nitrite and NO, have a long-standing role in blood flow regulation and thus in the treatment of diseases which are caused by vasoconstriction (narrowing of the blood vessels). Examples are angina pectoris and other ischemia-based tissue injuries. Infusions of inorganic nitrite can have comparable effects. The beneficial roles of sub-micromolar concentrations of NO in general, and nitrate/nitrite derived NO in particular, include protection of cells against harmful agents, referred to as cytoprotection. Harmful agents can be ROS, over-produced by mitochondrial signalling; the availability of NO contributes to the annihilation of ROS. An example is provided by Eq. (14.14).

$$NO + OH \rightarrow HNO_2 \tag{14.14}$$

Carbon monoxide has already undergone promising clinical tests in humans. CO is highly toxic when inhaled, mainly due to its high affinity for the $Fe^{2+}$ of haemoglobin (Hb): CO binds approximately 250 times more efficiently to Hb than $O_2$ and thus blocks off $O_2$ transport, resulting in $O_2$ undersupply (hypoxia) and thus tissue damage. But CO is also generated endogenously by the oxidative breakdown of the haem of senescent and injured red blood cells, Eq. (14.15), a process which is catalysed by a haem oxygenase. Along with CO and $Fe^{2+}$, biliverdin is formed, which is further reduced to bilirubin by nicotine-adenine dinucleotide phosphate, NADPH (Sidebar 9.2).

$$Hb(Fe^{2+}) + 3O_2 + 4(NADPH + H^+) \rightarrow biliverdin + CO + Fe^{2+} + 4NADP^+ + 3H_2O \tag{14.15}$$

Like NO, CO serves as a messenger molecule and thus can play a beneficial role in inflammatory and cardiovascular disorders [31, 32]. Given the toxicity of inhaled gaseous CO, the development of suitable compounds that can be administered orally or intravenously for the

---

[16] See also the gastroprotective effect of bismuth compounds, Section 14.3.4.

delivery of CO to diseased tissues, and which avoid toxic blood CO levels, is key. This can be achieved by prodrugs that release CO at the target site where CO can reach the endangered tissue before being scavenged by red blood cells. While CO has a high affinity for the ferrous site in the haem system of haemoglobin and myoglobin, the *kinetics* of CO binding are slow, which allows for a targeted application of a CO-releasing drug with minimized toxic effect.

Unlike compounds that deliver NO, no CO-releasing prodrugs have so far become clinically approved. Potential prodrugs for the liberation of CO in response of a bio-activation stimulus are transition metal (M) carbonyls that are soluble in aqueous media, and that at least partially survive circulation within the body before arriving at the target site, where a trigger mechanism takes care of the release of CO. Since metal-based carbonyls are commonly insoluble in water, and disposal of CO is hampered by the partial double-bond character of the M-CO bond ($\sigma+\pi$; cf. Sidebar 13.1), this is a non-trivial demand. In Fig. 14.21, three promising carbonyl complexes fulfilling the conditions of a CO-releasing prodrug are sketched. Of particular interest is the $Fe^0$ complex $Fe(CO)_3$(Ac-*cyclo*-hexadiene) with an acetyl (Ac) substituent at the *cyclo*-hexadiene ring: Under mild oxidative conditions and in the presence of an esterase, this complex releases CO [33].

Hydrogen sulfide ($H_2S$) is about as toxic as CO: 800 ppm of $H_2S$ in inhaled air are lethal within minutes. Unlike CO, $H_2S$ 'betrays' its presence by its foul odour; our olfactory sense detects $H_2S$ at concentrations as low as 5 ppb! In natural environments, where $H_2S$ is produced by sulfate reducing bacteria and by microbial degradation of organic matter, we are therefore warned of its presence in a timely way.

$H_2S$ is also produced by bacteria in the gastrointestinal tract, where it contributes to the relaxation of the intestinal muscles and thus helps to prevent obstipation. On the other hand, overproduction of $H_2S$ is counteracted by its oxidation to thiosulfate $S_2O_3^{2-}$ in the intestinal mucosa. Oxygen radicals can oxidize thiosulfate further to tetrathionate $S_4O_6^{2-}$, which then serves as a respiratory electron acceptor for *Salmonella*, a pathogen that causes intestinal inflammation [34]. The reaction sequence is summarized in Eq. (14.16).

$$2H_2S + 3H_2O \rightarrow S_2O_3^{2-} + 8e^- + 10H^+; \quad 2S_2O_3^{2-} \leftrightarrows S_4O_6^{2-} + 2e^- \tag{14.16}$$

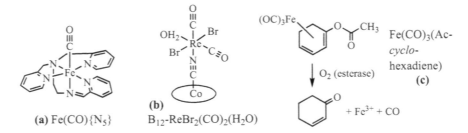

**(a)** $Fe(CO)\{N_5\}$

**(b)** $B_{12}$-ReBr$_2$(CO)$_2$(H$_2$O)

$Fe(CO)_3$(Ac-*cyclo*-hexadiene) **(c)**

Figure 14.21 Three transition metal carbonyl complexes that release CO under physiological conditions. (a) The five nitrogens of the pentadentate ligand in $Fe(CO)\{N_5\}$ are provided by three pyridines, an amine and an imine function. (b) The encircled cobalt in the rhenium(II) complex represents the corrinoid system of vitamin $B_{12}$ (Fig. 13.4 in Section 13.1). (c) The release of CO in the iron complex $Fe(CO)_3$(Ac-*cyclo*-hexadiene) (Ac=acetyl) is catalysed by esterases.

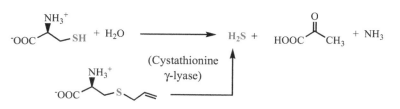

**Scheme 14.5** The endogenous formation of hydrogen sulfide from L-cysteine, and from the garlic component S-allyl-L-cysteine (bottom).

As in the case of CO and NO, trace amounts of $H_2S$ are also produced in our body tissues, where this gas contributes to the regulation of metabolism and of vasodilation. The latter effect goes back to the ability of $H_2S$ to open potassium ion channels in the outer membrane of the smooth muscle of the blood vessels, resulting in extrusion of $K^+$ and reduction of $Ca^{2+}$ influx, which in turn causes muscle relaxation, and thus vascular dilatation and improved blood flow. The starting material for the body's own $H_2S$ production is cysteine. The endogenous production of $H_2S$ is attributed to cystathionine γ-lyase, an enzyme that catalyses several trans-sulfuration pathways [35], and also the hydrolytic conversion of cysteine to pyruvic acid, $H_2S$ and $NH_3$, as illustrated in Scheme 14.5. The enzyme further produces $H_2S$ from S-allyl-L-cysteine, one of the sulfur-based constituents of garlic that are believed to be responsible for the cardio-protective and other beneficial health effects of garlic.

## ➕ Summary

Most of the inorganic prescriptions used in medicinal applications are based on metals and metalloids, but the importance of non-metals (e.g. fluorine and iodine) and molecules such as CO, NO, and $H_2S$ is also increasingly acknowledged.

Metallo-drugs introduced into an individual commonly undergo speciation, i.e. their genuine ligand systems are (partially) exchanged for ligands provided by the body. These can be simple ligands such as $H_2O/OH^-$ and $Cl^-$, more complex ones (e.g. porphyrins), or high-molecular mass binders. Examples for the latter are proteins and DNA. Typical binding sites in proteins are thiolate, amine, carboxylate, and hydroxide functions in amino acid side-chains; typical binding site in DNA is N7 of guanine. In the cytosol, glutathione is an important reductant and metal ion binder. Many metals are potentially toxic. The extent to which a metal is poisonous largely depends on its concentration, physiological availability, and interaction with the body's own substances. A common means to remove toxic metals from the body is chelation therapy, i.e. injection of ligands that effectively bind to and clear the metal from the body.

Iron and copper are examples of such essential elements, an overload of which causes severe malfunctions. Elevated absorption of ferrous iron in the small intestines can result in haemochromatosis. Accordingly, dysfunction of copper homeostasis leads to copper overload and development of Wilson's disease. However, undersupply of Cu (Menkes disease) and iron are equally harmful. People affected with thalassemia, a genetic disease that provides some resistance against malaria, suffer from undersupply of the $Fe^{2+}$-dependent oxygen carrier haemoglobin. Finally, Alzheimer's disease appears to be closely connected to copper dyshomeostasis in the brain.

Prescriptions of mercury, gold, and antimony have been used for millennia to cure syphilis (Hg), arthritis (Au), and skin infections (Sb). Gold in particular, along with bismuth, silver, platinum, and lithium, enjoys modern clinical applications. Gold-based anti-arthritic drugs contain $Au^+$ in a linear coordination environment of two thiolate ligands or thiolate plus phosphane. The drug myochrisine, with thiomalate as ligand, blocks cathepsin-K (a protease involved in arthritic bone degeneration) by linking a monomeric thiolatogold unit to a cysteinate residue of the protein.

Three platinum-based anticancer drugs are in clinical use: cisplatin, carboplatin, and oxaliplatin, all of which contain $Pt^{2+}$ in a square-planar coordination environment with two cisoid amino functions. During transport through the body, the drugs are subjected to (partial) speciation, and genuine copper transporters can take over in further transport and distribution of $Pt^{2+}$. The main coordination mode for $Pt^{2+}$ is to N7 of two adjacent, intrastrand guanosines. This causes a kink in the DNA, preventing DNA's proper functioning. Other metals with a prospective potential for cancer treatment are ruthenium, titanium, vanadium, gallium, and gold, in particular when having available aromatic ligands allowing for intercalation of the (pro-)drug into the DNA structure.

Other metals of use in medicine include vanadium (not yet in clinical application), lithium, bismuth, and silver. Vanadium compounds provide an anti-diabetic potential due to the ability of vanadate to compete with the active site phosphate in phosphate-dependent enzymes. Lithium is being successfully applied in the treatment of bipolar disorder. The basis for its mode of operation is the similarity (with respect to the ionic radius) between $Li^+$ and $Mg^{2+}$. $Li^+$ thus can down-regulate $Mg^{2+}$-dependent cell signalling. Bismuth prescriptions such as the colloidal subcitrates and subsalicylates of $Bi^{3+}$ have a long-standing role in the treatment of stomach ulcers caused by *Helicobacter pylori*. Silver, in the form of $AgNO_3$, $Ag^+$ in zeolites, or nanoparticulate Ag, is an efficient disinfectant and thus used in wound healing.

For the treatment of malignant tumours, *radio*pharmaceuticals are also in use. Examples are drugs with radionuclei such as $^{90}Y$, $^{153}Sm$, $^{186}Re$, and $^{188}Re$, all of which are short-lived nuclei (hours to days) that emit $\beta^-$ and $\gamma$ radiation, with $\beta^-$ causing tissue destruction. For application in the palliation of individuals suffering from metastasized bone cancers, these radionuclides are embedded in carriers that are functionalized with diphosphonates, allowing for the targeting of hydroxyapatatite associated with the affected osteoblast tissue.

Imaging methods are used to visualize soft body tissues. Well-established examples are $\gamma$-ray imaging and magnetic resonance imaging (MRI). In MRI, tissue water protons are excited by high-energy radiofrequency pulses, and the different relaxation times of the water protons in differing environments are detected and transformed into an image. Contrast improvement is achieved by paramagnetic contrast agents, in particular nanoparticulate ferric oxides, and gadolinium(III)-based coordination compounds. Common short-lived nuclei employed in $\gamma$-ray imaging (SPECT) are the metastable $^{99m}Tc$ ($\rightarrow ^{99}Tc + \gamma$), and positron emitters such as $^{123}I$ ($\rightarrow ^{123}Te + \beta^+$; $\beta^+ + e^- \rightarrow 2\gamma$). The latter is referred to as positron emission tomography, PET. In order to channel $^{99m}Tc$ to tissues to be imaged, e.g. a diseased bone structure, the technetium compound is functionalized so as to become recognized by the target tissue. In the case of bone, *bis*(phosphonates) have proved to be suitable. Hydroxy-*bis*(phosphonate) as a targeting function is presently also in clinical trial in coordination compounds for PET based on $^{68}Ga$ ($\rightarrow ^{68}Zn + \beta^-$).

The highly toxic gases NO, CO, and $H_2S$ are also produced *in vivo* in submicromolar concentrations. Here, they are essential regulatory molecules in signalling pathways, inducing vasodilation and cell protection. At hypoxic conditions, nitrate and nitrite serve as mammalian NO storage pools. NO is liberated from nitrite $NO_2^-$ enzymatically by denitrification or, in acidic compartments of the gastrointestinal tract, non-enzymatically by reduction and disproportionation. While NO-releasing agents are used medicinally to counteract vasoconstriction, prodrugs that release CO are still at the trial stage. CO has vasodilatory effects comparable to those of NO, and has specific potential in the treatment of inflammatory disorders. Vasodilation and thus control of the blood pressure is also a pivotal regulatory function of $H_2S$ that is produced endogenously from cysteine and its derivatives. $H_2S$ produced by bacteria in the intestines counteracts obstipation. On the other hand, infectious gut bacteria such as *Salmonella* can thrive by profiting from thiosulfate, an oxidation product of $H_2S$, as a respiratory electron acceptor.

## Suggested reading

**Berdoukas V and Wood J. In search of optimal iron chelation therapy for patients with thalassemia major.** *Haematologica* 2011; 96: 5–8.
A critical account on applications and further developments of iron chelation therapy in iron overload diseases.

**Faller P. Copper in Alzheimer disease: too much, too little, or misplaced?** *Free Radical Biol. Med.* 2012; 52: 747–748.
This brief account compares, in a clearly arranged manner, several of the main aspects of Wilson, Menkes, and Alzheimer diseases, focussing on novel findings on the maldistribution of copper in Alzheimer disease.

**Dabrowiak J.** *Metals in medicine.* **Chichester, UK: Wiley & Sons, 2009.**
The author carefully and comprehensively reviews the role of metal-based drugs and prodrugs in medicinal applications, emphasizing both chemical and pharmaceutical aspects.

**Jones CJ and Thornback JR.** *Medicinal applications of coordination chemistry.* **Cambridge, UK: The Royal Society of Chemistry, 2007.**
The book covers a plethora of aspects directed towards the application of coordination compounds in diagnostic and therapeutic medicine, including introductory chapters on general aspects of coordination chemistry.

**Sigel A, Sigel H, Sigel RKO. Interrelations between essential metal ions and human diseases. In:** *Metal ions in life science,* **vol. 13. New York: Marcel Dekker Inc, 2013.**
Covers aspects of metals and metalloids in diagnostic and therapeutic medicine.

**Das DK. Hydrogen sulfide preconditioning by garlic when it starts to smell.** *Am. J. Physiol Heart Circ. Physiol.* 2007; 293: H2629–H2630.
The article provides a brief introduction into organosulfur components present in garlic, and their apparent beneficial effect in cardio-protection.

## References

1. Leszcyszyn OI, Zeitoun-Ghandour S, Stürzenbaum SR, et al. Tools for metal ion sorting: in vitro evidence for partitioning of zinc and cadmium in *C. elegans* metallothionein isoforms. *Chem. Commun.* 2011; 448–450.

2. Schmidt M, Raghavan B, Müller V, et al. Crucial role of human Toll-like receptor 4 in the development of contact allergy to nickel. *Nat. Immunol.* 2010; 11: 814–819.

3. Evans RW, Rafique R, Zarea A, et al. Nature of non-transferrin-bound iron: studies on iron citrate complexes and thalassemia sera. *J. Biol. Inorg. Chem.* 2008; 13: 57–74.

4. Pantopoulos K. (2008): Function of the hemochromatosis protein HFE: lessons from animal models. *World J. Gastroenterol.* 2008; 7: 6893–6901.

5. (a) Müller A, Bögge H, Scimanski U. Molybdenum–copper–sulphur containing cage system and its bioinorganic relevance. *J. Chem. Soc. Chem. Commun.* 1980; 91–92; (b) Alvarez HM, Xue Y, Robinson CD, et al. Tetrathiomolybdate inhibits copper trafficking proteins through metal cluster formation. *Science* 2010; 327: 331–334.

6. Delange P and Mintz E. Chelation therapy in Wilson's disease: from D-penicillamine to the design of selective bioinspired intracellular Cu(I) chelators. *Dalton Trans.* 2012; 6359–6370.

7. Rubino JT, Riggs-Gelasco P, Franz K-J. Methionine motifs of copper transport proteins provide general and flexible thioether-only binding sites for Cu(I) and Ag(I). *J. Biol. Inorg. Chem.* 2010; 15: 1033–1049.

8. James SA, Volitakis I, Adland PA, et al. Elevated labile Cu is associated with oxidative pathology in Alzheimer disease. *Free Radical Biol. Med.* 2012; 52: 298–302.

9. (a) Faller P and Hureau C. Bioinorganic chemistry of copper and zinc ions coordinated to amyloid-β peptide. *Dalton Trans.* 2009; 1080–1094; (b) Hung YH, Bush AI, Cherny RA. Copper in the brain and Alzheimer's disease. *J. Biol. Inorg. Chem.* 2010; 15: 61–76.

10. Healy ML, Lim KKT, Travers R. Jaques Forestier (1890–1978) and gold therapy. *Int. J. Rheumatic Disease* 2009; 12: 145–148.

11. (a) Bhabak KP, Bhuyan BJ, Mugesh G. Bioinorganic and medicinal chemistry: aspects of gold(I)–protein complexes. *Dalton Trans.* 2011; 2099–2111; (b) Bernes-Price SJ and Filipovska A. Gold compounds as therapeutic agents for human diseases. *Metallomics* 2011; 3: 863–873.

12. Weidauer E, Yasuda Y, Biswal BK, et al. Effects of disease-modifying anti-rheumatic drugs (DMARDs) on the activities of rheumatoid arthritis-associated cathepsins K and S. *Biol Chem.* 2007; 388: 331–336.

13. Dabrowiak JC. Cisplatin. In: *Metals in medicine.* Chichester, UK: Wiley & Sons, 2009. ch. 3.

14. Du X, Wang X, Li H, et al. Comparison between copper and cisplatin transport mediated by human copper transporter 1 (hCTR1). *Metallomics* 2012; 4: 679–685.

15. Reedijk J. Increased understanding of platinum anticancer chemistry. *Pure Appl. Chem.* 2011; 83: 1709–1719.

16. Ummat A, Rechkoblit O, Jain R, et al. Structural basis for cisplatin DANN damage tolerance by human polymerase η during cancer chemotherapy. *Nat. Struct. Mol. Biol.* 2012; 19: 628–633.

17. (a) Hartinger CG, Jakupec MA, Zorbas-Seifried S, et al. KP1019, a new redox-active anticancer agent—preclinical development and results of a clinical phase I study in tumor patients. *Chem. Biodivers.* 2008; 5: 2140–2155; (b) Costa Pessoa J and Tomaz I. Transport of therapeutic vanadium and ruthenium complexes by blood plasma components. *Curr. Med. Chem.* 2012; 17: 3701–3738.

18. (a) Liu H-K and Sadler PJ. Metal complexes as DNA intercalators. *Acc. Chem. Res.* 2011; 44: 349–359; (b) Liu H-K, Berners-Price SJ, Wang F, et al. Diversity in guanine-selective DNA binding modes for an organometallic ruthenium arene complex. *Angw. Chem. Int. Ed.* 2006; 45: 8153–8156.

19. Tshuva EY and Ashenhurst JA. Cytotoxic titanium(IV) complexes: renaissance. *Eur. J. Inorg. Chem.* 2009; 2203–2218.

20. (a) Rehder D. The potentiality of vanadium in medicinal applications. *Future Med. Chem.* 2012; 4: 1823–1837; (b) Rehder D. The future of/for vanadium. *Dalton Trans.* 2013; 42: 11749–11761.

21. Zhompson KH, Lichter J, LeBel C, et al. Vanadium treatment of type 2 diabetes: a view to the future. *J. Inorg. Biochem.* 2009; 103: 554–558.

22. Dudev T and Lim C. Competition between $Li^+$ and $Mg^{2+}$ in metalloproteins. Implications for lithium therapy. *J. Am. Chem. Soc.* 2011; 133: 9506–9515.

23. Maimovich A, Eliav U, Goldbourt A. Determination of the lithium binding site in Inositol monophosphatase, the putative target for lithium therapy, by magic-angle-spinning solid-state NMR. *J. Am. Chem. Soc.* 2012; 134: 5647–5651.

24. Li H and Sun H. Recent advances in bioinorganic chemistry of bismuth. *Curr. Opin. Chem. Biol.* 2012; 16: 74–83.

25. (a) Sintubin L, De Gusseme B, Van der Meeren P, et al. The antibacterial activity of biogenic silver and its mode of action. *Appl. Microbiol. Biotechnol.* 2011; 91: 153–162; (b) Marambio-Jones C and Hoek EMV. A review of the antibacterial effects of silver nanomaterials and potential implications for human health and enviromment. *J. Nanopart. Res.* 2010; 12: 1531–1551; (c) Morones-Ramirez JR, Winkler JA, Spina CS, et al. Silver enhances the antibiotic activity against Gram-negative bacteria. *Sci. Transl. Med.* 2013; 5: 190ra81.

26. Ogawa K and Saji H. Advances in drug design of radiometal-based imaging agents for bone disorder. *Int. J. Mol. Imaging* 2011; article ID 537687.

27. Ogawa K, Kawashima H, Shiba K, et al. Development of [$^{90}$Y]DOTA-conjugated bisphosphonate for treatment of painful bone metastases. *Nucl. Med. Biol.* 2009; 36: 129–135.

28. Norek M and Peters JA. MRI contrast agents based on dysprosium and holmium. *Progr. Nucl. Magn. Reson. Spectrosc.* 2011; 59: 64–82.

29. Palma E, Correie JDG, Campello MPC, et al. Bisphosphonates as radionuclide carriers for imaging or systemic therapy. *Mol. BioSyst.* 2011; 7: 2950–2966.

30. Lundberg JO, Weitzberg E, Gladwin MT. The nitrate–nitrite–nitric oxide pathway in physiology and therapeutics. *Nat. Rev. Drug Discov.* 2008; 7: 156–167.

31. Romão CC, Blättler WA, Seixas JD, et al. Developing drug molecules for therapy with carbon monoxide. *Chem. Soc. Rev.* 2012; 41: 3571–3583.

32. Otterbein LO. The evolution of carbon monoxide into medicine. *Respiratory Care* 2009; 54: 925–932.

33. Romanski S, Kraus B, Schatzschneider U, et al. Acyloxybutadiene iron tricarbonyl complexes as enzyme-triggered CO-releasing molecules (ET-CORMs). *Angew. Chem. Int. Ed.* 2011; 50: 2392–2396.

34. Winter SE, Thiennimitr P, Winter MG, et al. Gut inflammation provides a respiratory electron acceptor for *Salmonella*. *Nature* 2010; 467: 426–429.

35. Huang S, Chua JH, Yew WS, et al. Site-directed mutagenesis on human cystathionine-$\gamma$-lyase reveals insight into the modulation of $H_2S$ production. *J. Mol. Biol.* 2010; 396: 708–718.

# Index

Plate 1 Left: Dome-shaped stromatolites from the Shark Bay, Western Australia. From http://www.heritage. gov.au/ahpi. Right: Cross section of a fraction of a typical stromatolite of cyanobacterial origin, showing light calcareous layers along with dark, silica-enriched carbonaceous layers. Reprinted from Chapter 2. Courtesy of Didier Descouens. This file is licensed under the Creative Commons Attribution-Share Alike 3.0 Unported license.

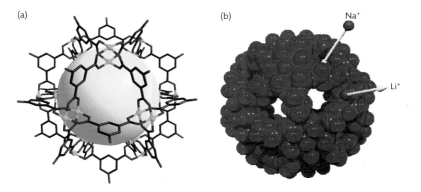

Plate 2 Organometallic/inorganic models for cation transporters. (**a**) MOP-18, viewed down one of the openings of threefold symmetry. Cu is green, O red, and C black; the cavity is highlighted in yellow. Reprinted from Chapter 3, ref. [5]; copyright © 2008 Wiley-VCH Verlag GmbH Co. KGaA, Weinheim. Part (a) reproduced with kind permission from Dr Kimoon Kim. (**b**) Space-filling representation of the polyoxidomolybdate $[Mo_{132}O_{372}(SO_4)_{30}]^{72-}$; Mo is blue, oxygen red (sulfate omitted). The counter transport of $Na^+/Li^+$ across one of the 20 channels of threefold symmetry connecting to the (water-filled) cavity is shown. From Chapter 3, ref. [6]; copyright © 2006 Wiley-VCH Verlag GmbH & Co. KGaA, Weinheim. Part (b) kindly supplied by Dr Achim Müller.

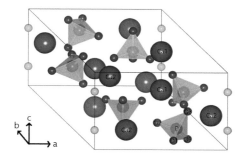

Plate 3 Crystal structure (unit cell) of apatite/fluorapatite A $Ca_5(PO_4)_3(OH/F)$. Purple: $PO_4^{3-}$ tetrahedra; red: $O^{2-}$; blue: $Ca^{2+}$; green: $F^-$. Reprinted from Chapter 3, ref. [18]; copyright © 2013, with permission from Elsevier. Reproduced with kind permission from Dr Barbara Pavan.

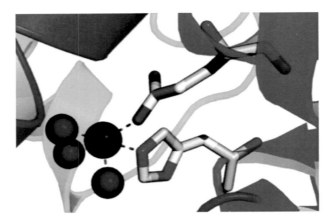

Plate 4 The coordination and protein environment of $Fe^{2+}$ (black sphere) in ferrochelatase, a protein that directs the insertion of $Fe^{2+}$ into the protoporphyrin IX system. The red spheres are water molecules; the coloured ribbon-like structures represent the protein matrix. Reprinted with permission from Chapter 5, ref. [2]; copyright © American Chemical Society.

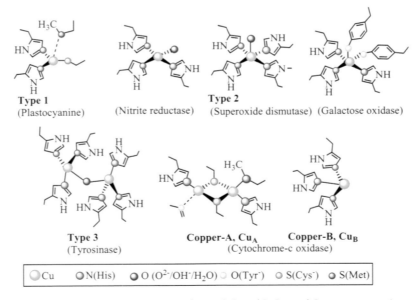

Plate 5 The typical coordination arrangements of types 1, 2, and 3, $Cu_A$ and $Cu_B$ copper proteins; sidebar 6.2.

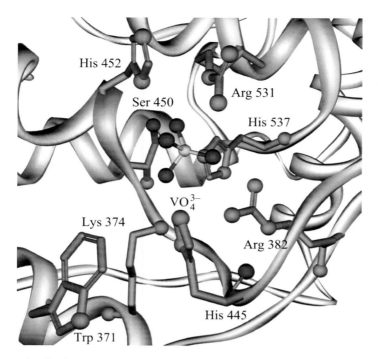

Plate 6 The vanadate binding pocket of vanadate-dependent bromoperoxidase from the marine macro-alga *Ascophyllum nodosum*. Colour code: yellow, vanadium; red, oxygen; blue, nitrogen. Reprinted from Chapter 7, ref. [9b], p. 29. Copyright © (2013), with permission from Elsevier. Kindly supplied by Dr Jens Hartung.

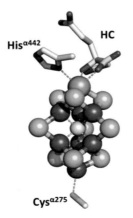

Plate 7 Structure of the FeMo-co; HC = homocitrate. Colour code: Fe, orange; Mo, light blue; S, yellow; C, grey; O, red; N, dark blue. Reprinted with permission from Chapter 9, ref. [8a]. Copyright © (2013) American Chemical Society. Kindly supplied by Dr Markus Ribbe.

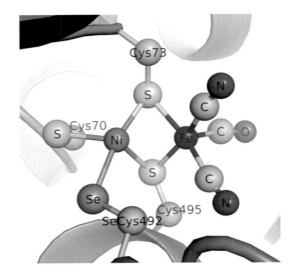

Plate 8 Active site of the {FeNiSe} hydrogenase from the sulfate reducing bacterium *Desulfomicrobium baculatum*. The ribbon structures represent the polypeptide chain environment. Reprinted from Chapter 10, with kind permission from Springer Science and Business Media; Baltazar CSA, et al. *J. Biol. Inorg. Chem.* 2012; 17: 543–555. Image kindly supplied by Dr Carla Baltazar.

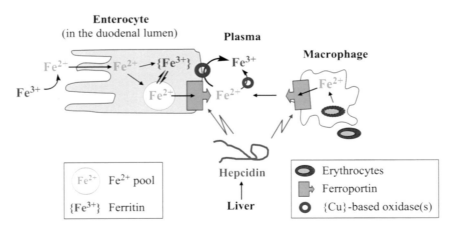

Plate 9 Dietary supply of iron (uptake by the intestinal lumen; left) and the body's own regulation of iron homeostasis (recycling of iron from haemoglobin; right). The peptide hormone *hepcidin*, produced in the liver, targets the ferrous ion transporter *ferroportin*. In a healthy individual, hepcidin levels and iron efflux increase in the case of low iron supply and hypoanaemia, and decrease in the case of excessive iron supply and hyperanaemia. Reprinted from Chapter 14.